Photopion Nuclear Physics

Photopion Nuclear Physics

Edited by
Paul Stoler
Rensselaer Polytechnic Institute
Troy, New York

Plenum Press · New York and London

Library of Congress Cataloging in Publication Data

Main entry under title:

Photopion nuclear physics.

"Based on lectures presented at the International Symposium on Photo-
pion Nuclear Physics at Rensselaer Polytechnic Institute, Troy, New York,
August 10–12, 1978."
QC793.5.M422P47   539.7'2162   78-31569
ISBN-13: 978-1-4684-3473-6   e-ISBN-13: 978-1-4684-3471-2
DOI: 10.1007/978-1-4684-3471-2

Based on Lectures presented at the International Symposium
on Photopion Nuclear Physics at Rensselaer Polytechnic Institute,
Troy, New York, August 10–12, 1978.

Preface

This volume consists of twelve review articles and a
variety of short contributions which are based on lectures
presented at the International Symposium on Photopion Nuclear
Physics. The review articles were submitted after the symposium, and there was considerable editing and cross referen-
cing with the aim of achieving a greater overall unity than
ordinarily found in conference proceedings.

Photopion Nuclear Physics, as the name suggests, combines
two of the most active areas of current intermediate energy
nuclear physics research - electromagnetic phenomena as
studied by means of electron scattering and photonuclear processes, and mesonic effects in nuclei. The potential value of
photopion studies as a complement to electron scattering for
the study of spin-isospin aspects of nuclear transitions has
been widely recognized for about a decade, thanks to the theoretical work at such institutions as Stanford and Catholic
Universities. In fact, some of this theoretical work was done
in anticipation of the advent of the new high duty cycle and
good energy resolution electron linacs. The potential for
using this reaction to study mesonic effects was not as broad-
ly addressed at that time. However, as seen in the following
pages of these proceedings, that situation has dramatically
changed. For example, within the past two years, theoretical
studies of experiments on D and ^{12}C, done respectively at the
Saclay and Bates electron linacs, have shown a strong sensitivity to different aspects of pion-nuclear interactions, so
that these problems have emerged into the forefront of this
field.

The symposium was conceived of at a time when most of the
initial experiments on charged pion production on light nuclei
and pion capture had been completed, and the subsequent agreement of the theoretical calculations gave a good degree of
confidence in the basic approach. Diverse higher energy data

and theoretical calculations (some speculative) were beginning
to appear from various laboratories. It was just about the
right time for a conference to see how it all fits together
and to define some coherent directions for this field.

The proceedings tend to follow the actual format of the
symposium. The first part contains three papers which give a
general review of the field from the theoretical and experimental points of view. Each of the next four sections begins
with a review of some specific area of interest, and contains
some contributed papers which are pertinent to that area.
Finally, the summary paper attempts to assess the present status of the field, and hints at some possible future directions.
There was a very large discussion time during the symposium.
These discussions were not reproduced in the Proceedings. However, as previously indicated, the final versions of all manu-
scripts were submitted after the symposium so that each author
could take all the other contributions and the discussions into account to some extent. Although there was considerable
editing, very little was substantive, and that only after complete consultations with the authors.

Thanks are due to H. A. Medicus and B. Schoch of the or-
ganizing committee, as well as to the editor's colleagues. A.
M. Bernstein, K. I. Blomqvist, K. Min, E. J. Winhold and P. F.
Yergin for their work on behalf of the symposium, and to the
International Advisory Committee (A. M. Bernstein, G. E. Brown,
K. M. Crowe, D. Drechsel, H. Feshbach, J. S. O'Connell, K.
Shoda, V. L. Telegdi, C. Tzara, H. Überall, J. D. Walecka, P.
F. Yergin) for their guidance. Special appreciation goes to
Phyllis Campbell and Clara Gorenstein for their work in over-
seeing the day-to-day organizational details, and to Deborah
Maniaci who bore the responsibility for most of the technical
typing in this volume. The editor and organizing committee
are grateful to the United States National Science Foundation,
the United States Department of Energy, the American Physical
Society and Rensselaer Polytechnic Institute for their sponsor-
ship. The Digital Equipment Corporation generously supplied
financial aid to help meet the travel expenses of some scien-
tists who otherwise could not have attended.

Troy, New York   Paul Stoler

November 1978

Contents

(A separate list of contributed papers follows on p. ix.)

INVITED PAPERS

INTRODUCTORY REVIEW   1
    G. E. Brown

PHOTOPION PHYSICS – AN EXPERIMENTAL
           OVERVIEW   19
    E. J. Winhold

RELATIONSHIP OF WEAK AND ELECTROMAGNETIC
           PROCESSES TO PHOTOPION
           REACTIONS   51
    T. W. Donnelly

RADIATIVE PION CAPTURE   69
    J. P. Perroud

    CONTRIBUTED PAPERS   101

TOTAL CROSS SECTION MEASUREMENTS FOR THE
           PHOTOPRODUCTION OF CHARGED
           PIONS FROM THRESHOLD TO
           350 MeV   129
    E. C. Booth

    CONTRIBUTED PAPERS   155

CHARGED PION SPECTROSCOPY   175
    K. Shoda

    CONTRIBUTED PAPERS   193

NEUTRAL PION PHOTOPRODUCTION ON NUCLEI
           NEAR THRESHOLD   219
    N. de Botton

    CONTRIBUTED PAPERS   239

CORRECTIONS TO THE QUASI-FREE PROCESS   255
    J. M. Laget

PHOTOPRODUCTION OF CHARGED PIONS
        FROM C^{12} USING A TAGGED
        PHOTON BEAM   275
    J. Arends, J. Eyink, H. Hartmann,
    A. Hegerath, B. Mecking, G. Nöldeke,
    and H. Rost

    CONTRIBUTED PAPER   295

PION-NUCLEUS INTERACTIONS AND
        PHOTOPRODUCTION   301
    F. Tabakin

    CONTRIBUTED PAPER   327

PION-NUCLEUS INTERACTIONS IN THE
        RESONANCE REGION   335
    E. J. Moniz

    CONTRIBUTED PAPERS   353

SUMMARY AND OUTLOOK   405
    A. M. Bernstein

LIST OF CONTRIBUTORS   425

LIST OF PARTICIPANTS   429

INDEX   433

CONTRIBUTED PAPERS

INVESTIGATION OF ANALOGS OF M1, E1 and M2 TRANSITIONS
IN ^{10}B, ^{12}C and ^{14}N THROUGH RADIATIVE PION CAPTURE. . .   101
    J. C. Alder, W. Dahme, B. Gabioud, C. Joseph,
    J. F. Loude, N. Morel, H. Panke, A. Perrenoud,
    J. P. Perroud, D. Renker, G. Strassner, M. T.
    Tran, P. Truöl and E. Winkelmann

RADIATIVE PION CAPTURE IN MEDIUM MASS NUCLEI: ^{32}S,   107
^{40}Ca . . . .
    J. C. Alder, W. Dahme, B. Gabioud, C. Joseph,
    J. F. Loude, H. Medicus, N. Morel, H. Panke,
    A. Perrenoud, J. P. Perroud, D. Renker, G.
    Strassner, M. T. Tran, P. Truöl and E.
    Winkelmann

BRANCHING RATIO FOR RADIATIVE PION CAPTURE WITH
RESPECT TO ABSORPTION OF STOPPED NEGATIVE PIONS   111
    Carl Werntz

GIANT RESONANCE EXCITATION IN VARIOUS PROCESSES.
EXAMPLE: ^{13}C.   117
    H. R. Kissener and R. A. Eramzhyan

NUCLEON SPECTRA FROM THE ^{16}O$(\pi-,\gamma)$ REACTION   121
    R. A. Eramzhyan, M. Gmitro, L. A. Tosunjan

THE UNIFIED DESCRIPTION OF DIRECT AND RESONANCE
PROCESSES IN THE RADIATIVE PION CAPTURE BY LIGHT
NUCLEI.   125
    V. V. Balashov and R. Wünsch

THEORY OF PION PHOTOPRODUCTION in ^{12}C and ^{16}O   155
    A. Nagl and H. Überall

NEGATIVE PION PHOTOPRODUCTION FROM ^{28}Si   161
    K. Srinivasa Rao and S. Susila

PHOTOPRODUCTION OF π^- FROM ^{14}N.   165
    V. DeCarlo, N. Freed, and W. Rhodes

A METHOD FOR MEASUREMENTS OF THE TOTAL ^{2}H(γ,π^-)PP
CROSS SECTIONS AT THRESHOLD.   171
    B. Schoch, F. Klein and G. Lührs

STUDY ON THE LOW-LYING STATES VIA THE PHOTOPION
PRODUCTION ON SEVERAL NUCLEI   193
    H. Ohashi, K. Nakahara, M. Yamazaki, K. Shoda
    and B. N. Sung

CHARGED PION ELECTROPRODUCTION ON LIGHT NUCLEI NEAR
THE THRESHOLD.   199
    S. Furui

PHOTOPION ENERGY DISTRIBUTIONS AND SPIN-ISOSPIN
GIANT RESONANCES   205
    K. Shoda, M. Yamazaki, K. Nakahara, H. Ohashi

PION SPECTRA TO INDIVIDUAL FINAL STATES IN THE
REACTIONS ^{12}C$(\gamma,\pi^{\pm})^{12}$B,^{12}N   211
    N. Paras, A. M. Bernstein, K. I. Blomqvist,
    G. Franklin, J. LeRose, K. Min, M. Pauli,
    D. Rowley, B. Schoch, P. Stoler, E. J.
    Winhold and P. F. Yergin

A MEASUREMENT OF π^0 PHOTOPRODUCTION ON ^{3}He and ^{4}He
NEAR THRESHOLD   239
    P. Argan, G. Audit, A. Bloch, N. de Botton, J. L.
    Faure, C. Schuhl, G. Tamas, C. Tzara and E.
    Vincent

NEUTRAL PION PHOTOPRODUCTION FROM COMPLEX NUCLEI
NEAR THRESHOLD   245
    F. L. Milder, E. C. Booth, B. L. Roberts,
    J. Comuzzi and H. Crannell

ON THE ^{6}Li$(\gamma,\pi^0$d$)^4$He REACTION   249
    R. Sridhar, K. Srinivasa Rao and S. Susila

QUASI-FREE PION PHOTOPRODUCTION FROM COPPER AND LEAD .   295
    K. Baba, I. Endo, M. Fujisaki, S. Kadota, Y.
    Sumi, H. Fujii, Y. Murata, S. Noguchi, A.
    Murakami

^{12}C$(\gamma,\pi^-)^{12}$Ng.s. FOR (0-50) MeV PIONS.   327
    G. N. Epstein, M. K. Singham and F. Tabakin

COHERENT π^o PHOTOPRODUCTION IN THE ISOBAR-HOLE
FORMALISM. 353
    J. H. Koch and E. J. Moniz

UNIFIED DESCRIPTION OF PION-PHOTOPRODUCTION AND
ELASTIC PION SCATTERING ON THE DEUTERON IN THE
ISOBAR-DOORWAY MODEL 361
    R. Händel, M. Dillig and M. G. Huber

A SUM RULE DESCRIPTION FOR THE INCLUSIVE ELECTROPRO-
DUCTION OF PIONS FROM NUCLEI 367
    R. Rosenfelder

INCLUSIVE ELECTROPRODUCTION OF LOW ENERGY PI MESONS. . 371
    M. Pauli, A. M. Bernstein, K. I. Blomqvist,
    G. B. Franklin, J. LeRose, K. Min, N. Paras,
    D. Rowley, B. Schoch, P. Stoler, E. J.
    Winhold and P. F. Yergin

PHOTOPION PRODUCTION IN THE FERMI GAS MODEL. 381
    J. S. O'Connell, W. M. MacDonald and E. T.
    Dressler

TOTAL PHOTONUCLEAR ABSORPTION CROSS SECTIONS FOR
Li and Be UP TO 350 MeV. 385
    J. Ahrens, H. Gimm, R. J. Hughes, R. Leicht, P.
    Minn, A. Zieger and B. Ziegler

ISOMERIC AND GROUND STATE FORMATION IN ^{197}Hg BY THE
^{197}Au$(\gamma,\pi^-)^{197}$Hg REACTION NEAR THRESHOLD 387
    P. H. Ballentine, J. K. Hersh, H. A. Medicus,
    S. Planeta, E. Potenziani and S. Rossdeutscher

INVESTIGATION OF THE CHARGE EXCHANGE PROCESS π^-d $\rightarrow$ 391
π^onn . . . .
    Il.-T. Cheon

PION-ELECTROPRODUCTION FROM SINGLE NUCLEONS. 395
    H. B. Miska

DEEP INELASTIC ELECTRON SCATTERING NEAR AND ABOVE
PION THRESHOLD. 399
    P. D. Zimmerman and C. F. Williamson

INTRODUCTORY REVIEW

G. E. Brown[*]

State University of New York, Stony Brook, NY

and NORDITA, Copenhagen

I. INTRODUCTION

Photopion physics is several decades old, but its application to nuclear structure is in its infancy. In this introduction I wish first to review what is known about photopion production on nucleons at threshold, where firm statements of the theoretical predictions can be made, and then to extend the discussion to higher-energy regimes, where models have to be used. In particular, we shall discuss the isobar-doorway model and its relationship to the optical model.

As examples of nuclear structure studies, we shall discuss π^- production by γ-rays on C^{12}, where good experiments and careful theoretical analyses have been carried out.

The use of photo- and electropion production to study nucleon-isobar interactions will be discussed.

II. LOW ENERGY THEOREMS

Current algebra gives very definite predictions for the photoproduction of pions near threshold. To lowest order in m_π/m_n the threshold production is given by the Kroll-Ruderman[1] result. This result also follows from gauge invariance in the following simple way.

The interaction of low-energy pions with nucleons is given by pseudo-vector coupling

1

Fig. 1.   Photoproduction through the gauge term.

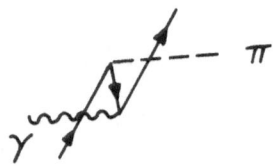

Figure 2. Pion photoproduction involving a virtual nucleon pair in pseudoscalar pion-nucleon coupling.

$$\delta L \;=\; \frac{f_\pi}{m_\pi}\, \bar\psi \,\vec\tau\, \vec\sigma \,\cdot\, \vec\nabla \vec\phi_\pi \psi \;.$$

$$(1)$$

Introduction of an electromagnetic field necessitates

$$\vec\nabla\phi_\pi \;\to\; (\vec\nabla - ie\vec A)\phi_\pi \;,$$

$$(2)$$

where e is the charge of the relevant pion, so that one is left with a term

$$\frac{-ief_\pi}{m_\pi}\, \bar\psi \,\vec\sigma\cdot\vec A\, \vec\tau\cdot\vec\phi_\pi\psi$$

$$(2.1)$$

in the Lagrangian. This term describes the photoproduction process shown in Fig. 1. In fact, the above process is equivalent to that which would proceed through virtual nucleon pairs, were one to employ pseudoscalar nucleon-pion coupling

$$\delta L \;=\; g\bar\psi\gamma_5\vec\tau\cdot\vec\phi_\pi\psi \;,$$

$$(3)$$

the process being shown in Fig. 2.

Current algebra[2] extends the above result to include the terms of order m_π/m_n in the amplitude. Using PCAC, the pion field can be related to the axial vector operator A_μ^j through

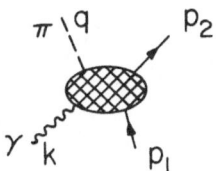

Figure 3. Kinematics of photopion production.

$$\phi_\pi^j(x) = c \frac{\partial A_\mu^j}{\partial x_\mu} , \tag{4}$$

where c is related to the pion decay constant. Here j labels the isospin. The matrix element for pion photoproduction is then given by

$$(M_{\gamma\pi})_{\nu\mu} = c<N(p_2)|\int d^4x e^{-iqx} T(\partial_\mu A_\mu^j(x)V_\nu^3(0))|N(p_1) >$$

$$= iq_\mu c<N(p_2)|\int d^4x e^{-iqx} T(A_\mu^j(x)V_\nu^3(0))|N(p_1) >$$

$$- c<N(p_2)|\int d^4x e^{-iqx}\delta(x_o)[A_o^j(x),V_\nu^3(0)]|N(p_1) >. \tag{5}$$

where V_ν is the isovector electromagnetic current operator, ν indicating the space-time component. In the above, T indicates the time ordering of the enclosed operators, and a partial integration has been performed to get to the second equation.

In the commutator term, the transition to q = 0 can now be made, and, upon integrating over x, one has

$$<N(p_2)|\int d^4x e^{-iqx}\delta(x_o)[A_o^j(x), V_\nu^3(0)]|N(p_1) >$$

$$\underset{q \to 0}{=} <N(p_2)|[\int d^4x A_o(x), V_\nu^3]|N(p_1) >$$

$$= <N(p_2)|i\epsilon_{j3\ell}A_\nu^\ell|N(p_1) > , \tag{6}$$

the last step following from the commutation rules of current algrebra. For q → 0 we can represent A_ν^ℓ by

$$A_\nu^\ell = g_A \frac{\tau_\ell}{2} \gamma_\nu\gamma_5 . \tag{6.1}$$

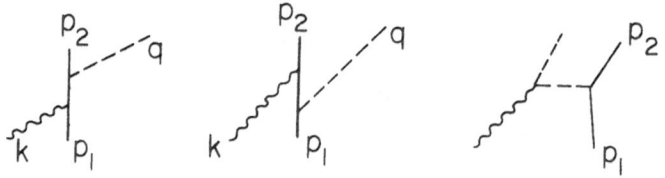

Figure 4.   Pole terms in pion photoproduction.

Remembering that $\vec{\sigma}$ is the nonrelativistic approximation for $\gamma_\nu \gamma_5$, we get back the result (2.1) derived from gauge invariance.

The other term in Eq. (5), which gives corrections of order m_π/m_n, can be written as

$$iq_\mu <N(p_2)| \int d^4x e^{-iqx} T(A_\mu^j(x) V_\nu^3(o))|N(p_1)>$$

$$= \text{Pole terms} + q_\mu N_{\mu\nu} \qquad (7)$$

where $N_{\mu\nu}$ is what is left after the pole terms are subtracted and well behaved as $q_\mu \to o$. By "pole terms" is meant the Born terms shown in Fig. 4. The point is that these have denominators such as $(p_1+q)^2 - m_n^2 = 2p_1 \cdot q + q^2$ which go to zero as $q \to 0$. Thus, they must be treated explicitly. These pole terms give the m_π/m_n correction.

Now, it turns out that the Kroll-Ruderman term (6) plus pole terms give a gauge invariant expression, so that the term $q_\mu N_{\mu\nu}$ must be gauge invariant by itself. We can express

$$q_\mu N_{\mu\nu} = q_\mu N_{\mu\nu}(q^2 = 0, q_\mu = 0, k_\mu = 0)$$

$$+ (\frac{q^2}{m_n}, \frac{q \cdot k}{m_\pi^2}) \ . \qquad (8)$$

Gauge invariance requires

$$q_\mu N_{\mu\nu} k_\nu = 0 \ , \qquad (8.1)$$

implying

$$N_{\mu\nu}(q^2 = 0, q_\mu = 0, k_\mu = 0) = 0 . \tag{8.2}$$

Consequently, gauge invariance tells us that, one the pole terms are evaluated, we have all of the amplitude up to terms of order q^2/m_n^2, $q\,k/m_n^2$.

The result of this calculation for $\pi^\pm$ photoproduction is:

$$\frac{\omega_k}{q_\pi} \frac{d\sigma}{d\Omega} \begin{matrix} \gamma + n \to p + \pi^- \\ \gamma + p \to n + \pi^+ \end{matrix}$$

$$= 2 \left(\frac{f}{m}\right)^2 \frac{1}{(1 + \frac{m_\pi}{m_n})^2} (1 \pm \frac{m_\pi}{2m_n})^2 . \tag{9}$$

Comparison of experimental results[3] and theoretical predictions is shown in Table I.

The preceding discussion applied only to charged pion photoproduction, where we have been strongly guided by gauge invariance [see esp. Eq. (2)]. Since to leading orders the processes do not contribute to π^0 production, this latter process is a much more delicate question, and effects which are small in charged pion photoproduction become important here. An example of such additional processes is shown in Fig. 5. In complex nuclei sizeable contributions to π^0 production arise from charged pion production, the amplitude for which is relatively large, and the subsequent charge exchange of the produced meson, with π^0 emerging. We will return in ¶ III to a discussion of π^0 production in the higher-energy domain, where it becomes an allowed process.

Table I. Reduced threshold differential cross sections (k/q) $d\sigma/d\Omega$ in µb/ster.

CHANNEL	EXPERIMENT	KROLL-RUDERMAN	SOFT PION Eq. (9)
$n\pi^+$	16.33 ± .16	23.4	15.2
$p\pi^-$	19.8 ± 1.3	23.4	20.5
$p\pi^0$	.065	0	
$n\pi^0$		0	

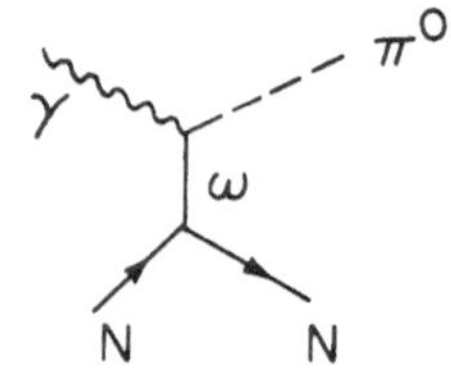

Figure 5. Example of a process which is important
in π^0 production.

Extensive and accurate photopion production experiments
near threshold have been carried out at Saclay on the two-and
three-body systems and are being extended to the four-body
system. I shall not go into detail, since these experiments
are discussed later in the proceedings.

The conclusion of these experiments is that π^+ photoproduction at threshold on very light nuclei seems to be correctly
described by impulse approximation, computed with accurate
conventional wave functions generated by an NN potential or by
a shell model. This is in contradistinction to the situation
with the magnetic dipole operator, which has significant corrections from exchange currents.

Summarizing the Saclay results on He^3, Tzara[5] says "Our
main and important conclusion is that the π^+ photoproduction
at threshold is insensitive to extranucleonic degrees of freedom in ^3He. In other words, this process is a faithful tool
for the exploration of the one body axial form factor, whereas
the magnetic spin dependent form factor reveals the presence
of exchange currents.[11]

In order to try to understand this situation, let us look
at the triton beta decay, which involves the axial vector
operator at essentially zero momentum transfer. Empirically,
the required matrix element is[6]

$$M_{GT} = <^3He|| \sum_{i=1}^{3} \sigma(i)\tau^+(i)||^3H> = - \sqrt{6} (0.973 \pm 0.010) . \quad (10)$$

Suppose we calculate this matrix element in impulse approximation (only one-body operators) using a simplified description
of H^3 as three nucleons in 1s shell-model states (no D-state

admixture). The result is

$$M_{GT} = - \sqrt{6} . \qquad (10.1)$$

From this one would say that there is near agreement. Inclusion of some mixed-symmetry s-state contribution would bring theory and experiment even closer. For a triton wave function

$$\psi = a\psi_s + c\psi_{s'} , \qquad (10.2)$$

where s' indicates mixed-symmetry component, the above M_{GT} is multiplied by $(a^2 - \frac{1}{3} c^2)$; c^2 is found to be small, .01 to .02 in detailed triton calculations. Yet the fact that the deuteron has a quadrupole moment tells us that the nucleon-nucleon interaction has a tensor force component.

The triton is known to have a D-state component of size roughly equal to that of the deuteron. If the D-state amplitude is b, then

$$M_{GT} = - \sqrt{6} [a^2 + \frac{1}{3} b^2 - \frac{1}{3} c^2] \qquad (10.3)$$

and we see that inclusion of the D-state throws agreement between theory and experiment a bit off. For $b^2 = .06$, $c^2 = .015$, we have, e.g., $M_{GT} = 0.94$. But we would not have noticed had we not included the D-state in the triton wave function.

Agreement with experiment is helped slightly by including the isobar component of the triton ψ_Δ.[7] For

$$\psi = a\psi_s + b\psi_D + c\psi_{s'} + d\psi_\Delta \qquad (11)$$

we find

$$M_{GT} = - \sqrt{6} [a^2 + \frac{1}{3} b^2 - \frac{1}{3} c^2 + \frac{8}{15} \sqrt{2} bd + \frac{1}{3} d^2]$$

with, of course, the normalization $a^2 + b^2 + c^2 + d^2 = 1$. For $b^2 = .06$ $c^2 = .015$, the value of M_{GT} remains unchanged at $-.95 \sqrt{6}$ as d^2 varies from .04 to .08. Thus, substantial isobar components can be present without introducing large changes in the β-decay, and, in fact the small changes that are introduced go in such a way as to improve agreement between theory and experiment.

Since relatively high momentum components are introduced by isobar admixtures, one would expect the same picture to apply to photopion production near threshold, where only low momentum would be required.

Thus we would expect it to be true that the axial-vector operator is relatively insensitive to extra-nucleonic components, this situation resulting, however, through an intricate interplay of tensor force and isobar admixture.

Going to heavier nuclei, I wish to single out the recent theoretical work by Epstein, Singham and Tabakin[8] on the $^{12}C(\gamma,\pi^-)^{12}N_{g.s.}$ reaction near threshold as an example of the well thought out and careful analyses which we need. Three principle ingredients enter the theoretical analysis:

1) the pion nucleus optical potential $V_{\pi N}$

2) the nuclear wave function

3) the elementary interaction operator $H_{\gamma\pi}$ which describes absorption of the photon and emission of the pion.

All of these ingredients appear to be reliably determined from independent analyses:

1) In the threshold region $V_{\pi N}$ can be obtained from atom data.

2) The ^{12}C and $^{12}N_{g.s.}$ wave functions are constrained by electron scattering data.

3) $H_{\gamma\pi}$ is fixed by data from photoproduction of pions from nucleons.

A number of improvements, some involving correction of errors in previous calculations are introduced, and with each improvement in the data, the theoretical results move closer to experiment, as shown in Fig. 6.

III. Coherent π_o - Production

Coherent π_o-production on the deuteron has been studied both experimentally and theoretically for decades. Of more

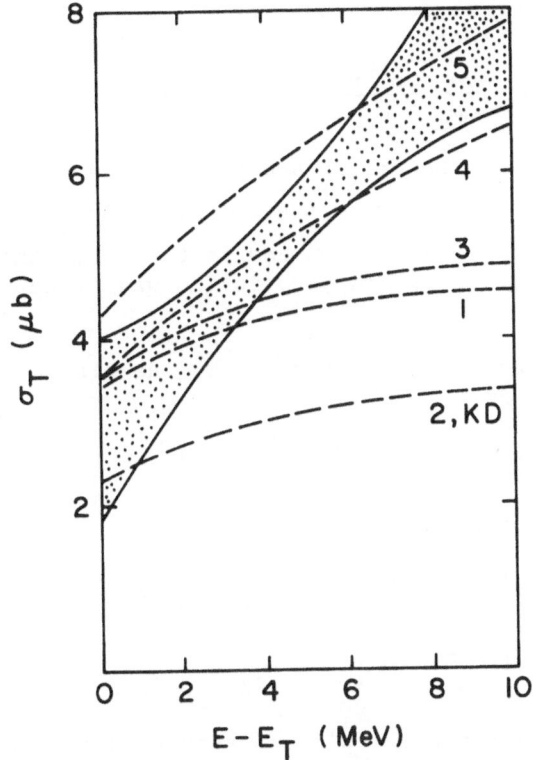

Figure 6.   Comparison of the theoretical results of Epstein,
Singham and Tabakin with the experimental results of Bernstein
et al.[9].   Curves 1 and 3 represent effects of correction of
errors in previous calculations (Curve 2), all within the context of S-wave pions.   Curve 4 is reached by including all
pion partial waves.   Curve 5 comes from increasing $H_{\gamma\pi}$ so as
to reproduce the threshold photopion production cross section
off free electrons.

immediate interest to us is a series of papers by Schrack, et
al.[10] showing how this process could be used to measure nuclear
charge distributions.

One can formulate the theory of this process within impulse approximation, but for the sake of connection with pion-
nucleus scattering, it may be useful to look at how it comes
out from isobar models.

The isovector magnetic coupling of γ-rays to nucleons and isobars can be written

$$\delta L_{N\Delta\gamma} = i \frac{e}{2m_n} \frac{\mu_p - \mu_n}{2} \bar{\Delta} T_3 \vec{S} \cdot [\vec{k}_\gamma \times \hat{\varepsilon}_\gamma] \psi(r) \qquad (12)$$

where μ_p and μ_n are the proton and neutron magnetic moments, respectively, and S and T are the spin and isospin transition operators,[11] connecting nucleon and isobar; Δ is the wave function of the (3,3) isobar and $\psi(r)$ is the wave function of the nucleon.

Within the same type of formalism, the pion coupling can be written

$$\delta L_{N\Delta\pi} = \frac{f^*}{m_\pi} \bar{\Delta} \vec{T} \vec{S} \cdot \vec{\nabla} \phi_\pi \psi(r) \; . \qquad (12.1)$$

The transition spins obey the relations[10]

$$\sum_{M_T} T_i | \Delta M_T \rangle \langle \Delta M_T | T_j = \delta_{ij} - \frac{1}{3} \tau_i \tau_j \qquad (12.2)$$

$$\sum_{M_S} S_i | \Delta M_S \rangle \langle \Delta M_S | S_j = \delta_{ij} - \frac{1}{3} \sigma_i \sigma_j \qquad (12.3)$$

where M_T and M_S are the isospin and spin projections of the isobars. These relations allow us to eliminate intermediate isobars.

We wish to consider the processes shown in Fig. 7.

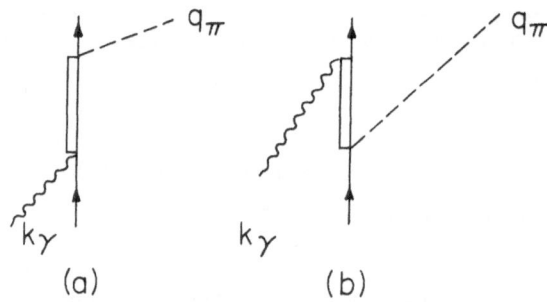

(a)   (b)

Figure 7.   Photoproduction proceeding through intermediate isobars.

For the moment, let us look at the process, Fig. 7a, involving the intermediate isobar. The matrix element is

$$M_{\gamma\pi_0} = \frac{ief^*}{2m_n m_\pi} \frac{\mu_p - \mu_n}{2} \sum_{M_T, M_S} <0|\vec{S}_i \cdot \vec{\nabla}\phi_\pi(\vec{x}_i)T_{3i}|\Delta M_S M_T>$$

$$\times <\Delta M_S M_T|T_{3i}\vec{S}_i \cdot [\vec{k}_\gamma \times \varepsilon_\gamma]|0>$$

$$M_{\gamma\pi_0} = \frac{ief^*}{2m_n m_\pi} \frac{\mu_p - \mu_n}{2} \sum_i \sum_{M_T, M_S} <0|\vec{S}_i \cdot \vec{\nabla}\phi_\pi^*(\vec{x}_i)T_{3i}|\Delta M_S M_T>$$

$$\times \frac{<\Delta M_S M_t|\vec{T}_{3i}\vec{S}_i \cdot [\vec{k}_\gamma \times \hat{\varepsilon}_\gamma] e^{i\vec{k}_\gamma \cdot \vec{x}_i}|0>}{\hbar\omega_k - \hbar\omega_\Delta} \qquad (12.5)$$

where the sum $\sum_i$ is carried out over all nucleons in the nucleus, $|0>$ represents the many-body ground state, and $\hbar\omega_\Delta$ is the (complex) energy of the Δ-isobar. We assume $0>$ to represent the wave function of a spin-zero nucleus, so that we can drop terms linear in the nucleon spin. This leaves us with the following expression which is essentially the same as that obtained by R. M. Woloshyn[12] in the isobar-doorway model, after carrying out the sums over M_S and M_T:

$$M_{\gamma\pi 0} = \frac{ief^*}{2m_n m_\pi} \frac{\mu_p - \mu_n}{2} \frac{4}{9} \sum_i \frac{<0|\vec{\nabla}\phi_\pi^*(\vec{x}_i) \cdot [\vec{k}_\gamma \times \vec{\varepsilon}_\gamma] e^{i\vec{k}_\gamma \cdot \vec{r}_i}|0>}{\hbar\omega_k - \hbar\omega_R} \qquad (13)$$

The strategy of Schrack[10] was to work only slightly above pion production threshold, where the pions would have low energies, so that the nuclear distortion of the pion wave function could be neglected. In this case,

$$\phi_\pi^*(x_i) = e^{-iq_\pi \cdot x_i} \qquad (13.1)$$

and his experiment measured the nuclear form factor at momentum $k_\gamma - q_\pi$. For higher pion energies, where p-wave pion nucleon interactions become important, the distortion of the pion wave function cannot be neglected.

In fact, the pion wave function inside the nucleus is a very interesting quantity, and present controversies relating to the role of short-range correlations, etc., bear directly on this quantity. George Bertsch[13] suggests turning matters around, and using the experiment to determine the distorted pion wave function $\vec{\nabla}\phi_\pi$ <u>inside</u> the <u>nucleus</u>. By varying $\vec{k}_\gamma$, one can determine the product

$$F = \rho\nabla\phi_\pi \qquad (14)$$

where ρ is the nuclear density, as a function of k_γ.

Now, ρ is very accurately determined from elastic electron scattering, so one would be able to extract $\vec{\nabla}\phi_\pi$, as a function of momentum. This affords a very exciting possibility for determining the distorted pion wave function.

An analysis of the $^{12}C(\gamma,\pi^0)^{12}C$ reaction has been carried out by Woloshyn[12] with the results shown in Fig. 8.

Woloshyn finds substantial differences between predictions from the isobar-doorway model and DWIA, both giving smaller cross sections than experiment. For these results, the Sopkovich[15] approximate treatment of distortion of the pion

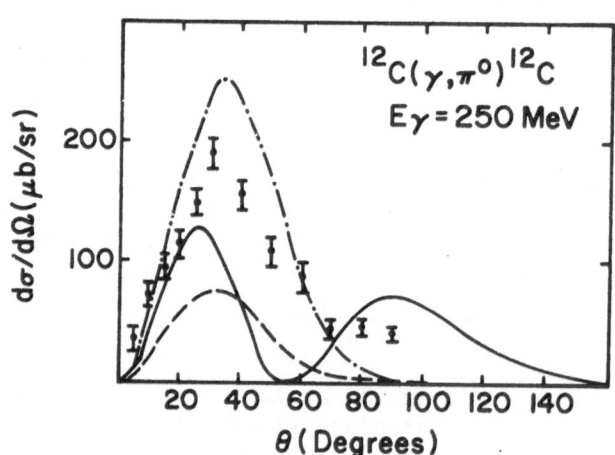

Fig. 8.   Angular distribution for $^{12}C(\gamma,\pi^0)^{12}C$ at photon energy 250 MeV in the isobar doorway (solid curve), plane-wave impulse approximation (dot-dashed) and DWIA (dashed) models. The data are taken from Saunders[14].

wave function was used. Later results[16] using the exact
optical potential wave functions for the DWIA gave results
even smaller than those shown.

Note that the data shows no evidence of a dip around $50°$.
In the isobar-doorway model, this dip reflects the first dif-
fractive minimum of the pion-nucleus elastic scattering differential cross section and is very difficult to avoid, since
the total amplitude (all partial waves) is parametrized by a
single form factor.

The isobar-doorway model shows some tendency to peak
around $90°$, which is understandable from Eq. (13) if
ϕ_π^* is approximated by a plane wave, Eq. (13.1). The factor
$\vec{q} \cdot [\vec{k}_\gamma \times \vec{\varepsilon}_\gamma]$ enters into the amplitude. The fact that the DWIA
does not show this peak must mean that the absorption acting
on the pion was terribly strong.

At lower energies the lowest-order pion-nucleus optical
model potential gives much too strong absorption; introduction
of short-range correlations between nucleons coming from ω-
and ρ-meson exchange effectively cut down the multiple scattering greatly[17]. I believe that Woloshyn's results are telling
us that with the standard optical-model potential, the absorption is so strong that the pion simply doesn't get through the
nucleus sufficiently to make a large-angle scattering in the
DWIA calculation.

Putting in short-range correlations will bring the isobar-
doorway and optical-model approaches closer together, as I
shall now discuss. This is important, if we want to concentrate on getting out the distorted pion wave function in the
nucleus; the result should not be model-dependent.

There is a large overlap between photonuclear and photopion research workers, so the isobar-hole model should be highly palatable to them, because of its similarity to the giant
dipole resonance. Unperturbed isobar-hole excitations are
mixed through the pion exchange interaction, which is complex,
describing the fact that the isobar can drop into the nucleon
hole, emitting an on-shell pion. The mixing produces, in
first approximation[18] a highly collective state carrying most
of the strength:

Fig. 9.   Isobar-particle, nucleon-hole configuration mixing.

$$|J> = N\Sigma C_i |(\Delta_i \phi_i^{-1})^J> \qquad (15)$$

where

$$C_i \simeq <(\Delta_i \phi_i^{-1})^J|\delta L_{N\Delta\pi}|0> \qquad (15.1)$$

with $\delta L_{N\Delta\pi}$ given by Eq. (12.1) and N is a normalization constant. This state is analogous to the giant dipole state, except that the particle state has been replaced by the Δ, and
$\delta L_{N\Delta\pi}$ replaces the dipole interaction.

Yet there are major differences between the isobar-hole
and dipole state, in addition to the great difference in energies. An isobar Δ has a width of $\sim$ 110 MeV; thus its lifetime
is

$$\tau = \frac{\hbar}{110 \text{ MeV}} = \frac{140}{110} \frac{(\hbar/mc)}{c} \qquad (16)$$

written in a way so as to show that a particle moving with the
speed of light can travel only a distance $\sim$ h/mc = 1.4 x 10^{-13}cm
during its lifetime. How far can an isobar travel? Typical
isobar velocities are $\sim$ 1/3 c. Configuration mixing adds a
width of the order of the unperturbed width for the collective
state. Thus, a typical isobar would travel a distance

$$\ell \sim \frac{1}{6} \frac{140}{110} \times (\hbar/mc) = 0.3 \text{ fm} \qquad (16.1)$$

during the lifetime of the coherent state.

Two questions arise:

(1) How can coherence be developed when each individual
    isobar is highly localized?

(2) How can the high nonlocality, of the type discussed
    early on by Kisslinger and Wang[19] develop?

My personal feeling is that:

(1) The coherence does not develop.

(2) The nonlocality is not there for reasons which I now develop.

First of all, introduction of short-range correlations[20] and, especially, of ρ-meson exchange cuts down considerably the interaction between isobar-hole states, so the collective state for each j is not displaced so far from its imperturbed position. The short-range correlations and ρ-meson exchange do not, however, affect the imaginary part of the interaction, so the widths of states with low j-values will still be large. Since these widths come in the denominator, they will cut down the role of the low j states. This is simply an expression that pions do not like to go through the center of the nucleus.

Once the strength for the relevant pionic process is not displaced far from its unperturbed position (more precisely, less than $\Gamma/2 \sim 55$ MeV, the isobar half width), we can put the different j values together again in the following way. The resonant part of the photoproduction amplitude can be expressed as

$$M_{res} = \frac{2M}{\hbar^2} \sum_j \frac{<0k'| \sum_i \delta L_{N\pi\Delta}(x'_i)|j><j| \sum_\ell \delta L_{N\Delta\gamma}(x_\ell)|0k>}{\hbar\omega_k - \hbar\tilde{\omega}_j} \qquad (17)$$

where j labels the collective state with angular momentum j, $\hbar\omega_h$ is the (highly) complex energy of that state, the sum over i and j is over particle label, k represents the wave function of the incoming pion and k' that of the outgoing one[*], and $|0>$ represents the ground state of the initial nucleus. Now

---

[*] Parenthetically, I would remark that the states $\vec{\phi_k}$ and $\vec{\phi_{k'}}$ should be states distorted in the optical-model potential which reproduces the average scattering amplitude. See Ch. IX of G. E. Brown, Unified Theory of Nuclear Models and Forces, North-Holland Publ. Co., Amsterdam, 1971. This bears on whether one should use $\nabla\phi_\pi^*$ or $-iq_\pi$ in Eq. (13).

$$\hbar\omega_j = E_j - i\,\frac{\Gamma_j}{2}\,. \tag{17.1}$$

To the extent that for the states $|j>$ important for the scattering, the E_j's are close to their unperturbed position,

$$(E_j - E_j^{(0)}) < \frac{\Gamma_j}{2} \tag{17.2}$$

we can use closure, obtaining*

$$M_{res} = Const \sum_{i,\ell} \frac{<0k'|\delta L_{N\pi\Delta}(x_i)\delta L_{N\Delta\pi}(x_\ell)|0k>}{\hbar\omega_k - \hbar\omega_j} \tag{18}$$

and the effective operator becomes completely local. Terms $i = \ell$ enter in lowest order in an expansion like that used to derive the optical-model potential.

Theorists will shout that I've done nothing but derive, in a circuitous fashion, the distorted wave impulse approximation. This is correct. But I've also shown that it's plausible that the isobar doorway model, using properly distorted pion wave functions, should lead to the same results.

After this involved theoretical discussion of isobars, let me bring matters down to earth by briefly discussing Laget's proposals for finding out about isobar-nucleon interactions experimentally. If a photopion process involves virtual Δ's close to threshold (zero kinetic energy) in a two- or few-body system, then the process will depend crucially on the Δ interaction with the other nucleons. In Fig. 10 is shown Laget's curve[21] for $\gamma D \rightarrow pp\pi^-$, with curves for two different values ($a = -1$ and $a = -1.2$) for the assumed scattering length of the Δ-nucleon system. Note that the present experimental data seems to demand an attractive interaction. Laget has many other studies in the few-body system, but I believe this work to most beautifully illustrate the possibilities of well chosen experiments.

---

* This expression is somewhat symbolic, because summations over the spins and isospins of the intermediate isobars produce numerical factors, as in Eqs. (12.2) and (12.3), which have been suppressed.

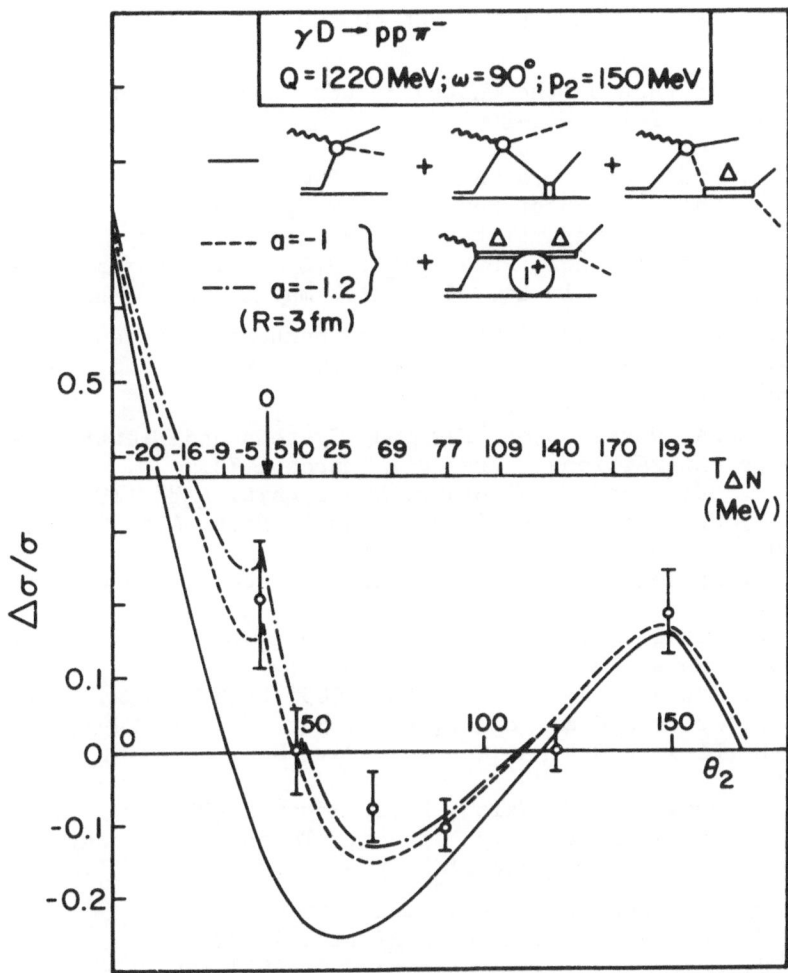

Fig. 10.   Laget's Curve.

Acknowledgments

I would like to thank George Bertsch for pointing out the possibilities of coherent π^0 production, Wick Haxton and Mannque Rho for instruction on low-energy theorems. I enjoyed much interaction with the Saclay linear accelerator group, and wish especially to thank C. Tzara and J. M. Laget for useful discussions.

REFERENCES AND NOTES

* Supported in part by the United States Department of Energy, Contract No. EY-76-S-02-3001.

1. N. M. Kroll and M. A. Ruderman, Phys. Rev. $\underline{93}$ (1954) 233.
2. G. W. Gaffney, Phys. Rev. $\underline{161}$ (1967) 1599.
3. N. de Botton, report, this conference.
4. P. De Baenst, Nucl. Phys. $\underline{B24}$ (1970) 633, keeping terms to order (m_π/m_n) in Eq. (9), which may be more consistent, obtained 13.0 and 19.9 for these numbers. It can be seen from this that the deviations between theory and experiment are of order (m_π/m_n).
5. C. Tzara, Workshop on "Few Body Systems and Electromagnetic Interactions," Frascati, March 1978.
6. R. C. Salgo and H. H. Staub, Nucl. Phys. $\underline{A138}$ (1969) 417.
7. A. M. Green and T. H. Schucan, Nucl. Phys. $\underline{A188}$ (1972) 289.
8. G. N. Epstein, M. K. Singham and F. Tabakin (1978) preprint.
9. A. Bernstein, N. Paras, W. Turchinetz, B. Chasan and E.C. Booth, Phys. Rev. Lett. $\underline{37}$ (1976) 819.
10. R. A. Schrack, Phys. Rev. $\underline{140}$ (1965) B 897; R. A. Schrack, J. E. Liess and S. Penner, Phys. Rev. $\underline{127}$ (1962) 1772; J. E. Liess and R. A. Schrack, Rev. Mod. Phys. $\underline{30}$ (1958) 456.
11. G. E. Brown and W. Weise, Phys. Reports $\underline{22C}$ (1975) 279.
12. R. M. Woloshyn, to be published. To obtain Woloshyn's expression, we replace $\nabla\phi_\pi^*$ by $-iq_\pi$.
13. George Bertsch, private communication.
14. L. M. Saunders, Nucl. Phys. $\underline{B7}$ (1968) 293.
15. N. J. Sopkovich, Nuovo Cimento $\underline{26}$ (1962) 186.
16. R. M. Woloshyn, private communication
17. G. E. Brown, B. K. Jennings and V. I. Rostokin, Phys. Reports, to be published.
18. M. Hirata, J. H. Koch, F. Lenz, E. J. Moniz, Phys. Letts. $\underline{70B}$ (1977) 281.
19. L. Kisslinger and W. Wang, Phys. Rev. Lett. $\underline{30}$ (1973) 1071.
20. E. Oset and W. Weise (1978) preprint.
21. J. M. Laget, private communication.

PHOTOPION PHYSICS - AN EXPERIMENTAL OVERVIEW

E. J. Winhold

Physics Department

Rensselaer Polytechnic Institute

Troy, New York   12181

1.   INTRODUCTION

This review attempts to survey recent and current photo-pion experiments giving some attention also to future directions over the next couple of years. In doing this we need to keep in mind the experimental limitations on the measurements being made, as well as the physics motivation for the experiments. Our task will be to provide an overview of the field without intruding too much on the more detailed reviews of the various subfields which will follow.

Figure 1 gives a schematic view of experimental activity in the field. In the lower energy region at photon energies within, say, 50 MeV of the photopion threshold, much of the activity involves the study of transitions between well-defined specific nuclear states. Several general comments can be made on work in this region:

1. The study of gamma ray spectra associated with radiative π^- capture by nuclei has been carried on now for nearly a decade, first at Berkeley, and more recently at SIN and LAMPF. The initial results from this work were really the first to demonstrate the potential of photopion studies for nuclear spectroscopy, as well as their selectivity in exciting M1 and spin-flip giant dipole analogs; more recently the 1p shell nuclei have been rather thoroughly studied with a resolution

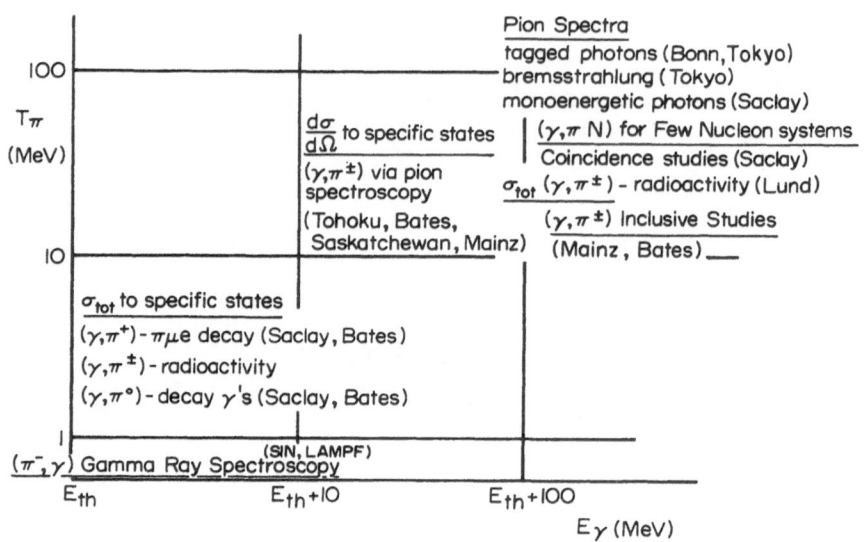

Figure 1. Schematic view of current experimental work.

of 0.9 MeV and some attention has been paid to the 2s-1d shell nuclei.

2.   Turning to total photopion cross section measurements, there now exists a fair body of precise work in light nuclei within several MeV of threshold. This is a special region in that the elementary photopion production amplitude for charged pions is dominated by the momentum-independent term proportional to $\underline{\sigma} \cdot \underline{e}$ (where $\underline{\sigma}$ is the nucleon spin and $\underline{e}$ is the photon polarization), s-wave pions predominate, and the effects of the pion-nucleus final-state interactions are relatively small, at least for (γ, π^+) reactions on the lightest nuclei.

3.   A significant development in the past two years has been the emergence of magnetic spectrometer studies of photopion spectra as a viable high resolution spectroscopic tool alongside radiative pion capture. These studies involve double differential cross section measurements with respect to angle and pion energy in the general pion energy range between 10 and 50 MeV.

The development of these latter two programs has, of course, depended for its feasibility on the good beam properties (intensity, duty cycle, and beam energy spread) of current-generation electron linacs.

There continues to be less overall activity in the (3,3) resonance region than at lower energies. However, there are some new developments here too. The Saclay program of ($\gamma,\pi N$) coincidence studies on few-nucleon systems continues to produce detailed and incisive results. In addition, an important new development in this energy region is the use of monoenergetic photon beams, either via the tagging of bremsstrahlung photons, as at Bonn, or via positron annihilations, as at Saclay.

One final general remark can be made. The results for all the experiments listed in Fig. 1 will reflect the combined effect of four factors: (a) the elementary amplitude for $\gamma N \rightarrow \pi N$, (b) the pion production mechanism, (c) the pion wave function, (d) the nuclear wave functions. Then we can hope that if (a) and (b) are well understood, the experiments can be focussed, by a suitable choice of experimental parameters, to yield information on either nuclear structure questions or on the pion-nucleus interaction.

2.   TOTAL CROSS SECTIONS NEAR THRESHOLD

The region near threshold in light nuclei, where factors (c) and (d) are relatively well understood, is a good initial testing ground for our understanding of photoproduction in nuclei. In all but a few cases the total (γ,π) cross section measurements are limited to an energy region within a few MeV of threshold in order that individual final states are selected. The measurement of (γ,π^+) cross sections by detection of the pion decay chain positrons was pioneered by the Saclay-Louvain collaboration[1]. Their experimental arrangement is shown in Fig. 2; they require a triple coincidence in either set of counters during a gate following each beam burst. This class of measurement has been most important because it provided the first demonstration that precise photopion measurements giving cross sections to a few percent were possible, and it encouraged theoretical interest in making detailed calculations.

We will first briefly review the benchmark case of ^{6}Li(γ,π^+)^{6}He(g.s.). This case received a lot of publicity

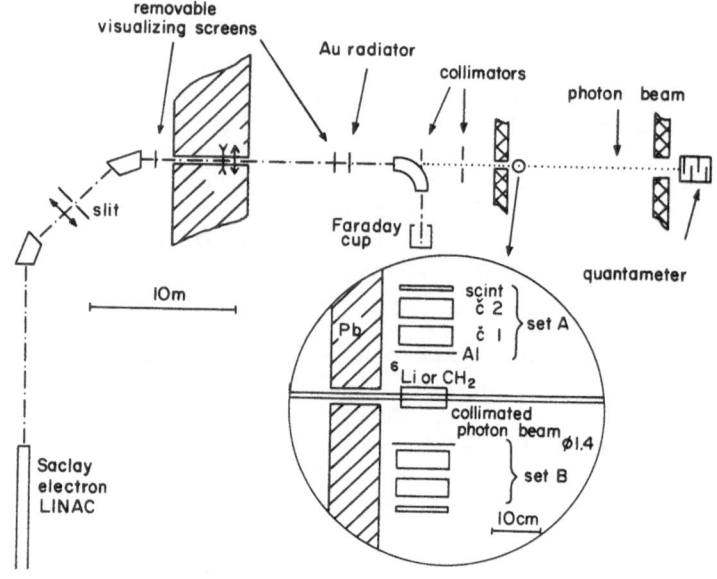

Figure 2. Experimental arrangement for (γ, π^+) cross section measurements at Saclay.

in recent years because of the vicissitudes of both the experimental and theoretical results before they finally settled down. As seen in Fig. 3, the ^{6}He ground state is the analog of the giant M1 state in ^{6}Li at 3.56 MeV, and one expects strong excitation in (γ, π) of this class of spin-flip transition via the spin-dependent terms in the photopion operator. The ground state (γ, π^+) transition can be studied over a range of almost 2 MeV before the onset of transitions to the first excited state. The final experimental result from Saclay[1], measured relative to hydrogen, appeared a year ago and was quoted with an error of $\pm$ 4%. This is an especially favorable case because the nuclear structure factors are well known -- the analog M1 transition in ^{6}Li has been well studied in (e, e') -- and pion-nucleus interaction effects are small. There are several theoretical calculations which calculate ^{6}Li(γ, π^+) in impulse approximation obtaining their nuclear structure information from the experimental data on electron scattering and

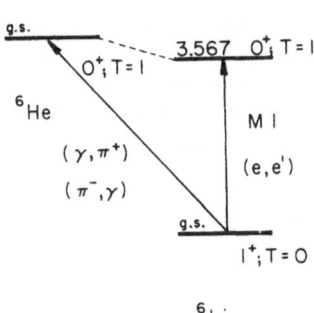

Figure 3

weak interaction data[2,3]. They differ among each other to
within 10% or less, with the differences primarily due to nuclear structure uncertainties. They agree with experiment to
about 10%.

In addition to (γ,π^{+}), the (π^{-},γ) transition between
these same two states has been studied more recently at SIN by
the Lausanne-Munich-Zurich collaboration[4] with 0.9 MeV resolution. The ground state peak in their gamma ray spectrum is
well resolved. From the gamma ray counting rate N_{γ}, the
branching ratio R_{γ} for capture per stopped pion can be obtained
if the detector acceptance η_{γ} (efficiency x solid angle) is
known. This depends on the transition rates λ_{γ} and λ_{a} for
radiative capture and absorption as well as on the correspon-
ding capture probabilities $\omega(n\ell)$:

$$N = N_{\pi}\,\eta_{\gamma}\,R_{\gamma} = N_{\pi}\,\eta_{\gamma}\,[\,\frac{\lambda_{\gamma}(1s)}{\lambda_{a}(1s)}\,\omega(1s) + \frac{\lambda_{\gamma}(2p)}{\lambda_{a}(2p)}\,\omega(2p) + \ldots]$$

A theoretical calculation of λ_{γ} depends on the same nuclear
matrix elements as for $\sigma(\gamma,\pi)$. However, a comparison of theory
and experiment also depends on a knowledge of the λ_{a}'s and
ω's from pionic atom data, and for the ^{6}Li the uncertainty

in the ω's gives rise to $\simeq 20\%$ uncertainty in R_γ. To avoid
this uncertainty, the ^{6}Li gamma spectrum corresponding to
capture from the 1s orbit has been meausred at SIN^5 in a
coincidence experiment where the 2p→1s x-rays are detected
in thin NaI counters. In this case the ratio of coincidence
counts $N_{\gamma x}$ to x-ray singles counts N_x is independent of the
ω's and of the x-ray counter efficiency:

$$\frac{N_{\gamma x}}{N_x} = \frac{\eta_\gamma \lambda_\gamma(1s)}{\lambda_a(1s)}$$

The price paid in the coincidence experiment is a reduction
in counting rate of two orders of magnitude compared to the
singles experiment, so that counting statistics become an
important factor. The uncertainty in the experimental determination of $\lambda_\gamma(1s)$ and the consistency of the experimental
value with theory are both not far from the 10% level. At
any rate ^{6}Li is a special case in that we have precise measurements of both (γ,π^+) and (π^-,γ) combined with quantitative
theoretical understanding, with agreement at the 10% level.

We will next consider briefly a second case, namely
(γ,π^+) on the deuteron, which has been studied at both Bates[6]
and Saclay[7]. In Fig. 4 the datapoints are from Saclay. The
Bates data extend to higher energies than the Saclay data but
both experiments agree in their overlap region within their
errors, which are at the few percent level. This case is of
great interest because we have a simple system with a known
wave function where the rescattering effects in the final state
are calculable. The solid and dashed lines in Fig. 4 represent two independent impulse approximation calculations[8,9].
Their good agreement with the data implies that meson exchange
current effects in this process are near the few percent level
or lower. The calculations ignore pion-nucleon rescattering;
neutron-neutron rescattering effects, on the other hand, are
very important. Nevertheless this total cross section measurement, which does not look at the energy breakup in the π^+ nn
state, is not a sensitive way to determine a_{nn}, the n-n
scattering length. This specific case is also discussed by
E.C. Booth in these Proceedings. In contrast, ^{2}H(π^-,γ)nn

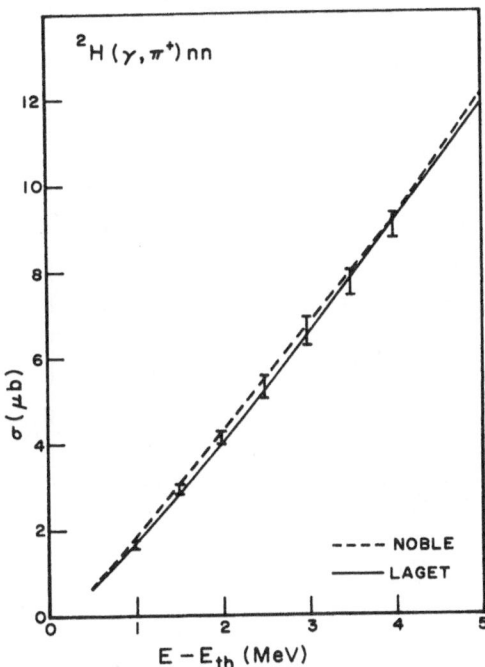

Figure 4. The experimental points are from Ref. 7; the solid and dashed curves are the results of two independent calculations (Refs. 8 and 9).

the gamma spectrum shape is sensitive to the n-n interaction, especially at its high energy end, which corresponds to the case of a correlated pair of neutrons recoiling in the opposite direction to the photon. The later paper by Dr. Perroud reports in more detail results from SIN on this experiment.

In these (γ, π^+) cross sections on very light nuclei near threshold, pion rescattering effects appear to be quite small. As pointed out by Koch and Woloshyn[10], this is not necessarily the case for low energy (γ, π°) cross sections because the elementary production amplitude for π°'s is an order of magnitude smaller than for charged π's. The case of (γ, π°) on the deuteron measured at Saclay[11] (Fig. 5) is particularly clean. The calculations[12,13] show the importance of the process where a charged pion is produced on one nucleon and is

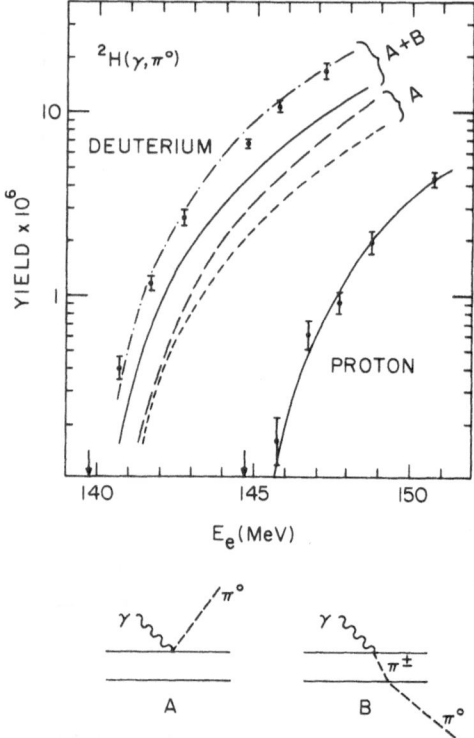

Figure 5. The experimental points are from Saclay (Ref. 11).
The lines are from two independent calculations (Refs. 12
and 13). The curves A include direct π^0 production only,
while curves A + B also include charge-exchange rescattering.

charge-exchange scattered off the other. The two curves labelled A are the result of two independent calculations which
include direct (γ, π^0) production only, while the curves labelled A + B result from inclusion of charge-exchange rescattering in these calculations. These two calculations differ
primarily in their treatment of the deuteron wave function.

     Turning to (γ, π^-) cross section measurements by detecting
the radioactivity of the residual nucleus, we find substan-

tially more serious background problems than in the (γ, π^+) $\pi\mu e$ studies. There is one case of special interest: $^{12}C(\gamma, \pi^-)\ ^{12}N(g.s.)\ (\beta^+, T\frac{1}{2} = 11\ ms.)$. Here ^{12}N has no particle-bound excited states, which allows the measurement in principle to be carried well up in energy. Moreover the ^{12}N ground state is the analog of the well-studied 15 11 MeV 1^+ giant M1 state in ^{12}C. The Bates data[14], (Fig. 6) covers the energy range up to 11 MeV above threshold. The backgrounds, which are due to (γ, p) followed by (p, n) reactions in the target, are rather large as indicated by the large sub-threshold yields and have to be estimated in the region of interest by extrapolating the sub-threshold data. The cross section values extracted from these yield data are shown in Fig. 7 and Fig. 6 of the preceding paper by G.E. Brown, where the shaded area shows the uncertainty in the cross sections. The curves represent several different theoretical calculations. Epstein et al.[15] and Haxton[16] differ in their treatment of the nuclear wave functions. The Helm model calculation of Nagl and Überall[17] is not shown but is also consistent with the data. The two Haxton curves use two different sets of pion optical model parameters, both derived from pionic atom data. The dashed curve is obtained by Haxton if only s-wave pions are kept. This has the wrong slope and misses the data badly, so one sees that the higher partial waves are important. These data are not of the quality of some of the (γ, π^+) results. Even so, they are clearly useful. With a factor of two or so improvement, they would begin to select between different treatments of the nuclear structure factors or the pion-nucleus interaction. At higher energies we can expect a greater sensitivity to momentum-dependent terms in the production amplitude and to pion-nucleus interactions. This is seen, for example, in the calculations of Nagl and Überall[17] and Epstein et al.[18], presented at this symposium. The region up to 25 MeV pion energy or so is not easily covered in pion elastic scattering experiments, and photopion production may well provide important information on the pion-nucleus interaction at these energies.

3. PION SPECTROSCOPY

Charged pion spectroscopy has emerged in the past two years as a promising tool for studies of $(\gamma, \pi^\pm)$ reactions to

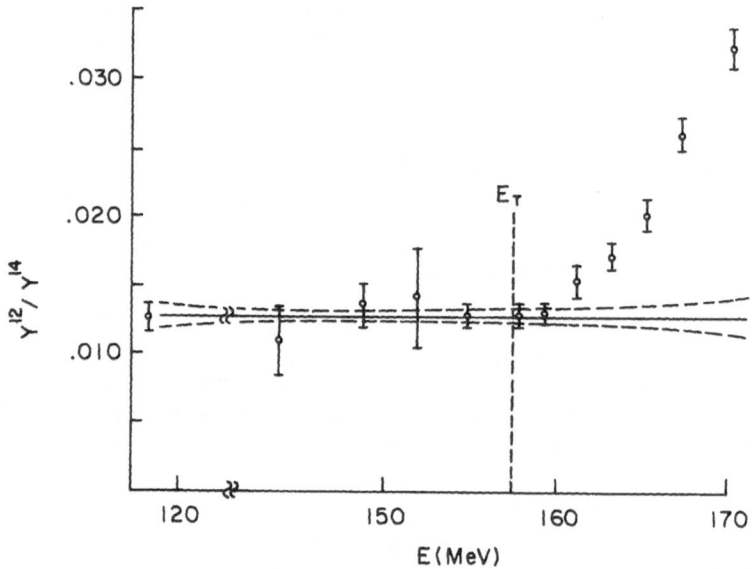

Figure 6. The ratio of the yield for $^{12}C(\gamma,\pi^-)$ $^{12}N(g.s.)$ to that for $^{14}N(\gamma,2n)$ $^{12}N(g.s.)$, as a function of electron energy (Ref. 14). E_T is the expected threshold for $^{12}C(\gamma,\pi^-)$ and the yield below this energy is due to two-step reaction background.

specific final nuclear states. Magnetic spectrometers have been set up at Tohoku, Bates, and more recently at Saskatchewan and Mainz, and the first results from Tohoku[19] were presented two years ago. Pion photoproduction has the advantage over (π^-,γ) that both π^+ and π^- can be observed. Moreover, the momentum transfer[9] can be varied by varying the reaction angle and/or the energy (Fig. 8), unlike (π^-,γ) for which q is fixed near 0.6 fm^{-1}. For 50 MeV pions, for example, one can cover a range of q from about 0.5 to 1.7 fm^{-1}, which for $^{12}C(\gamma,\pi^+)^{12}N(g.s.)$ covers the range from near the first maximum of the M1 form factor to past the first minimum. One has the option of varying either momentum or energy transfer sepa-

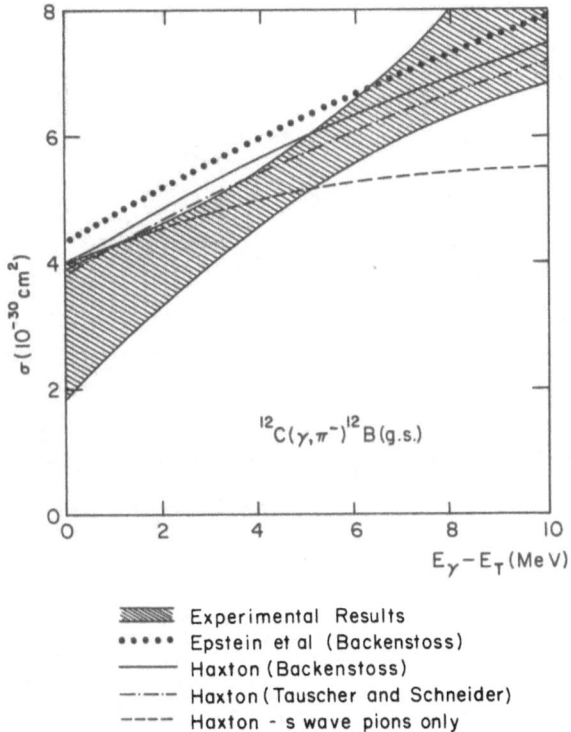

$$^{12}C(\gamma,\pi^-)^{12}B(g.s.)$$

░░░░ Experimental Results
••••• Epstein et al (Backenstoss)
——— Haxton (Backenstoss)
—·—·— Haxton (Tauscher and Schneider)
— — — Haxton - s wave pions only

Figure 7

rately. Thus for investigating nuclear structure effects while keeping pion final state interactions constant one can vary q at fixed T_π by varying θ, while for studying the pion-nucleus interaction one can instead vary T_π at fixed q.

As an example of the experimental arrangements, Fig. 9 shows the low energy pion spectrometer being used by the RPI-MIT collaboration at Bates. Photo- or electroproduction can be studied depending on whether or not the radiator upstream from the target is inserted. The spectrometer uses a quadru-pole-dipole combination with a multi-wire proportional chamber in the focal plane to determine particle momentum. Backup counters (three thin scintillators in coincidence together with a Cerenkov veto counter) with pulse height requirements select pions from electrons and other background events. The

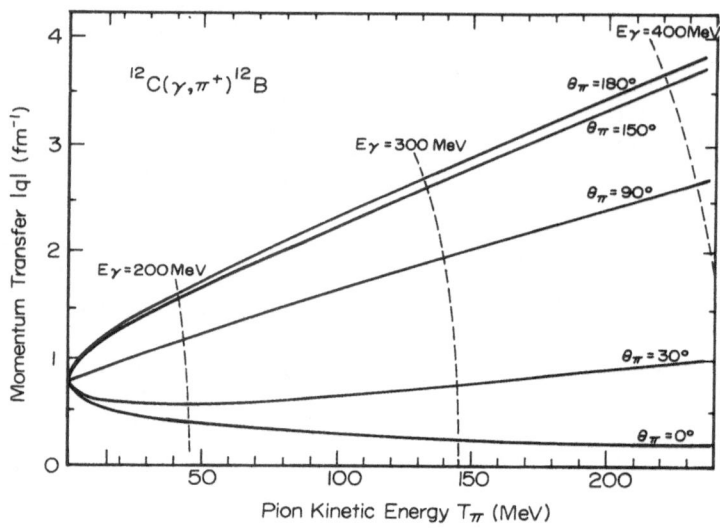

Figure 8

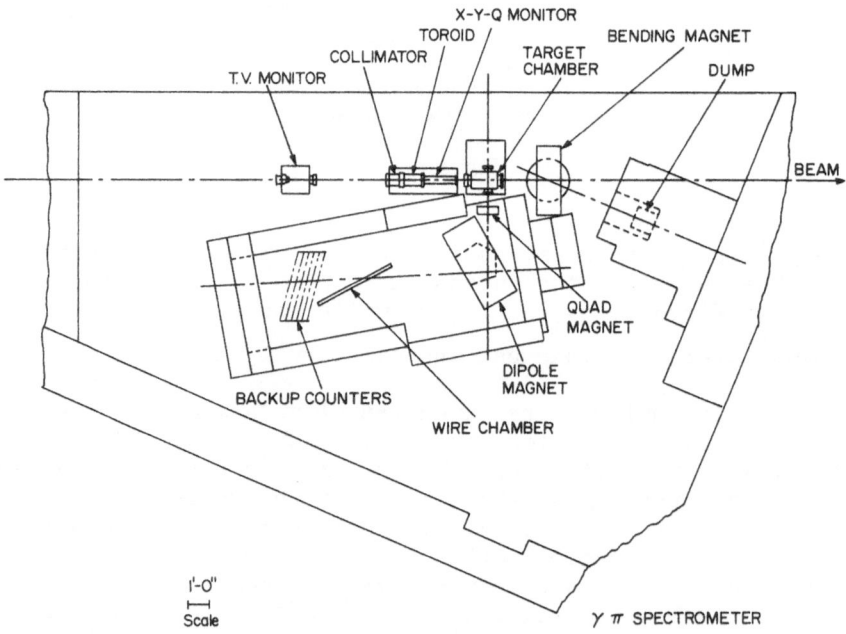

Figure 9. Low energy photopion spectrometer at Bates.

TABLE I
RPI–MIT Magnetic Pion Spectrometer Parameters

Magnetic Configuration	QD
Solid Angle	17 msr
Angular Acceptance: (in radial plane)	+ 64, -70 mr.
(in transverse plane)	71 mr.
Path length (central ray, target to wire chamber)	2.76 m
Dipole: (Radius, central trajectory	0.43 m
Maximum B	9.0 kilogauss
Dynamic Range $\frac{\Delta p}{p}$	15 %
Maximum momentum	125 MeV/c
Maximum pion-energy	45 MeV >
Minimum pion-energy	15 MeV
Energy resolution	1.4×10^{-2}
Angular range	Fixed at 90°

TABLE II
Tohoku Magnetic Spectrometer Parameters

	Present Pion Spectrometer	New Spectrometer (tentative specifications)
Magnetic Configuration	double focussing (169.8° deflection)	dipole (Enge type)
Solid angle	6 msr.	5 ~ 10 msr.
Angular Acceptance: (in horizontal plane)	± 20 mr.	± 30 mr.
(in vertical plane)	± 57 mr.	± 70 mr.
Path length (central ray, target to central detector)	4.6 m.	1.3 m.
Radius		
Maximum B (*)	1.0 m. 10 kilogauss	~ 0.2 – 0.4 m. 10 kilogauss
Dynamic Range $\frac{\Delta p}{p}$	3.3 %	~ 50 %
Maximum momentum	300 MeV/c	150 MeV/c
Maximum pion-energy	200 MeV	65 MeV
Momentum resolution	5×10^{-4}	2×10^{-3}
Angular range	Variable (25°-155°)	Variable (40°-140°)

* Available up to ~ 13 Kilogauss

spectrometer is fixed at an angle of 90°. Table I shows
the specifications for this spectrometer. The accessible
pion energy range is from ~ 15 MeV to 45 MeV, the pion energy
bite at 30 MeV kinetic energy is about 7 MeV, and the overall
energy resolution including incident electron beam spread,
at a solid angle of 17 msr, is about 0.6 MeV at T_π = 30 MeV.

In Table II the specifications for the Tohoku systems are
listed. The present Tohoku spectrometer has the great advantage over the Bates system of permitting angle variation.
Its main limitation is the narrow 3% pion momentum range

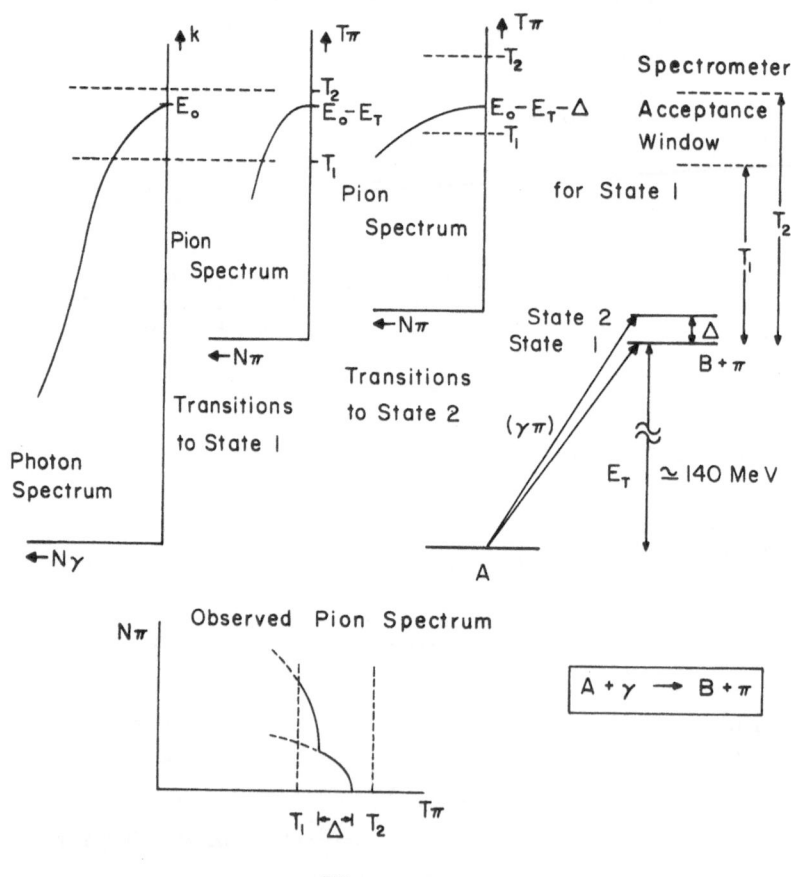

Figure 10

covered at a single setting. This will not be the case for their new spectrometer which goes into service around the end of the year and which will be a compact, broad-range instrument. The data-taking methods used in the spectrometer experiments require some explanation. The situation is shown schematically in Fig. 10. It can apply either to photoproduction where the continuous photon spectrum shown consists of real bremsstrahlung photons, or to electroproduction in which case the curve will represent the virtual photon spectrum associated with the electrons. We suppose the pion spectrometer is set to accept pions and measure their momentum in the energy interval T_1 to T_2, and we suppose a photon spectrum of end point E_o is incident on a target nucleus A. For transitions to a single specific final state, the observed pion spectrum will have a sepctral line shape which reflects the known photon line shape, provided the (γ,π) cross section is effectively constant over the photon energy range corresponding to the pion energy acceptance interval $T_2 - T_1$. This will be true if the cross section is slowly-varying with photon energy and if $T_2 - T_1 << E_o - E_\pi$. If we have transitions to two final states, the observed spectrum will be the sum of two individual spectra separated in energy by Δ, the energy difference of the two states. Since the line shape for each transition is known, such a spectrum can be analyzed to obtain the cross section for individual transitions. To evaluate the dependence on photon energy of the cross section to particular states, or to extend the range of photon and excitation energy covered, the spectrometer acceptance interval $T_2 - T_1$ and/or the end point energy E_o can be varied.

As an example of how this works in a simple case, Fig. 11 shows the $^{16}O(\gamma,\pi^+)^{16}N$ spectrum in the ground state region as measured at Bates. In this plot T_π increases to the left. The ^{16}N ground state quartet of odd parity levels spans 400 KeV and is not resolvable with the Bates spectrometer. Then there is a gap of over 3 MeV before one encounters higher levels. The measured spectrum is indeed well fit by a single level spectrum shape in this energy region. The deviation from this shape between 3 and 4 MeV corresponds to the onset of transitions to higher states.

We will now consider in some detail the case of ^{12}C for which there are π^- capture spectra from SIN as well as some

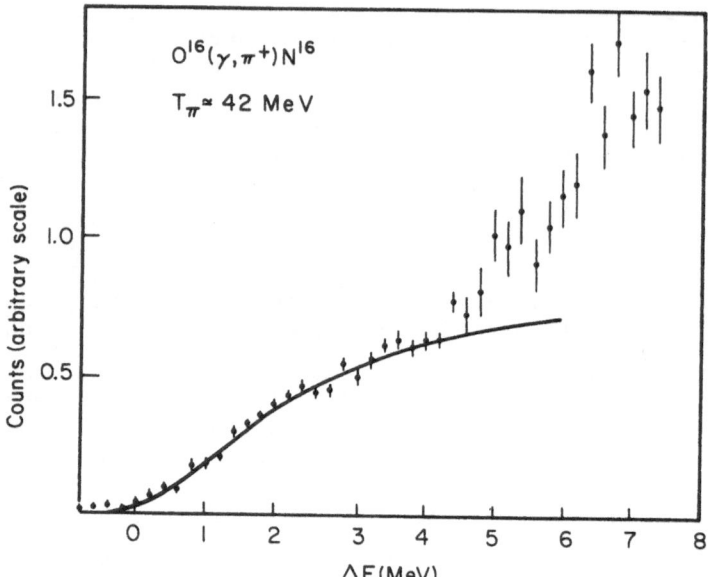

Figure 11. Bates data for $^{16}O(\gamma,\pi^+)^{16}N$. Solid curve is single level fit corresponding to transitions to the ^{16}N ground state quartet of levels. The deviation of the experimental spectrum from this shape above 3 MeV is due to transitions to higher states in ^{16}N.

photopion spectrum and angular distribution data from Tohoku and Bates. The (π^-,γ) spectrum, which is presented in the contribution by J.C. Alder et al.[20], shows (Fig. 12) why ^{12}C is an attractive and interesting case for study. The right-hand peak is dominated by spin-flip transitions to the ground state of ^{12}B, which is the analog of the 15·11 MeV giant M1 state in ^{12}C. In addition weaker transitions to the three lowest-lying bound excited states of ^{12}B lie under the tail of this peak; even though these transitions are not resolved, their strengths can be extracted because the line shape for an individual transition is known well. At higher energies the spectrum is dominated by the two prominent peaks at

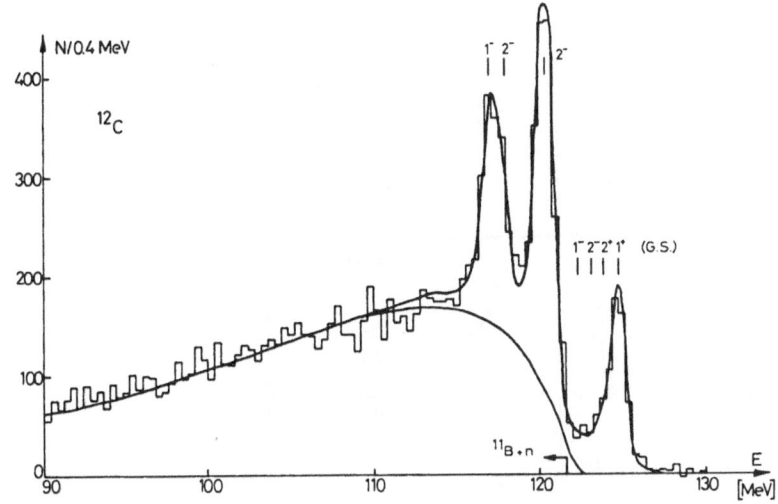

Figure 12.   The gamma ray spectrum from ^{12}C(π^{-},γ) measured at SIN (Ref. 20) (see text).

$E_x \sim 4\frac{1}{2}$ and $7\frac{1}{2}$ MeV, which probably correspond to the strong excitation of 2^- and 1^- spin-isospin states. These peaks were first seen in the earlier Berkeley data[21]; this case was in fact the first experimetal demonstration of the predicted strong selective excitation of spin-isospin giant dipole analog states by the (γ,π) reaction.

Turning to the (γ,π) results, the spectra taken at three angles at Tohoku[22] is shown in Fig. 13. Especially in the 90° spectrum, one can see the presence of strong transitions to the state or states near $4\frac{1}{2}$ MeV. The end points for the ground state and the 0.95 MeV 2+ first excited state are shown by the arrows. One sees at 30° ($q \simeq 0.5$ fm^{-1}) that the ground state contribution is strongly present while the first excited state is weak. However, the reverse is true at 90° ($q \simeq 1.1$ fm^{-1}): the ground state transition is weak compared to the first excited state. These trends are consistent with the form factor shapes for the analog states in ^{12}C as seen in (e,e'). This is seen more clearly in the more detailed an-

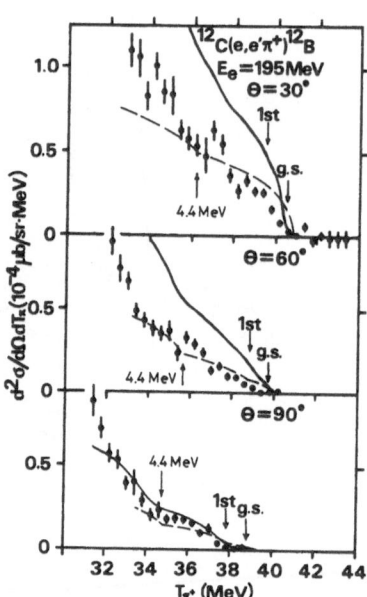

Figure 13. ^{12}C pion spectra measured at three angles at Tohoku (Ref. 22).

gular distribution data obtained at Tohoku[23] for these two states (see Fig. 4 of the contribution by H. Ohashi et al.). The two angular distributions are quite different in shape. The solid and dotted curves are Helm[24] and shell model[25] calculations respectively. For the ground state, the M1 form factor has its minimum showing up at back angles, while the E2 form factor for the first excited state peaks at higher q, and the corresponding angular distribution peaks at back angles.

To extract the strength for weakly excited higher states is more difficult. At Bates 90° spectra have been measured for π^+ and π^- at pion energies centered at 15, 28, and 40 MeV. These results are discussed in detail by Paras et al. in a symposium contribution. Figure 14 shows the π^+ spectrum at T_π = 28 MeV. The solid line is a fit to the data assuming transitions to states in ^{12}B at excitation energies of 0.0, 0.95, 1.67, 2.62, 3.4, 4.4, and 5.6 MeV, adjusting the strength of each transition to optimize the fit. The quality of fit is

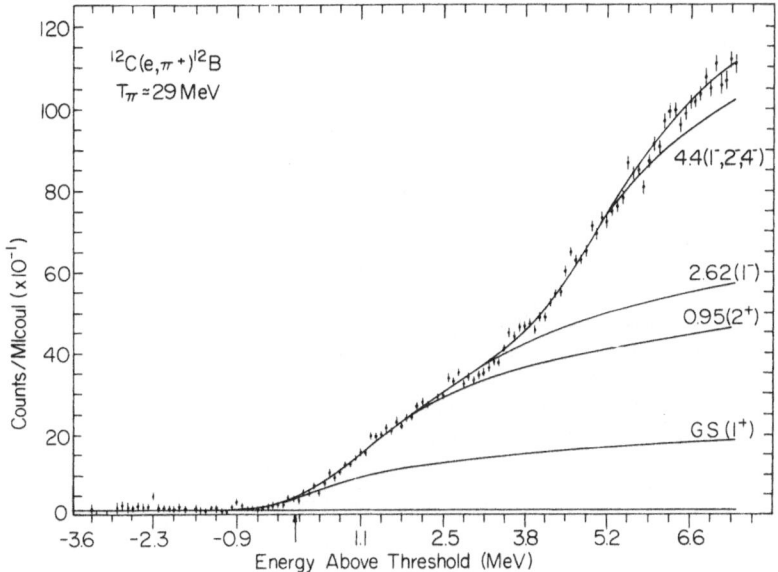

Figure 14. Bates pion spectrum for $^{12}C(e,e'\pi^+)$ at T = 28 MeV. The solid line fit assumes transitions to ^{12}B states at 0.0, 0.95, 1.67, 2.62, 3.4, 4.4, and 5.6 MeV.

seen to be good (χ^2 per degree of freedom = 1.1). For each fitted line one can extract a cross section value. The presence of excited state transitions is evident from the data, and indeed one can extract good cross section values for the ground, 0.95, and 4.4 MeV states. However, the cross section values obtained for the more weakly excited states between 0.95 and 4.4 MeV have large uncertainties. The bar graph in Fig. 15 shows the cross section values from an earlier Bates run on ^{12}C at 90° under slightly different experimentla conditions. The results are in satisfactory agreement with the newer data of Paras et al. in these Proceedings. The rectangular boxes are the results of calculations by Nagl and Uberall[26]. The height of each box represents the spread in the calculated cross section values obtained by varying the pion optical model parameters used. The nuclear structure input for these Helm model calculations is obtained by parameerizing $^{12}C(e,e')$ data. In fact, the uncertainties in the

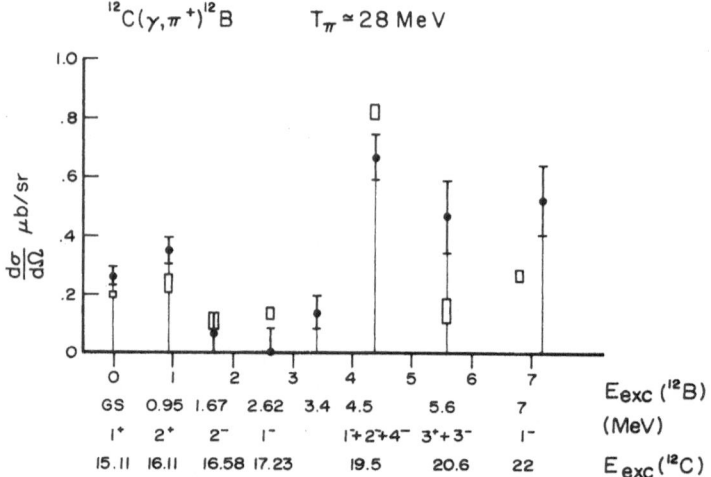

Figure 15.   See comments in text.

experimental form factors from electron scattering give rise
to a major uncertainty in the calculated cross sections,
typically ≃ 25%.

These (γ, π^+) results can be compared to the branching
ratios for (π^-, γ) to the same ^{12}B states, obtained from the
spectrum of Fig. 12, and plotted in Fig. 16. When account is
taken of the differences in momentum transfer between the two
experiments ($q \simeq 0.6$ fm^{-1} for (π^-, γ); $q \simeq 1.1$ fm^{-1} for (γ, π^+)),
there appears to be no gross discrepancy between the two sets
of results.

One final remark about ^{12}C: a definitive $(\gamma, \pi^\pm)$ study of
the giant dipole analog states above 4 MeV excitation has yet
to be done, but should be rewarding. A combination of good
statistics and moderate energy resolution, together with angular distribution measurements to help identify states, will be
needed to obtain the strength distribution in this region*.

Pion spectrometer systems have also been set up recently

*Editor's note: the contribution by K. Shoda et al. begins to
address this question.

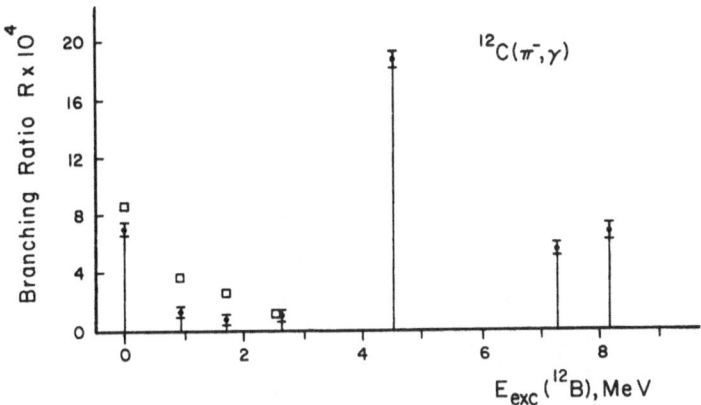

Figure 16. Experimental branching ratios for radiative capture on ^{12}C to states in ^{12}B (Ref. 20). The square boxes are
the results of the calculation of Maguire and Werntz (Ref.
42).

at Mainz and Saskatchewan, and at Mainz a program of studies
to specific states in light nuclei is under way[27]. The initial measurements already completed have however been inclusive studies of low energy π^+ and π^- yields on carbon in the
energy range between 5 and 16 MeV for E_e = 200 MeV at Saskatchewan[28], and at 21.6 MeV for E_e = 280 MeV at Mainz[29]
Direct interpretation of these experiments is difficult because there are many accessible final states extending to
high excitation energies. It is hoped, however, that π^+/π^-
ratios at these energies will be relatively insensitive to
nuclear structure factors but sensitive to pion final-state
interactions. There are several contributions to the·sympos-
ium in this area.

To assess the sort of spectrometer performance one can
hope to bring to bear in pion spectroscopy in the immediate
future, we can consider the pion spectrometer being planned
by the RPI-MIT collaboration for deployment in the new Bates
experimental hall in early 1980. The system (Fig. 17) uses
two quadrupoles and a split dipole of radius 0·7 meters, with
pole edges shaped to minimize aberrations. The pion energy
range covered is from about 20 MeV to 260 MeV. The optics

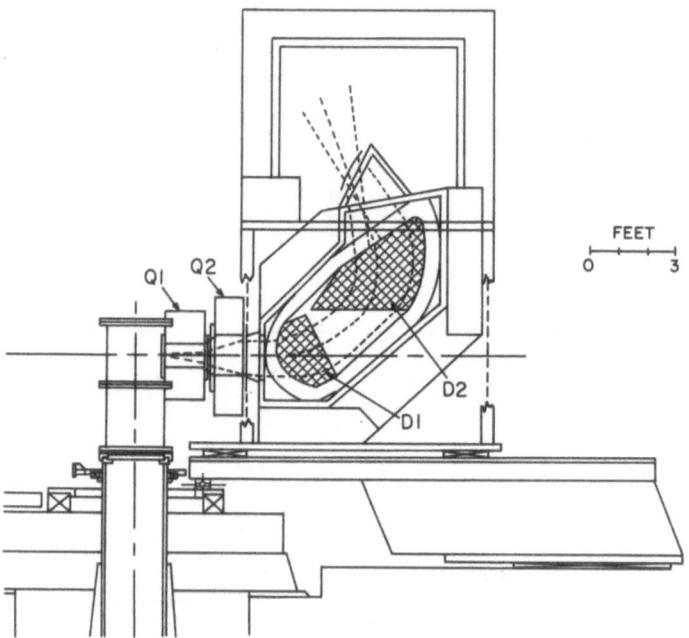

Figure 17.   New QQSD pion spectrometer planned for Bates.

calculations by Ingvar Blomqvist for this system have resulted in a design with excellent properties. This combines a
solid angle of 35 msr with momentum resolution of $\simeq 5 \times 10^{-4}$
and a momentum bite of 20% at a given setting. Two angles
and two coordinates will be measured in the focal plane for
each event. The reaction angle has to be determined to about
$\frac{1}{2}^{\circ}$. For the best resolution, the spectrometer will have to
be operated in the energy loss mode. In this mode the incident electron beam is dispersed vertically on the sample with
its dispersion matched to that of the spectrometer in order
to minimize contributions to the resolution from the incident
electron beam energy spread.

Figure 18 shows the results of a computer simulation of
experimental data from this spectrometer by Gregg Franklin to
demonstrate its spectroscopic capabilities. The upper spectrum is for $^{12}C(\gamma,\pi^{+})$ to the three lowest ^{12}B states. The

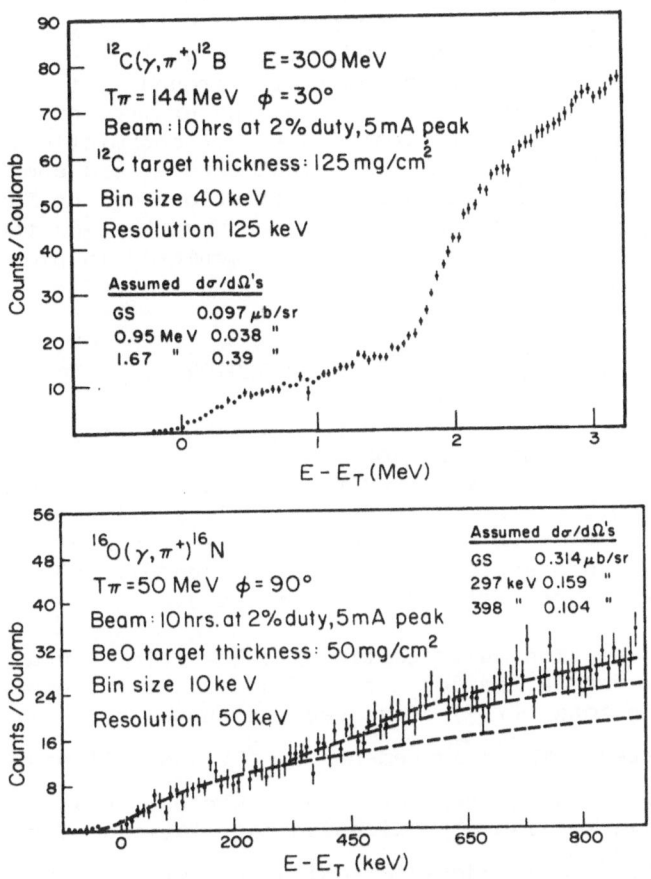

Figure 18. Computer simulated pion spectra, with scatter of points and error bars determined by statistics only.

simulated run is for 10 hours at 30° at an electron beam energy of 300 MeV. The cross sections assumed are those calculated by Nagl and Überall for these conditions. The points are what would be measured under these conditions with the scatter determined by statistics alone. If one goes through the exercise of analyzing these mocked-up data to extract cross sections, one finds the ground and second excited states are determined to a few percent while the weak first excited state can be obtained to about 30%. The lower example is a

severe test: it is for $^{16}O(\gamma,\pi^+)$ to the ^{16}N ground state
quartet. With the best spectrometer resolution (50 keV)
at a pion energy of 50 MeV, and again assuming only statistical errors are present, one finds one can just about resolve
the four states. In this exercise, the ground state cross
section is determined well ($\simeq$ 5%) and the excited states to
$\simeq$ 50%. Of course these estimates are over-optimistic in that
they ignore effects of non-statistical errors, and one should
not take these results as literal predictions of the future
capabilities of this new spectrometer. Nevertheless they do
provide some indication of what can be achieved in pion spectroscopy in the near future.

4. THE (3,3) RESONANCE REGION

The (3,3) resonance region is of great potential interest
from the point of view of studying Δ production and the Δ-nucleus interaction. There has been relatively little detailed
experimental work here so far, with the exception of the
Saclay coincidence studies on very light nuclei, and what we
have to look at are mostly various bits and pieces.

Our information on transitions to low-lying bound states
comes mainly from a limited group of radioactivity experiments
measuring the total (γ,π) cross section summed over all bound
residual states. The measurements on ^{51}V and ^{27}Al made at
Lund[30] a few years ago generated some interest because the
distorted wave impulse approximation calculations of Freed
and Ostrander[30] were order-of-magnitude lower than experiment
in the (3,3) resonance region. One difficulty in these calculations involves the nuclear structure input – for both
nuclei there are many bound states which must be included.
The case of $^{14}N(\gamma,\pi^-)^{14}O$ recently measured at Lund[31] does not
have this uncertainty because there is only one bound final
state and the nuclear wave functions are fairly well known.
Although the calculations of DeCarlo[32] for this case do not
fit the data well below 300 MeV, there are not as big discrepancies between experiment and calculation at higher energies as for ^{27}Al or ^{51}V.

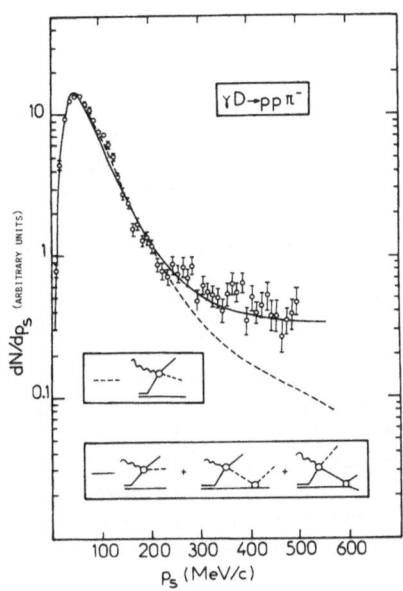

Figure 19.   Proton momentum distribution in the $\gamma D \rightarrow pp\pi^-$ reaction.   Experimental data is from Ref. 34.   Dashed curve is
from quasi-free model; solid curve includes single rescattering
effects.   This figure is from Ref. 33.

These radioactivity experiments give no information on
the transitions to higher-lying continuum states which dominate (γ,π) reactions in the (3,3) resonance region.   Single-arm
measurements of pion spectra are limited in scope so far but
they do give us some infromation on these continuum transitions.   The measurements appear consistent with the dominance
of a quasi-free process in which photoproduction occurs on a
single nucleon in the nucleus acting independently, but with
the results colored by the momentum distribution of nucleons
in the nucleus as in (e,e') quasielastic scattering, as well
as by the final state interaction of the pions.

Looking first at $D(\gamma,\pi^-)$ pp, Fig. 19, which is due to
Laget[33], shows the experimental data from a DESY bubble chamber experiment[34] on the momentum distribution of recoil protons.   The dotted curve is a calculation based on the quasi-
free process, with no final state rescattering effects in-

cluded, while the solid curve includes pion and nucleon
single rescattering effects. We see that the lion's share of
the events are well described by the quasi-free process but
that rescattering effects become important at high momenta.
At Saclay a series of double-arm coincidence experiments
have been carried out to study more directly these rescattering effects at high momentum values of the recoil proton.
We will return to that work below and comment briefly on it.

At the Tokyo synchrotron in the past several years[35,36],
spectrometer studies of charged pions from ^{12}C have been
carried out for a somewhat limited range of pion energies and
forward angles. The spectra correspond to differences between
two runs with bremsstrahlung end-point energies 50 MeV apart,
so that the photon energy is effectively defined to $\simeq$ 50 MeV.
The solid curves (Fig. 20) correspond to a quasi-free model.
Broadening due to nucleon momentum distribution and to experimental resolution effects is included. One sees that under these kinematic conditions there is good agreement between experiment and this simple theory. Even for copper at
forward angles[36] there appears a well defined peak in the pion
spectrum.

More recently, tagged photon beams with comparable energy
resolution have been set up at both the Tokyo[37] and Bonn[38]

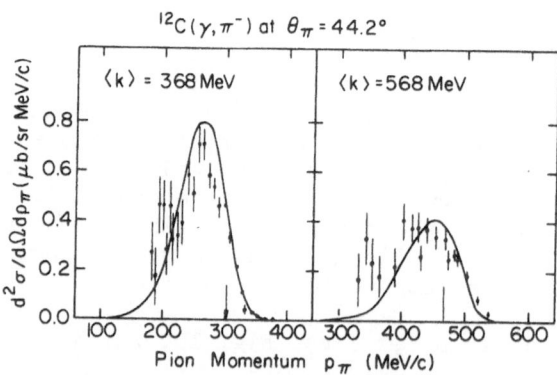

$^{12}C(\gamma, \pi^-)$ at $\theta_\pi = 44.2°$

Figure 20. Representative forward-angle pion momentum spectra
measured at Tokyo (Ref. 35). The solid curves are a quasi-
free model calculation. The arrow is at the momentum expected for a free nucleon.

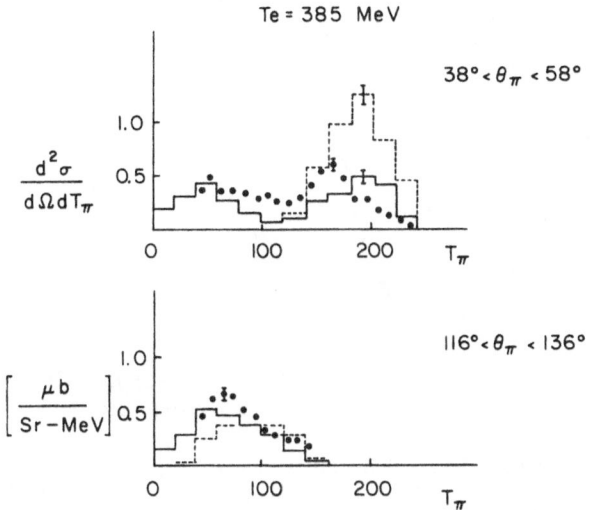

Figure 21.   Representative pion spectra from Ref. 38.   See comments in text.

synchrotrons and used for studies of pion spectra.   The tagged photon experiment at Bonn has now produced considerably more complete results for ^{12}C than the earlier bremsstrahlung work.   This work has been done with the photon slices with an effective energy width of 10 MeV, and covers the angular range from 38° to 136°, at photon energies from 200 to 390 MeV.   Pions are detected by the spectrometer down to 40 MeV.   Mecking describes these results in detail in a later contribution.

Representative results from Bonn are shown here in Figs. 20 and 21.   The two plots are of pion spectra at forward and back angles for a mean photon energy of 385 MeV.   The histograms are from Monte Carlo calculations based on the quasi-free model with (solid histogram) and without (dotted histogram) inclusion of pion interactions.   One sees that there is a fair fit between the model and data; the spectra appear to be significantly reduced in energy by pion-nucleus interactions.

One additional experimental note on spectrum measurements: the Saclay monochromatic photon beam obtained from positron

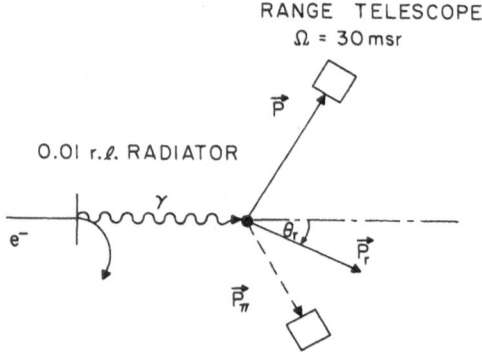

RANGE TELESCOPE
Ω = 30 msr

0.01 r.ℓ. RADIATOR

MAGNETIC SPECTROMETER "400"
Ω = 3.4 msr   ΔP/P = 9%

Figure 22. Saclay coincidence arrangement for study of
$D(\gamma, \pi^- p)p$ (Ref. 39). In more recent work (Ref. 40), the two
protons were detected rather than the pion-proton pair, using
the "700" spectrometer and range telescope.

annihilation is now operational, with beam energies monochromatic to about 1%. This promises to be an excellent tool for
single-arm studies in the (3,3) resonance region.

Double-arm coincidence experiments are being carried out
principally at Saclay, with emphasis on a detailed study of
$\gamma D \to \pi^- pp$ [39,40] As indicated schematically in Fig. 22, they
detect the pion or one proton in a magnetic spectrometer and
the other proton in a counter telescope, measure their directions and momenta, and hence completely determine the reaction kinematics. To study deviations from the simple quasi-
free picture, they look at large values of recoil momenta.
One interesting example of their results is shown in Fig. 23.
They measure the angular distribution of recoil protons and
they plot the ratio of measured yield to that expected on the
simple quasi-free picture with no rescattering. At a recoil
momentum of 400 MeV/c they see a strong peak near 50° which
is well accounted for by the sum of two rescattering effects.
The dominant effect is pion-nucleon single rescattering (the
upper diagram; a calculation of this yields the solid line).
In addition one appears to see effects of a two-pion term
(lower diagram). This involves meson exchange between the
nucleons modifying the single pion process. This two-pion

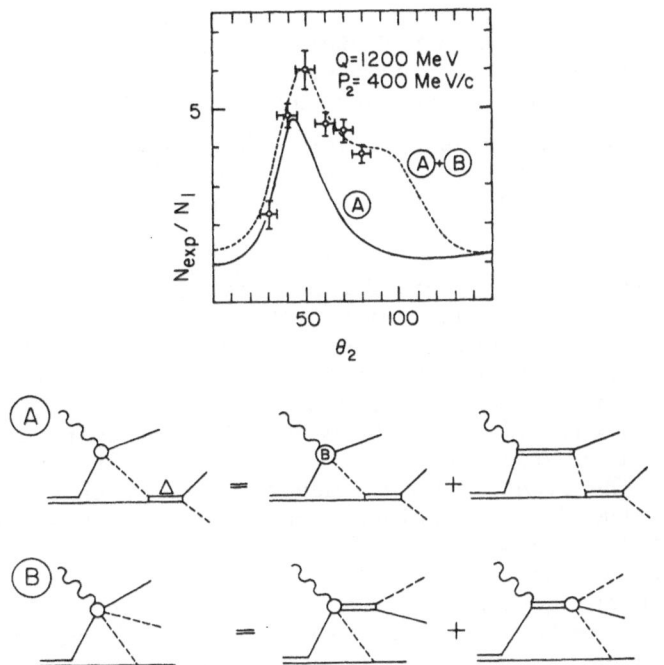

Figure 23. Saclay data on the angular distribution of protons
emitted in $\gamma D \rightarrow \pi^- pp$ with recoil momentum of 400 MeV/c when Q =
1200 MeV. The ratio of measured yield to that expected from
the quasi-free process is plotted. The calculated curve A
includes pion-nucleon single rescattering; curve A + B also
includes the effects of meson exchange between the nucleons
(diagram B) (Ref. 40).

term should come in more strongly at higher proton energies
and this expectation is borne out by data at higher recoil
momenta. In these studies of rescattering effects, we are
really looking at the important pieces of the Δ-N interaction
in relatively direct experiments. In his report Laget dis-
cusses these very pretty experiments in greater detail.

Aside from this program at Saclay, very little double-
arm coincidence work has been done. One class of such experiments of considerable potential is $(e,e'\pi)$ where the outgoing pion and electron are detected in coincidence. To see
the difficulty in these experiments with current generation
linacs with few-percent duty factors, one can look at the

(e,e'π) experiment on the proton[41] done at Saclay several
years ago. With a beam duty factor of 1 - 2%, the beam current had to be held below 5 a to keep the accidental rate
manageable (true coincidence rate accidentals $\simeq$ 1), so that

coincidence counting rates were very low (4 - 40 h^{-1}) and
six weeks of accelerator time were needed for the expeirment.
This experience points up the need for high duty factor beams
for coincidence experiments.

There are some prospects for new high duty factor facilities. The Amsterdam linac, with a 10% duty factor beam at
600 MeV, and with two spectrometers on the same pivot post
(one a 550 MeV/c high resolution device with Ω = 5 msr, the
other a more compact larger solid angle instrument [Ω = 18
msr at 750 MeV/c]), will have the capability for coincidence
work when it is operational at full energy, possibly in a year
and a half. The Stanford superconducting microtron will be
pushed up to the 200 MeV region, perhaps by late 1979, and
will be useful for near-threshold photopion studies. The
Illinois superconducting microtron and the room temperature
facility at Mainz both have plans to add high energy stages
which would bring them into the pion-producing energy range.
But these stages are yet to be funded and are certainly
several years off at the best, for both Mainz and Illinois.

So high duty factor facilities, while not a present
reality, are indeed on the horizon. But if we come back to
the present situation and, in scanning current activity in
the field, ask ourselves which areas of experimental activity
represent specially significant developments in the past year
or so, two areas come to mind. There is, first, the use of
"monoenergetic" photon beams in single arm experiments in the
(3,3) resonance region, and second, the use of magnetic spectrometers to study individual nuclear transitions, so far at
photon energies within 50 MeV of threshold, though the technique is surely not limited to that region.

REFERENCES

1. G. Audit et al., Phys. Rev. C15, 1415 (1977).
2. J. Bergstrom et al., Nucl. Phys. A251, 401 (1975).
3. J.B. Cammarata and T.W. Donnelly, Nucl. Phys. A267, 365
    (1976).

4. J.C. Alder et al., Meson-Nuclear Physics, 1976 (Carnegie-Mellon Conference), AIP Conf. Proc. No. 23, p. 626-7.

5. J.C. Alder et al., Contribution C14, 7th International Conference on High Energy Physics and Nuclear Structure, Zurich, 1977.

6. E.C. Booth et al., Phys. Lett. 66B, 236 (1977).

7. G. Audit et al., Phys. Rev. C16, 1517 (1977).

8. J.M. Laget, Nucl. Phys. A296, 388 (1978).

9. J.V. Noble, Phys. Lett. 67B, 39 (1977).

10. J.H. Koch and R.M. Woloshyn, Phys. Lett. 60B, 221 (1976).

11. P. Argan et al., to be published.

12. J.H. Koch and R.M. Woloshyn, Phys. Rev. C16, 1968 (1977).

13. P. Bosted and J.M. Laget, Nucl. Phys. A296, 413 (1978).

14. A.M. Bernstein et al., Phys. Rev. Lett. 37, 819 (1976).

15. G.N. Epstein et al., Phys. Rev. C17, 702 (1978).

16. W.C. Haxton, LASL preprint LA-UR-77-2898.

17. A. Nagl and H. Überall, this symposium.

18. G.N. Epstein et al., this symposium.

19. K. Shoda et al., Meson-Nuclear Physics-1976 (Carnegie-Mellon Conference), AIP Conf. Proc. No. 23, p. 604-5.

20. J.C. Alder et al., this symposium.

21. J.A. Bistirlich et al., Phys. Rev. Lett. 25, 689 (1970); Phys. Rev. C5, 1872 (1972).

22. K. Shoda et al., Proc. Int. Conf. Nucl. Structure, Tokyo, 1977, J. Phys. Soc. Japan 44, Suppl., 482 (1978).

23. K. Shoda et al., Phys. Rev. Lett. 39, 1131 (1977).

24. F. Cannata et al., Can. J. Phys. 52, 1405 (1974).

25. J.B. Seaborn et al., Nucl. Phys. A219, 461 (1974).

26. A. Nagl, private communication.

27. B. Schoch, private communication.

28. H.S. Caplan, private communication.

29. F. Borkowski et al., Phys. Rev. Lett. 38, 742 (1977).

30. I. Blomqvist et al., Phys. Rev. C15, 988 (1977).

31. V. DeCarlo et al., to be published.

32. V. DeCarlo, this symposium.

33. J.M. Laget, Phys. Lett. 68B, 58 (1977).

34. P. Benz et al., Nucl. Phys. B65, 158 (1973).

35. K. Baba et al., J. Phys. Soc. Japan 42, 1049 (1977).

36. K. Baba et al., this symposium.

37. I. Arai et al., J. Phys. Soc. Japan 45, 1 (1978).

38. J. Arends et al., to be published.

39. P. Argan et al., Nucl. Phys. A296, 373 (1978).

40. P. Argan et al., Phys. Rev. Lett. 41, 86 (1978).

41. G. Bardin et al., Nucl. Phys. B120, 45 (1977).

42. W. Maguire and C. Werntz, Nucl. Phys. A205 211 (1973).

RELATIONSHIP OF WEAK AND ELECTROMAGNETIC PROCESSES IN NUCLEI TO PHOTOPION REACTIONS*

T.W. Donnelly

Institute of Theoretical Physics
Department of Physics
Stanford University
Stanford, California   94305

Within the context of this symposium, the present discussion is focused on how weak and electromagnetic processes in nuclei relate to photopion reactions. In a wider sense the ideas summarized here provide a general framework for describing any process which is "weak" in strength and proceeds through few-body transition operators. Nuclear matrix elements of such operators may be factored conveniently into a linear combination of terms which contain the elementary particle matrix elements of the relevant operators multiplied by numbers which contain the complexity of nuclear dynamics - the nuclear density matrix elements. It is via the density matrix that such "weak" processes can be interrelated as discussed below. First the general problem is summarized, after which two simple (but interesting) examples are presented to illustrate the method.

We begin our discussion by considering the weak and electromagnetic processes shown schematically in Fig. 1. These include the electromagnetic processes (Fig. 1a), electron scattering involving the exchange of a virtual photon and real photon processes such as gamma decay or photo-excitation. We have in mind here processes in which no real pion is present (emitted or absorbed), that is, where $|i>$ and $|f>$ are nuclear states. Photopion processes are discussed a little later. In Fig. 1b, the conventional weak interaction processes of β-decay and μ-capture are diagrammed. Here the interaction vertex denotes the current-current interaction which results from the $W^{\pm}$ exchange of gauge theory models of

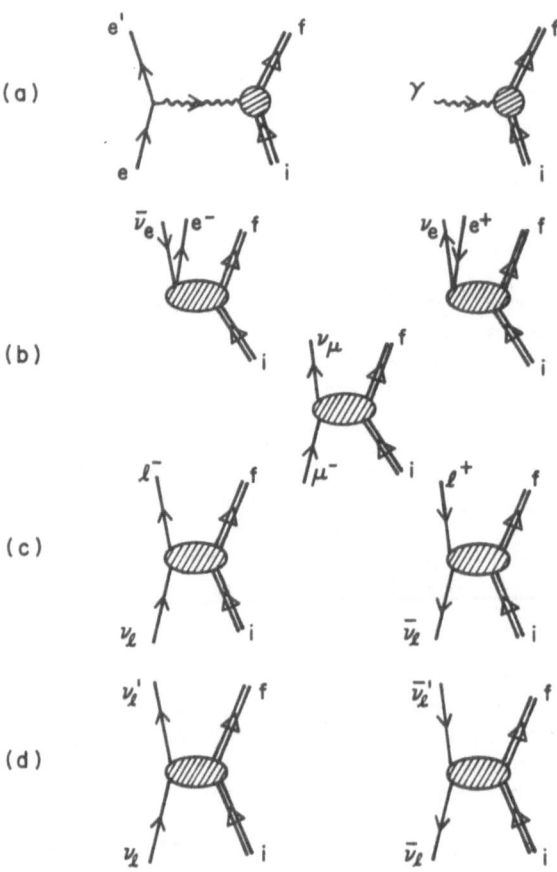

Figure 1. Weak and electromagnetic processes: (a) electromagnetic, (b) conventional weak, (c) charge-changing neutrino
reactions and (d) neutral current neutrino scattering.

the process when taken to the low energies relevant to nuclear
physics. Somewhat less conventional, but potentially richer
are the neutrino reactions (Fig. 1c) which also proceed via
$\underline{W}\pm$ exchange and neutrino scattering (Fig. 1d) which involves
neutral heavy bosons, Z^0, etc. All of these processes (except for the real photon reaction, which is just a special
case contained as a simple point on the electron scattering

form factor) are semi-leptonic in that they involve the product of leptonic and hadronic currents. We assume that the leptonic side is fully described by quantum electrodynamics or its extension via the gauge theories to the unified weak/electromagnetic interaction. Thus, cross sections for all of these processes can be written[1-4]

$$\frac{d\sigma}{d\Omega} \propto \left| \int d\vec{x}\ e^{-i\vec{q}\cdot\vec{x}} \ell_\mu <f|\hat{J}_\mu(\vec{x})|i> \right|^2, \qquad (1)$$

where ℓ_μ is the leptonic matrix element, $\vec{q}$ is the 3-momentum transferred between the leptons and hadrons and $\vec{J}_\mu(\vec{x})$ is the nuclear current density operator.† So we require the Fourier transforms of the nuclear matrix elements of the relevant nuclear current operators to attain a description of the processes in Fig. 1. Let us suppose that we are dealing with discrete nuclear states and that isospin is a good quantum number. Then the states are labelled with angular momentum, parity and isospin: $i \rightarrow J_i^{\pi i} T_i; J_{Ji}, M_{Ti}, f \rightarrow J_f^{\pi f} T_f; M_{Jf}, M_{Tf}$. It is convenient to decompose the current operators into multipoles (see Refs. 1-4 for details) which we may label $\hat{T}_{JM_J;TM_T}^{X\lambda}(q)$ where J and T refer to the angular momentum and isospin nature of the multipole, X refers to the elementary nature of the current (vector, V or axial-vector, A for the processes in Fig. 1) and λ refers to the space-time components of the current ($\lambda=0,1,2,3$ here). This is just alternative, more compact notation for the multipoles used in the past[1-4]: $(V)\hat{M}, \hat{L}, \hat{T}^{el}$ and $\hat{T}^{mag}$ and $(A)\hat{M}^5, \hat{L}^5, \hat{T}^{el5}$ and T^{mag5}. Upon using the Wigner-Eckart theorem we see that the required quantities are the nuclear reduced matrix elements $<J_f^{\pi f}\ T_f\ \vdots\vdots\ \hat{T}_{J;T}^{X\lambda}(q)\ \vdots\vdots\ J_i^{\pi i}\ T_i>$, where the symbols $\vdots\vdots$ denote reduction both in angular momentum and isospin.

To learn something about these quantities let us first consider the elementary nature of the processes under dis-

†We use carets to indicate second-quantized operators operating in the nuclear Hilbert space.

cussion.   Take the electro-magnetic interaction (Fig. la)
to be specific: the nuclear current operator may be decom-
posed into 0-body, 1-body, 2-body, etc. operators.   That is,
there may be 0-body or c-number pieces (for example the total
charge Z), 1-body pieces which behave as $a_\alpha^+ a_{\alpha'}$, (annihilating
a nucleon with quantum numbers α and creating one with
quantum numbers α'), 2-body pieces which behave as $a_\alpha^+ a_\beta^+ a_{\beta'} a_{\alpha'}$,
3-body pieces, etc.   Some examples are shown in Fig. 2 for
a photon (real or virtual) interacting with nucleons.   The
top diagram (1) is the (usually dominant) 1-body interaction
of a photon with a single nucleon, whereas the lower diagrams
(2) provide examples of meson exchange currents (MEC) where
the photon interacts with the subnuclear structure of the
nucleus (for example, see Ref. 5).   Let us denote the operators which result from analyses of such diagrams by

$$[1]_{\hat{T}} = \sum_{\alpha\alpha'}, \ <\alpha| \ ^{[1]}T|\alpha > a_\alpha^+ a_{\alpha'}, \tag{2a}$$

for 1-body operators, and

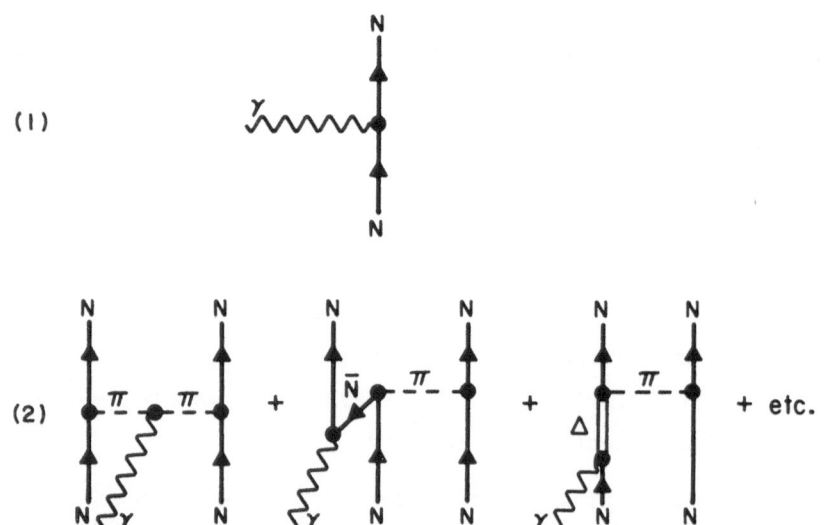

Figure 2.   Elementary electromagnetic interactions: (1) 1-
body contribution, (2) 2-body meson exchange current contributions.

$$[2]_{\hat{T}} = \sum_{\alpha\beta\alpha'\beta'} <\alpha\beta|[2]_{T}|\alpha'\beta'> a_{\alpha}^{+} a_{\beta}^{+} a_{\beta'} a_{\alpha'} \qquad (2b)$$

for 2-body operators. We neglect n-body operators with
n $\geq$ 3 and at present we have suppressed the quantum numbers
which characterize the operators for simplicity. Here
$<\alpha|[1]_{T}|\alpha'>$ and $<\alpha\beta|[2]_{T}|\alpha'\beta'>$ are single-particle and two-
particle matrix elements of the operators respectively. If
we apply the Feynman rules to the diagrams in Fig. 2, we
obtain such matrix elements (for example, see Refs. 2 and
5). Thus, for a given set of diagrams describing the elementary interaction of a photon with a collection of nucleons
all in free (plane-wave) states, we have the single-particle
and two-particle matrix elements required in Eqs. (2).

So far we are not describing nuclear physics, but are
addressing the problems of one plane-wave nucleon, two plane-
wave nucleons, etc., with no correlations between the nucleons. However, a nucleus is a self-bound, highly correlated
system of A nucleons, bound together via exchange of a vari-
ety of mesons. That is, the nuclear states |i> and |f>,
between which the transition of interest takes place, are
represented by A-particle nuclear wave functions. For a
general transition mediated by an arbitrary operator we
would require all of the information contained in such wave
functions, something rarely if ever attained to much accuracy. On the other hand, for "weak" transitions such as we are
considering here, we require the nuclear matrix elements
only of a few-body operators; that is, we find that the matrix
elements of $[1]_{\hat{T}}$ are the dominant ones, with $[2]_{\hat{T}}$ generally
less important, $[3]_{\hat{T}}$ very much less important, etc. Thus
we require specific nuclear matrix elements,
$<f|a_{\alpha}^{+} a_{\alpha'}|i>$, $<f|a_{\alpha}^{+} a_{\beta}^{+} a_{\beta'} a_{\alpha'}|i>$, etc, in which reside the nuclear dynamical content of these transition matrix elements
for 1-body, 2-body, etc., contributions.

To be specific, let us consider 1-body operators. The
single-particle matrix elements for plane-wave nucleons are
taken as given; for example, for electromagnetic interactions[1-4]

$$\langle \vec{k}m_s | J_\mu | (0) | \vec{k}'m_s' \rangle = i\bar{u}(\vec{k}m_s)[F_1\gamma_\mu + F_2 \,\sigma_{\mu\upsilon}q_\upsilon]u(\vec{k}'m_s'), \qquad (3)$$

where the plane-wave states are labelled with momenta $\vec{k}$, $\vec{k}'$ and spin projections m_s, m_s', the 4-momentum transfer is $q_\mu = (k' - k)_\mu$ and F_i, i-1,2 are the electromagnetic form factors (functions of q_μ) of the nucleon (we have suppressed isospin here). Information on the single-nucleon form factors may be obtained by scattering electrons from the proton and deuteron. Given the single-nucleon matrix elements we may make a non-relativistic reduction of Eq. (3) and change to a set of single-particle basis states more appropriate to nuclear physics, for example to a set of harmonic oscilla-tors where now $\alpha \to n\ell j m_j$; $\frac{1}{2} m_t$, a complete set of single-parti-cle quantum numbers (this is not an approximation as long as we sum over the complete set). Thus we have the single-particle matrix elements of the current (and its multipole projections). In nuclei we then have, using the Wigner-Eckart theorem on Eq. (2a),

$$\langle f \vdots [1] \hat{T}^V_{J;T}(q) \vdots i \rangle = \sum_{aa'} \langle a \vdots [1] T^{V\,\lambda}_{J;T}(q) \vdots a' \rangle \times \psi^{(fi)}_{J;T}(aa'), \qquad (4)$$

where $a \to n\ell j;1/2$, $\alpha \to a; m_j, m_t$ and the sums run over a complete set of single-particle quantum numbers. The numbers $\psi^{(fi)}_{J;T}(aa')$ are the 1-body density matrix elements. Eq. (4) is basic to our analysis. It expresses the many-body nuclear matrix element of a 1-body oeprator in terms of (presumed known including their q-dependence) single-particle matrix elements $\langle a \vdots [1] T^{V\,\lambda}_{J;T}(q) \vdots a' \rangle$ multiplied by constants $\psi^{(fi)}_{J;T}(aa')$, and the 1-body density matrix elements in which the complexities of nuclear structure are buried. These density matrices are reduced matrix elements of $a_\alpha a_{\alpha'}$ (coupled to form a tensor of rank J,T and projection M_J, M_T) be-tween states $|i\rangle$ and $|f\rangle$ (see Refs. 1,4). As long as we sum over complete sets of single-particle quantum numbers a,a' in Eq. (4), we have been quite general in our treatment of 1-body electromagnetic operators. However, we usually must truncate the sums to a small model space to obtain a tractable,

but approximate problem to address. In this case we have
only a very small set of numbers, $\psi_{J;T}^{(fi)}(aa')$ to specify
the nuclear physics content of Eq. (4). In other words, the
vast amount of information contained in the overlap of two
A-body nuclear wave functions is condensed down to a small
set of numbers which are relevant for 1-body operators.
Of course we could repeat the analysis for the 1-body contributions to the weak interactions shown in Fig. 1, but we
note that the <u>same</u> density matrices are involved, only the
single-particle matrix elements differ. In fact, any 1-body
operator of the same rank J,T requires this same set of
density matrices. For example we may apply this analysis
to a description of photopion processes (Fig. 3) such as
(γ,π), (π,γ) or $(e,e'\pi)$. Having the required single-particle
matrix elements for the process (see Refs. 6-8) and the 1-
body density matrices described above, it is straightforward
to compute the cross section for these processes using Eq.
(4).

A similar approach can be followed for 2-body operators.
Their many-body reduced matrix elements will then be expressed as linear combinations of two-particle reduced matrix
elements (assumed known from the elementary nature of the
process, see above) multiplied by 2-body density matrices
$\psi_{J;T}^{(fi)}((ab)J_1;T_1,(a'b')J_1';T_1')$ (defined in analogy to the
1-body density matrices). Then, given the 2-body density
matrices within some model space, the matrix elements of any
2-body operator can be computed. Clearly the sequence continues to 3-body operators and beyond, but we restrict our
attention only to 1- and 2-body operators.

There are several ways to proceed from here. One is to

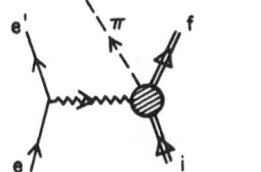 

Figure 3. Photo- and electro-pion processes.

attempt a solution to the A-particle Schrödringer equation
with a given nuclear potential. With the resulting nuclear
wave functions it is straightforward to calculate the 1-
and 2-body density matrices and hence to obtain cross sections for the processes of interest using Eq. (4). Although
quite successful in certain cases (e.g. the few-body systems
or closed shells plus or minus a few particles) there are
many instances where the cross sections obtained (having
made the approximations which must inevitably be made, such
as the RPA) are not in particularly good agreement with
experiment. Thus we are led to follow a different approach
in which we work directly with the density matrices, attempt-
ing to deduce something about them from one process (say
electron scattering) and then to employ this knowledge to
predict cross sections for another process (say the weak
interactions). Of course there are generally too few experimental data of insufficient accuracy to constrain the density
matrix fully and so we must use a hybrid of the straight
density matrix analysis with a shell model analysis. Specif-
ically, all of the 2-body density matrices might be calculated using shell model wave functions, while the 1-body density matrices could be obtained directly from analysis of
electron scattering. Since generally the 2-body MEC contributions are small corrections to the dominant 1-body contributions, we can tolerate the reduced accuracy in the 2-body
calculation. These general ideas are now applied to two
simple examples for illustration.

We consider first the A=6 system[9] composed of the 1^+0
ground state of ^{6}Li, the 0^+1 (3.562 MeV) second excited state

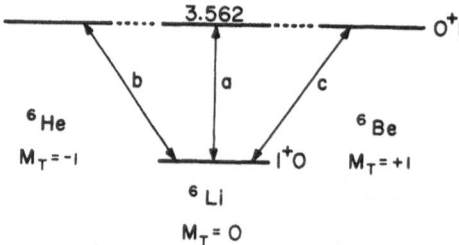

Figure 4. The A=6 system indicating transitions considered:
(a) neutral currents, (b) and (c) charge-changing currents.

of ^{6}Li and its isospin analogs the ground states of ^{6}He and ^{6}Be. The levels and transitions of interests are shown schematically in Fig. 4. Neutral current transitions (line a in Fig. 4) include electron scattering (e,e'), γ-decay, photoexcitation and neutrino scattering (ν,ν') or anti-neutrino scattering $(\bar{\nu},\bar{\nu}')$. Charge-changing transitions (lines b and c) include $\beta\pm$-decay, μ-capture, neutrino and anti-neutrino reactions (ν_ℓ,ℓ^-), (ν_ℓ,ℓ^+) (where ℓ = e or μ) and the photo- and electro-pion reactions $(\gamma,\pi\pm)$, $(\pi-,\gamma)$, (e,e'$\pi\pm$). As a model space for the nuclear wave functions we choose the ^{4}He shell to be closed, $(1s_{1/2})^4$ and take the remaining two nucleons to be in the 1p-shell. The states can then be written[9]

$$|1^+0;M_J M_T> = A|(1p_{3/2})^2 1^+0;M_J,M_T> + B|(1p_{3/2}1p_{1/2})1^+0;M_J,M_T>$$

$$+ C|(1p_{1/2})^2 1^+0;M_J,M_T> \tag{5a}$$

$$|0^+1;M_J,M_T> = D|(1p_{3/2})^2 0^+1;M_J,M_T> + E|(1p_{1/2})^2 0^+1;M_J,M_T>, \tag{5b}$$

where A,B,C,D, and E are constants which satisfy the normal-ization requirement $A^2+B^2+C^2 = D^2+E^2 = 1$. The core config-uration $(1s_{1/2})^4 0^+0$ has been suppressed for clarity. This system is especially simple in that we have only two active nucleons and can readily obtain the 1- and 2-body density matrix elements in terms of the few parameters A,...,E. For example, the 1-body density matrices are given by

$$\psi_{1;1}^{(fi)} (1p_{3/2} \, 1p_{3/2}) = 1 \frac{1}{\sqrt{2}} AD$$

$$\psi_{1;1}^{(fi)} (1p_{1/2} \, 1p_{3/2}) = - \frac{1}{2} BD$$

$$\psi_{1;1}^{(fi)} (1p_{1/2} \, 1p_{3/2}) = \frac{1}{\sqrt{2}} BE$$

$$\psi_{1;1}^{(fi)} \; (1p_{1/2} \; 1p_{1/2}) = -CE \tag{6}$$

where "f" refers to the 0^+1 states and "i" to the 1^+0 state.
The two-body density matrix can be obtained likewise. Thus
at this stage we have six parameters (A,...,E plus the harmonic oscillator parameter b if we choose such a basis) with
two normalization conditions. To fix the remaining parameters we do the following[9]: first we neglect all 2-body MEC
contributions and consider only the 1-body currents. We
then use the magnetic dipole and electric quadrupole moments
of the ^{6}Li ground state (electromagnetic data) to obtain two
relationships involving A,B,C, and b. Next we fit the elastic magnetic and inelastic ($1^+0 \to 0^+1$) electron scattering data
to fix b and to obtain a relationship between D and E. This
determines the entire set of parameters. Finally we can
check the results in several ways: we can compute the
^{6}He$\to$^{6}Li β-decay rate and the ^{6}Li$\to$^{6}He μ-capture rate and compare with experiment (the agreement is excellent; see Ref.
9 and the subsequent discussion in Refs. 7 and 8) and we can
compute the MEC corrections to the above 1-body analysis.
This latter calculation for electron scattering was reported
in Ref. 5 and some of the results are reporduced in Figs.
5 and 6. In Fig. 5 the inelastic form factor is shown as
a function of momentum transfer q both with and without MEC
included (the solid and dashed curves respectively). It is
clear that for momentum transfers below q $\simeq$ 500 MeV the MEC
corrections to the 1-body calculation are small ($\sim$ 10%) and
can safely be neglected. For q > 500 MeV there appears to
be large, even dominant MEC contributions (however, see the
discussion of Fig. 8 below). In Fig. 6 the same information
is presented as a fucntion

$$p(y) \equiv F_T(q)/f_{SN}f_{CM}(q/M_N)e^{-y}, \tag{7}$$

where h = $(bq/2)^2$, M_N is the nucleon mass, F_{SN} is the single
nucleon electromagnetic form factor and f_{CM} is the harmonic
oscillator center-of-mass correction[9]. If harmonic oscilla-
tors are sufficient for representing the radial dependence
of the 1p-shell wave functions then p(y) versus y should
yield a straight line. With the (e,e') data available when

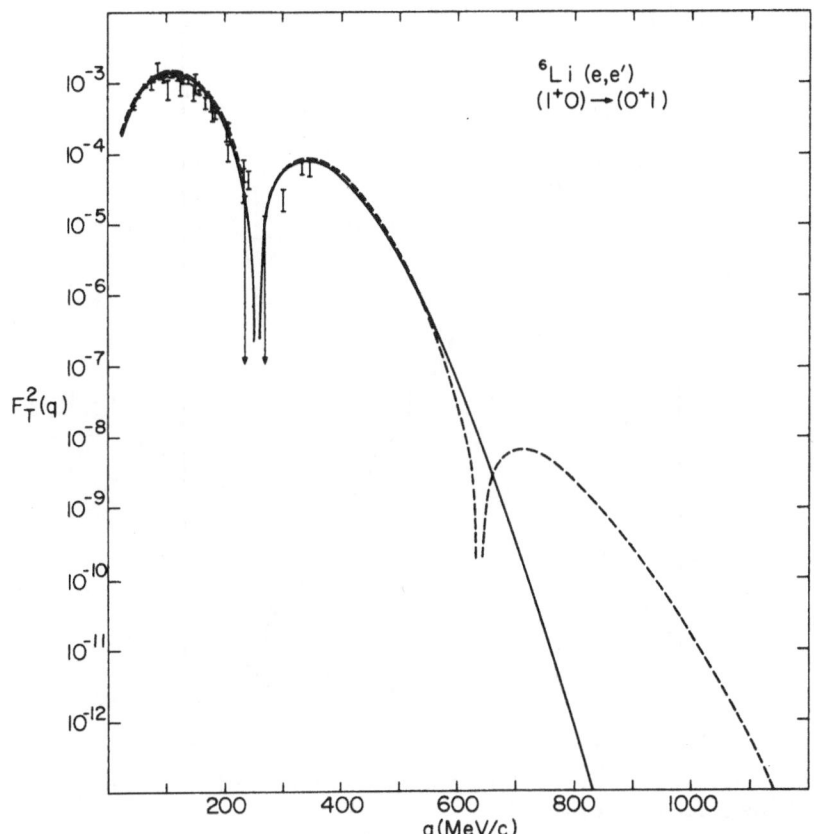

Figure 5. Transverse electron scattering form factor for the
$1^+0 \to 0^+1$ transition in ^{6}Li, with (dashed curve) and without
(solid curve) meson exchange current contributions. Sources
of experimental data are given in Ref. 9.

the analysis of Ref. 9 was undertaken the straight line fit
was adequate. However, it was interesting even then to see
how much the predictions would change with different radial
wave functions. In Ref. 9 Woods-Saxon 1p-shell wave functions
were employed and they yielded the modification $\Delta p(y) =$
$p(y)_{WS} - p(y)_{HO}$ shown in Fig. 7. Subsequently, when more pre-

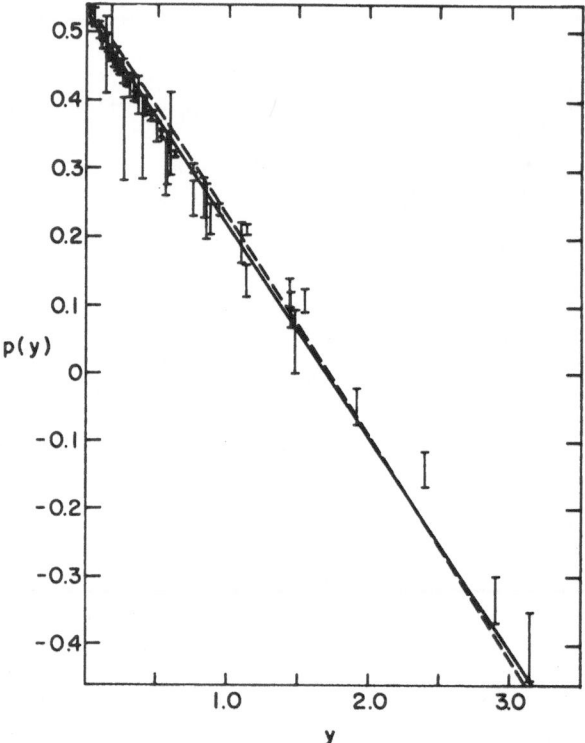

Figure 6. As for Fig. 5, but shown in terms of the function p(y) defined in Eq. (7).

cise data became available, it was seen that the oscillation about the straight line shown in Fig. 7 was in agreement with experiment (see the discussion in Ref. 8, where still other radial wave functions were employed, and also Ref. 7). Clearly the effect is not large, particularly for low to medium values of q. As can be seen in Fig. 8, the effect is large at high q and makes it hard to separate clearly the MEC effects from the choice of single-particle radial wave functions. Ideally all of these effects should be incorporated from the start into our analysis in terms of density matrices, however it is apparent that for low to medium values of momentum transfer we can use the basic 1-body density approach and hope to attain accuracies in our predictions at about

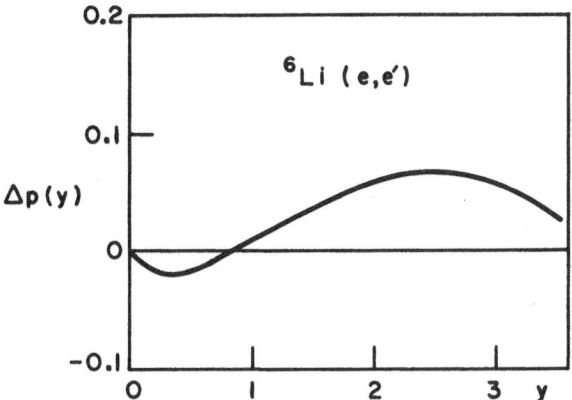

Figure 7. Modification of p(y) for Woods-Saxon wave functions compared to harmonic oscillators, $\Delta p(y) = p(y)_{WS} - p(y)_{HO}$.

10-15% level.

We are then in a position to calculate the cross-section for the photopion reactions (Fig. 3). With the checks on the analysis mentioned above we have confidence that the nuclear physics aspects of the problem (the density matrices) should be reliable to perhaps the 10-15% level and so comparisons with experimental photopion cross sections may tell us something about the elementary reaction mechanism. Such comparisons were carried out for $^6Li(\gamma,\pi^+)^6He$ near threshold (see Refs. 7 and 8 in the most recent data given in Ref. 10 and show excellent agreement between theory and experiment. Only the leading 1-body $\vec{\sigma} \cdot \vec{\epsilon}$ term was included in the elementary amplitude and we may conclude that near threshold this is the dominant elementary process.

Let us now turn briefly to a discussion of another system, the A=12 nuclei shown schematically in Fig. 9. Let us ignore the 2-body effects completely in this case and proceed with a 1-body density matrix analysis. We shall take the ^4He core to be a closed 1s-shell and allow the remaining eight active nucleons to be in all allowed configurations in the 1p-shell. Despite the complexity of the nuclear wave functions there are in fact only four 1-body density matrix

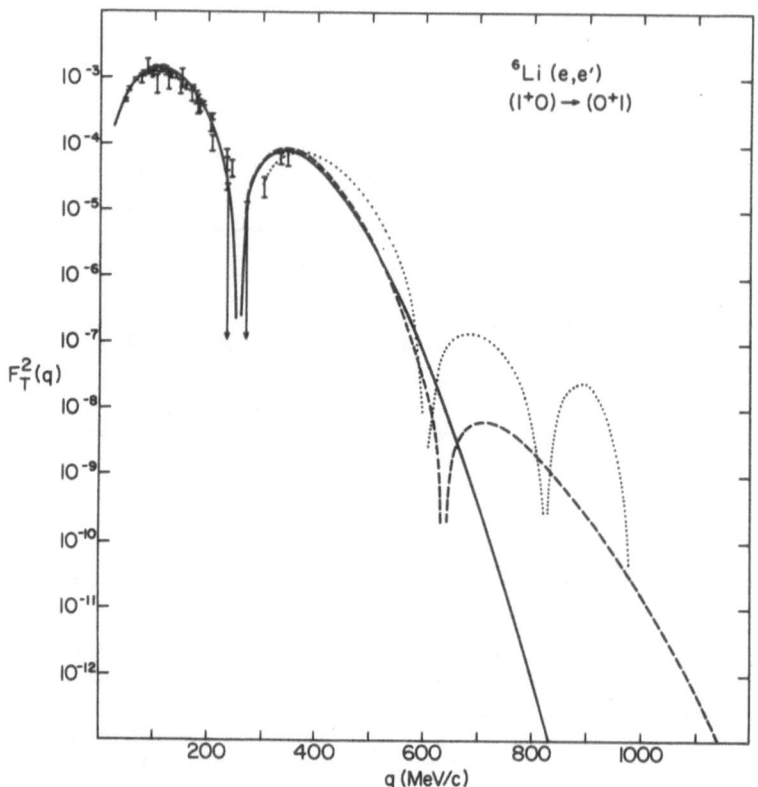

Figure 8. As for Fig. 5, but now including as well the 1-body result using Woods-Saxon wafe functions (dotted curve).

elements to consider:

$$\psi_{1;1}(3/2) \equiv \psi_{1;1}^{(fi)} (1p_{3/2} \ 1p_{3/2})$$

$$\psi_{1;1}(1/2) \equiv \psi_{1;1}^{(fi)} (1p_{1/2} \ 1p_{1/2})$$

$$\psi_{1;1}^{(\pm)} \equiv \psi_{1;1}^{(fi)} (1p_{3/2} \ 1p_{1/2}) \pm \psi_{1;1}^{(fi)}(1p_{1/2} \ 1p_{3/2}), \quad (8)$$

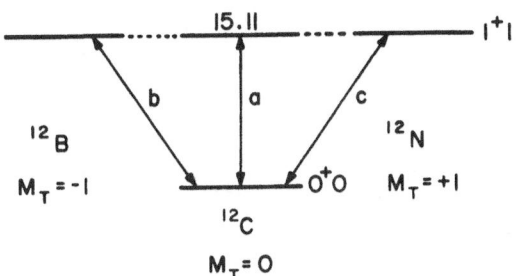

Figure 9.   The A=12 system indicating transitions considered:
(a) neutral currents, (b) and (c) charge-changing currents.

where "f" refers to the 1^+1 states and "i" to the 0^+0 state.
We employ harmonic oscillator wave functions again and so
have the oscillator parameter b in addition to the four density matrix elements. By obtaining a best fit to the inelastic electron scattering form factor (the details are given
in Ref. 4) we may extract the oscillator parameter and two
relationships amongst the ψ's which may be used to eliminate
$\psi_{1;1}(3/2)$ and $\psi_{1;1}(1/2)$ in terms of $\psi_{1;1}^{(-)}$ (there is no dependence on $\psi_{1;1}^{(\pm)}$ for the electromagnetic current):

$$\psi_{1;1}(3/2) = C_1\ \psi_{1;1}^{(-)} + C_2$$

$$\psi_{1;1}(1/2) = C_3\ \psi_{1;1}^{(-)} + C_4, \tag{9}$$

where the constants $C_1, \ldots, C_4$ are determined by the fit to
the (e,e') data. The conventional weak interaction processes
(Fig. 1b) may be then used to determine the last two un-
knowns, $\psi_{1;1}^{(\pm)}$. In Fig. 10 is shown the functional relationships between $\psi_{1;1}^{(\pm)}$ and $\psi_{1;1}^{(-)}$ which result when successively
the β^--decay rate, the β^+-decay rate and the μ-capture rate
are employed. These are the curves shown in Fig. 10, with
the maximum and minimum curves shown for μ-capture as there
is a relatively large error bar in this case. Somewhere
whinin the region of intersection of these curves is the
appropriate pair of values of $\psi_{1;1}^{(\pm)}$ and, through Eqs. (9),

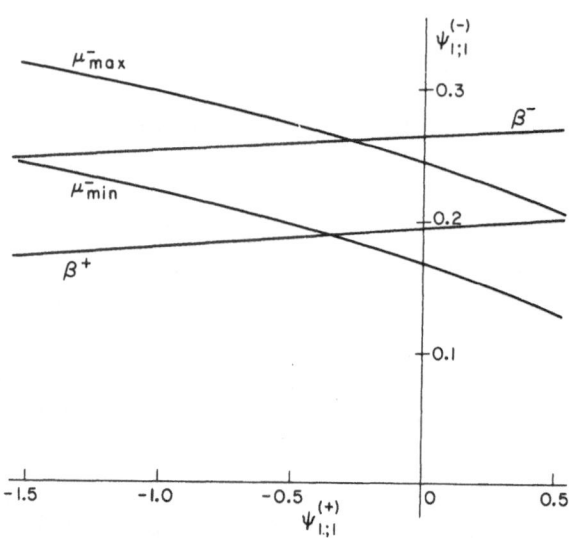

Figure 10. Acceptable regions of 1-body denstiy matrix
space using electron scattering data plus β^--decay rate,
plus β^+-decay rate and plus μ-capture rate (see text).

the appropriate full set of density matrix elements. We can
now proceed to compute cross sections for the other more
exotic weak interaction processes (Figs. 1c and 1d) or for
the photopion reactions (Fig. 3). Although there appears
to be considerable uncertainty left in the determination of
the ψ's in Fig. 10, such is not actually the case: when the
density matrix elements are used to predict these other
cross sections it is found that for a generous region of
parameter space the results change only on the 10-15% level.
Thus we can hope to test the nature of the elementary
processes by comparing with experiment. It may be found that
it is impossible to get consistently good agreement with all
of the data and that this may signal the entrance of other
effects such as the necessity of a larger model space or the
requirement that MEC effects be included. In fact, it appears
in the light of recent data on (e,e') from Bates/MIT[11] that

this may be the case. The possibility of learning such information is precisely the reason for pursuing the density matrix approach.

Finally, we may compute the photopion cross sections for comparison with experiment. This was done by Haxton[12], who performed an analysis similar to the one sketched above. He found good agreement with the threshold (γ, π) cross sections where, as in the A=6 case, the simple Kroll-Ruderman amplitude was employed.

In summary, we have been pursuing the following procedure:

1) The 1-body density matrix elements for a given nuclear transition are deduced from electromagnetic interaction data (i.e. electron scattering) by using the known q-dependence of the single-particle matrix elements in the basic equation (4).

2) 2-body MEC effects are computed using shell model wave functions. As these are frequently small corrections to the 1-body contributions we may tolerate reduced accuracy here. If these effects are sufficiently small they may be neglected; ideally they should be incorporated with the 1-body analysis.

3) **Weak interaction cross sections may be predicted.** Using the conventional weak processes (β-decay and μ-capture) the analysis may be checked.

4) Cross sections for other processes, such as the photopion reactions, may be calculated and the elementary nature of these processes investigated.

REFERENCES

*Research supported by the National Science Foundation, Grant NSF PHY 77-16188.
1. T.W. Donnelly and J.D. Walecka, Ann. Rev. Nucl. Sci. <u>25</u> (1975), 329.
2. J.D. Walecka, In <u>Muon Physics</u>, Vol. 2, ed. V.W. Hughes and C.S. Wu (Academic Press, New York, 1975), p. 113.

3. J.S. O'Connell, T.W. Donnelly and J.D. Walecka, Phys. Rev. C6 (1972), 719.

4. T.W. Donnelly and R.D. Peccei, Phys. Repts. (to be published).

5. J. Dubach, J.H. Koch and T.W. Donnelly, Nucl. Phys. A271 (1976), 279.

6. J.H. Koch and T.W. Donnelly, Nucl. Phys. B64 (1973), 478; Phys. Rev. C10 (1974) 2618 .

7. J.B. Cammarata and T.W. Donnelly, Nucl. Phys. A267 (1976) 365.

8. J.C. Bergstrom, I,P. Auer and R.S. Hicks, Nucl. Phys. A251 (1975) 401.

9. T.W. Donnelly and J.D. Walecka, Phys. Lett. 44B (1973) 330.

10. G. Audit et al., Phys. Rev. C15 (1977) 1415.

11. J. Dubach and W.C. Haxton (private communication).

12. W.C. Haxton (to be published).

RADIATIVE PION CAPTURE

J.P. Perroud*

Institute of Nuclear Physics

BSP Dorigny

University of Lausanne

CH 1015 Lausanne, Switzerland

1. INTRODUCTION

The field of radiative pion capture was opened in 1950 by Panofsky et al.[1] by measuring with a pair ospectormeter what is now called the Panofsky ratio. In 1970 a measurement of $^{12}C(\pi^-,\gamma)^{12}B$ by Crowe et al.[2] showed sharp lines in the photon spectrum around 125 MeV establishing the sensitivity of the process to the nuclear structure. In the past three years, a pair spectrometer[3] with a resolution of 900 KeV (FWHM) at 130 MeV was used at SIN to measure high energy photon spectra throughout the periodic table with many new levels resolved. Furthermore, the high pion intensity available at meson facilities allowed to collect a high statistic ^2H spectrum which led to a new measurement of the neutron-neutron scattering length a_{nn}, and to isolate for the first time 1s radiative capture by a coincidence experiment in ^6Li. Except for very light nuclei ^3H, ^3He and ^4He which have been already extensively reviewed[4] almost all of the Berkeley group's 2 MeV resolution spectra have been remeasured at SIN in a collaboration between the Universities of Lausanne, Munich and Zurich (LMZ). I will therefore concentrate essentially on these new data and try to make some comparisons with measurements in other fields.

2.   THEORETICAL BACKGROUND

Except for the so-called elementary particle (EP) treat-ment[5] based on current algebra techniques, which will be discussed below, all theories use the shell model and impulse approximation to describe the process ${}_{N}^{Z}A(\pi^{-},\gamma){}_{N+1}^{Z-1}A$. In this framework, the radiative capture rate for a transition from a pionic orbit $(n\ell)$ leading from an initial nuclear state J_i M_i to a final state J_f M_f may be written as[6]:

$$\Lambda_\gamma(n\ell;\ i{\to}f) = \frac{1}{2J_i+1}\frac{1}{2\ell+1}\frac{k}{m_\pi}\frac{1}{1+k/M_A}\int d^3k\,|<J_fM_f|H_{eff}|J_iM_i>|^2$$

$$H_{eff} = i(1+\frac{m_\pi}{m_n})\sum_{j=1}^{A}e^{-i\vec{k}\cdot\vec{r}_j}F_{j\lambda}\,\phi_{n\ell}^m(\vec{r}_j)t_j^+ \tag{1}$$

where $\vec{k}$ is the photon momentum, t_j^+ the isospin operator changing proton j into a neutron, $\phi_{n\ell}^m$ the pion wave function and m_π, m_n, M_A the pion, nucleon and target nucleus masses. $F_{j\lambda}$ is the elementary amplitude for the process $\pi^-p{\to}n\gamma$ and is given by:

$$F_{j\lambda} = A\vec{\sigma}_j\cdot\hat{e}_\lambda + B(\vec{\sigma}_j\cdot\hat{e}_\lambda)(\vec{q}\cdot\vec{k}) + C(\vec{\sigma}_j\cdot\vec{k})(\hat{e}_\lambda\cdot\vec{q}) - iD\hat{e}_\lambda\cdot(\vec{q}x\vec{k})$$

$$+ E(\vec{\sigma}_j\cdot\vec{q})(\vec{q}\cdot\vec{e}_\lambda) \tag{2}$$

where $\vec{q}$, $\vec{\sigma}_j$ and $\hat{e}_\lambda$ are the pion momentum, the spin operator acting on nucleon j and the photon polarization: A,B,C,D,E are linear combinations of photoproduction multipoles discussed in Ref. 4.

For 1s capture only the first term is important, but for 2p capture already sizable in ^{3}He, the momentum dependent terms contribute significantly and should be retained. The expansion of the photon wave function in terms of vector spherical harmonics gives for the rate[8]

$$\Lambda_\gamma (n\ell; i \to f) = 16\pi^2 \frac{k}{m_\pi} \frac{1}{1+k/M_A} \frac{1}{2J_i+1} \frac{1}{2\ell+1} (1+ \frac{m_\pi}{m_n})^2$$

$$\times \sum_{JL} |<J_f||A(JL\ell)||J_i>|^2 + |<J_f||B(JL\ell)||J_i>|^2 \qquad (3)$$

$A(JL\ell)$ and $B(JL\ell)$ being spherical tensors of rank J with respective parities $(-1)^{L+\ell}$ and $(-1)^{L+\ell+1}$, which can be written as:

$$\sum_{j=1}^{A} \sum_\alpha F_\alpha^{JL} (k,r_j) [\vec{\sigma}_j x Y_\alpha(\hat{r}_j)]^J t_j^+ + \sum_{j=1}^{A} G^{JL} (k,r_j) Y_J(\hat{r}_j) t_j^+$$

$$(4)$$

The contribution of all spin dependent terms (A,B,C,E) is given by the first part, where the sum runs over a few number vales of α. The second part arises from the D spin independent term. F and G functions contain only radial coordinates including pion and photon wave functions as well as Racah coefficients.

The Gamow Teller (GT) operator $\sum_{j=1}^{A} \sigma_j t_j^\beta$, a generator of the SU (4) group, appears in the first term and will lead to nuclear structure sensitivity involving spin flip. In fact, the search for the spin isospin type of excitation, not seen in photoabsorption, motivated the Berkeley group for their experiment on $^{12}C^2$. In the SU (4) classification, starting with a $J^\pi = 0^+$ target, these states belong to the L = 1 S = 1 $J^\pi=0^-,1^-,2^-$ T=1 $T_z=0\pm1$ supermultiplet. The GT operator is also the link between radiative pion capture and weak processes like β decay and muon capture and also with electron scattering. For the latter, the GT dominance is explicit in the isovector part of the MI operator, responsible for $|\Delta T| = 1$ transitions in inelastic scattering. The analogs of the strongly excited M1 state will be excited in radiative capture which will act as an analyzer of the SU (4) purity of the nuclear states involved.[9]

Either hydrogenic π^- wave functions are used, introducing a correction coefficient R to account for the mean distortion in the nuclear volume[10] or solutions of the Klein-Gordon equation in the appropriate optical potential[8,11]. The wave function distortion manifests itself as level shifts

and widths measured in π^--atom studies[12]. Since π^- are in general not captured from a single atomic orbit, the measured branching ratio R_γ is connected to the calculated

rates through capture schedule ω_{nl} and total absorption widths Γ_a (nl). Including some approximations[4] for p shell nuclei,

one has $R_\gamma = \omega_{1s}\Lambda_\gamma(1s)/\Lambda_a(1s) + (1-\omega_{1s})\Lambda_\gamma(2p)/\Lambda_a(2p)$. No matter how precise the branching ratio is measured, the uncertainties in π^- atom studies will spoil the comparison with theories. The effect of these uncertainties are somewhat reduced if one compares ratios between different states[13].

It is worth to note here the similarity of radiative capture and the reverse process of photoproduction near threshold (γ,π). Except for some minor differences in pion wave function distortion and different values of A,B,C,D, and E for π^+,π° and π^-, the physics involved is quite the same. In (γ,π) process one has the attractive possibility of varying the momentum transfer $q = |\vec{k}_\gamma - \vec{p}_\pi|$, e.g., from .4 to 1.5 fm^{-1} for $T_\pi < 50$ MeV, whereas in (π^-,γ) this quantity is fixed for a transition to a specific level (typically $q = k_\gamma = .65$ fm^{-1}).

An easy quantity to measure is the total branching ratio which has not found an interpretation. A calculation by Werntz[7] predicts linear variation with (A-Z)/2A and a strong isotopic effect for this quantity.

3. VERY LIGHT NUCLEI

3.1 Panofsky Ratios for ^{1}H, ^{2}H, ^{3}He and ^{6}Li

The ratios for charge exchange (π^-,π°) to radiative capture (π^-,γ) at rest in ^{1}H, ^{2}H, ^{3}He and ^{6}Li have been measured recently at Triumph by an UBC-Montreal-McGill collaboration.

The very low value of the Panofsky ratio in deuterium requires a good separation between deuterium and hydrogen
charge exchange events (3% [1]H contamination in the liquid
target). The π° decay 2γ were measured in coincidence by
two NaI crystals at 180°, restricting detection to π° travel-
ing towards the detectors. In one of the detectors mono-
energetic (2.9 MeV)π° from π^- p$\rightarrow\pi^\circ$n produces two peaks
whereas the continuum spectrum π° ($T_{\pi^\circ} < 1$ MeV) from π^- d$\rightarrow\pi^\circ$nn
fills the central region from 61.7 to 73.9 MeV, allowing a
nice separation as seen in Fig. 1. This technique has produced for the first time a value for the ^{2}H Panofsky ratio
(Table 1). It was also applied to ^{6}Li, where a small upper
limit was obtained. In hydrogen, accuracy is improved by a
factor 2 over Cocconi's result[14] and leads to an accurate
value for $|a_1 - a_3|$, the $T = \frac{1}{2}$ and $\frac{3}{2}$ S-wave π^- n scattering
length difference. For ^{3}He, the Panofsky ratio agrees with
the earlier Berkeley result, but is inconsistent with E.P.
treatment (see Ref. 15 for a discussion). In the context of
mass-3 wave functions calculated by Phillips and Roig[16], the
percentage of D state is limited by the high Panofsky ratio.

3.2 π^-d$\rightarrow$nnγ and the Effective Range Theory Parameters a_{nn} and r_{nn}

The effective range expansion of the ^{1}S$_0$ n-n phase
shift $\delta_0(p)$ at a relative momentum p is pcotg $\delta_0(p) = -1/a_{nn}$
$+ p^2 r_{nn}/2$. The numerous experiments (about 50) done so far
yield an average value[20] of $-16.6 \pm .6$ fm for the scattering
length a_{nn}. Amongst all reactions, ^{2}H (π^-,γ)nn and ^{2}H(μ^-,ν)nn
are unique for studying the two neutrons final state inter-
action (FSI) since no other strong interacting particle is
present, a fact which makes the theoretical calculations
necessary to extract a_{nn} more reliably. The commonly admitted
value of Henley[21]: $-16.4 \pm .9$ fm is heavily weighed by the
complete radiative pion capture experiment of Haddock[22]
($a_{nn} = -16.7 \pm 1.3$ fm) and has not changed these last years[20].

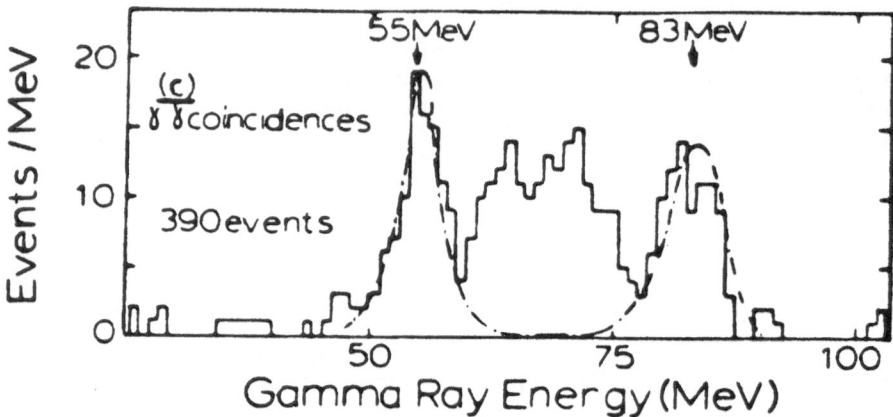

Figure 1.   Energy spectrum of one of the detectors for coincidence events showing the separation between hydrogen events
(two peaks) and deuterium events (continuum between the
peaks).

Bander's[23] theory was used to analyze the experiment. It is
based on impulse approximation and a dispersion relation
method is used to describe the FSI of the two neutrons.   The
theoretical uncertaintly is less than 1 fm, provided the
relative momentum p is limited to 24 MeV/c.   Two more recent
theories give lower theoretical uncertainties.   Both are
based on the impulse approximation but differ in the FSI
description.   Gibbs[24] et al. use a semi-phenomenological potential, whereas de Teramond's[25] theory is a continuation of
Bander's approach, solving the Omnes-Muskheleshvili[26] equation
exactly.   In both cases careful studies of the approximation
made, in the range $p < 35$ MeV/c, resulted in an uncertainty
of less than .3 fm for the extraction of a_{nn}.   In a complete
coincidence experiment one can easily select events with
$p < 35$ MeV/c but the neutron detection will introduce experimental difficulties and uncertainties associated with the
efficiency calibration of the neutron detectors.   On the
other hand, if the theory is not too sensitive to the limit

TABLE 1
Panofsky Ratios in Light Nuclei

Target	Panofsky Ratio	Ref.	Interpretation
^1H	$1.546 \pm .009$	17	$\|a_1 - a_3\| = (.263 \pm .005)m_\pi^{-1}$
	$1.533 \pm .021$	14	
^2H	$(5.76 \pm .76) \times 10^{-4}$	18	$\lim q \to 0 \; \|q\sigma(\pi^- d \to \pi^\circ nn)\|$ $= .0358$ MeV mb
^3He	$2.28 \pm .18$	44	
	$2.68 \pm .13$	15	Inconsistent with EP value of $1.9 - 2.1$
			Mass-3 w.f. $P(S^1) = 1.6\%$ $P(D) = 4-7.5\%$ (Ref. 16)
	$2.83 \pm .07$	19	
^6Li	$< 2 \; 10^{-6}$	19	

on p, the last 2 MeV (p $\sim$ 45 MeV/c) of a huge **statistics photon** spectrum obtained with a resolution below 1 MeV should provide a good measurement of a_{nn}.

To extract a_{nn} the theoretical spectrum is folded with the experimental response curve of the spectrometer and fitted with the measured spectrum. The deuterium spectrum and the line shape ($\pi^- p \to \gamma n$) are measured in exactly the same conditions by filling alternatively the liquid target with hydrogen and deuterium. The critical point is the stability of the spectrometer since a 10 KeV energy shift between hydrogen and deuterium runs will result in a .7fm error in a_{nn}. To achieve this stability, temperature, pressure (affecting energy losses in the chambers) and magnetic field were recorded and the energy of each event corrected accordingly. The stability was checked and a fluctuation of less than 2 KeV was observed. The fitted theoretical curves[24,25] and the measured spectrum are shown in Fig. 2 for a fit in the energy domain between 130 and 132 MeV. In Fig. 3 we have a typical χ^2 dependence on a_{nn} with $\Delta a_{nn} = .34$ fm for $\chi^2 - \chi^2_{min} = 1$.

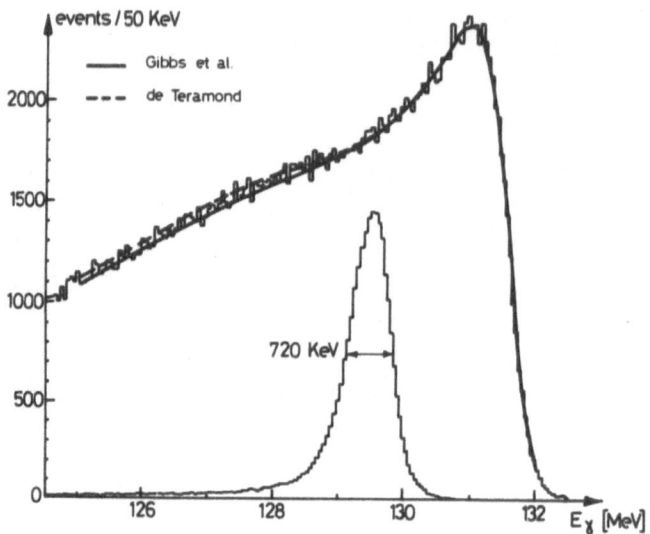

Figure 2.   Higher part of the γ spectrum from $\pi^- d \rightarrow nn\gamma$, the response of the sepctrometer ($\pi p \rightarrow \gamma n$) is also shown.   The fit to the theoretical spectra of Gibbs[24] and de Teramond[25] for $E_\gamma > 130$ MeV lead to $a_{nn} = -18.3 \pm .55$ fm.

Splitting the data in 8 runs gives no variation of a_{nn} with time.   However, the data showed a fluctuation (.4 fm) a little larger than the .34 fm quoted above which was associated with slight fluctuations in the shape of the spectra.   Figure 4 gives the variation of a_{nn} with the energy range used for the fit.   For $E_{\gamma min} > 120$ MeV (p < 50 MeV/c) no difference between the two theories is seen.   For all these fits, r_{nn} was fixed at the r_{pp} equivalent value of 2.8 fm.   The final value of a_{nn}, obtained with $E_\gamma > 130$ MeV, is not sensitive to the value of r_{nn}.   The latter is then determined from a fit in the energy range from 127 to 132 MeV with a fixed value for a_{nn}.   The results are

$$a_{nn} = -18.3 \text{ fm} \quad \Delta a_{nn} = .55 \text{ fm} \quad r_{nn} = 2.85 \quad \Delta r_{nn} = .35 \text{ fm}$$

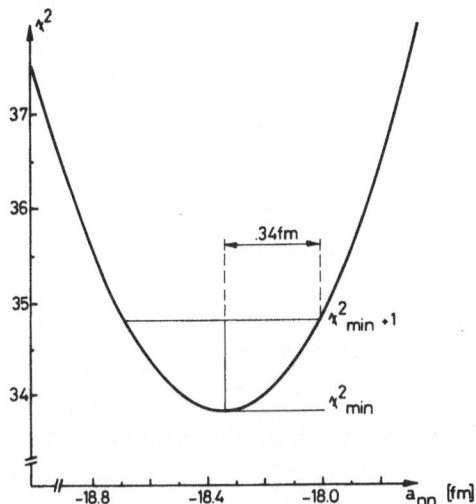

Figure 3. Sensitivity of the shape of the higher part of the spectrum to a_{nn}.

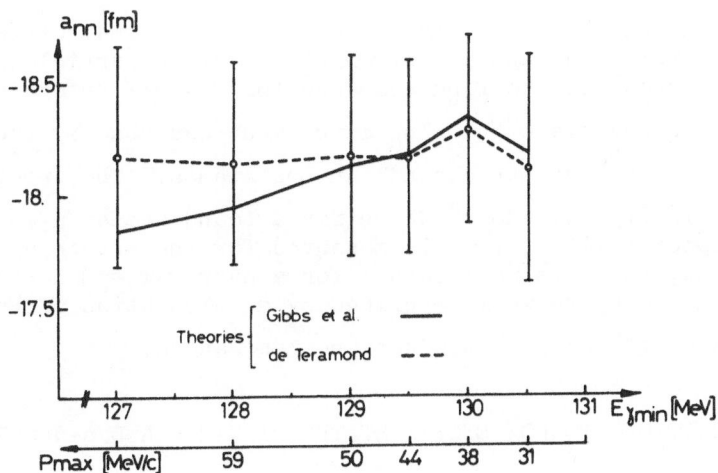

Figure 4. Stability of the extracted value of a_{nn} versus the range of the fit for the two theories[24,25].

The uncertainty in a_{nn} includes fluctuations of the recorded parameters (.2 fm), statistical error including the shape fluctuation (.4 fm) and the theoretical uncertainty of .3 fm. The uncertainty in r_{nn} is kept at a conservative value to account for unknown theoretical behavior.

Although more precise, the value for a_{nn} differs significantly from the 1972 value of Henley and Wilkinson: a_{nn} = -16.4 ± .9 fm. Neglecting the Sauer[27] difficulty associated with the Coulomb correction, the p-p scattering parameters given by the same authors are: a_{pp} = -17.1 ± .3 fm and r_{pp} = 2.84 ± .03 fm. The difference $|a_{nn}| - |a_{pp}| \simeq 1$ fm support the calculations of the electromagnetic indirect corrections due to ω-ρ mixing[28,29] and agree with the binding energy difference in the A = 3 mirror nuclei suggesting a short range potential a little more attractive in the n-n case.

3.3 Radiative Capture on Tritium and 3-Neutron Excitation Spectrum

As in the deuterium case, the 3 neutrons final state is not disturbed by another strong interacting particle and affords a good case to find evidence for T = 3/2 resonances or even a bound state ^{3}n. The experiment was done by the Berkeley group[30] at LEP channel in Los Alamos. The spectrum is given in Fig. 5. No evidence for a bound state ^{3}n is seen and an upper limit of .3 % is measured for the branching ratio. Also, there is no evidence for a narrow T = 3/2 resonance, the data are well described by a calculation of Phillips and Roig[32] which introduces no resonances.

4.   TRANSITIONS TO LOW ENERGY STATES WITHIN p AND s-d SHELLS

4.1 1s Capture in ^{6}Li

The LMZ collaboration isolated for the first time radiative capture from a single pionic orbit in ^{6}Li[31]. The

high energy photon was observed in coincidence with the
(25 to 30 KeV) x-rays feeding the ls orbit by means of four
properly shielded NaI crystals. The shape of the coincidence spectrum (Fig. 6b) is not dramatically different from
the singles spectrum (Fig. 6a). The branching ratios to the
well resolved 0^+ ^{6}He ground state and 2^+ first excited state
were extracted after substraction, below the latter, for a
small contribution due to the direct transition to the four
body final state (^{4}He,n,n,). The ratio $\Lambda_\gamma^{ls}/\Lambda_a^{ls}$ is directly given by the ratio for the same amount of beam: γ-x co-
incidences/ x events, divided by the acceptance of the spectrometer. The results are given in Table 2. The errors in
R_{2p} and Λ_γ^{2p} reflect the uncertainties in π^- atom data[38,46]

The well documented ^{6}Li presents an ideal case to test
the elementary particle treatment which link weak and electromagnetic processes like (π^-,γ) or (γ,π^+) at threshold.
In EP description the nucleus as a whole is described by a
set of form factors. For ls radiative capture or photoproduction near threshold, using PCAC and the pion pole domi-

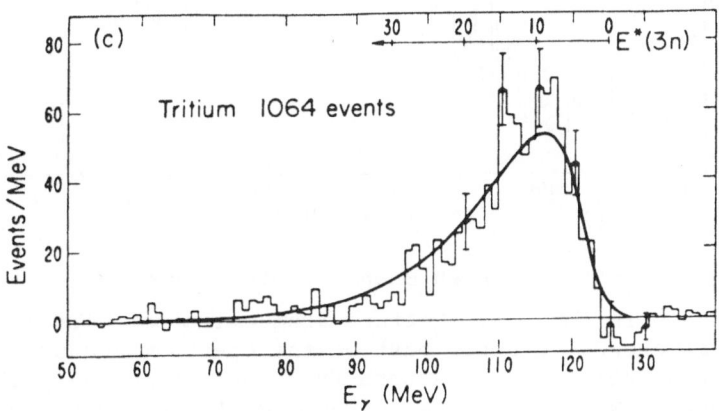

Figure 5. Photon spectrum from ^{3}H(π^-,γ)nnn. The solid curve
is the result of a calculation by Phillips and Roig[32] with
no resonances introduced.

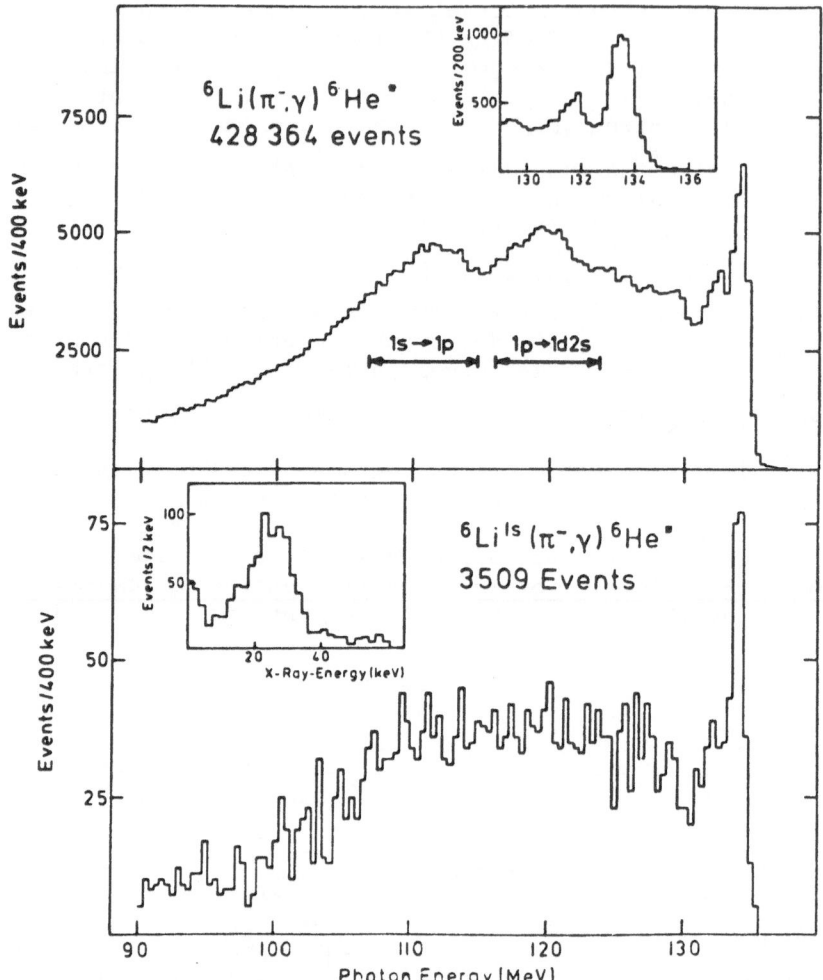

Figure 6. Spectrum of high energy photons from radiative pion capture in ^{6}Li. a) Capture from all atomic levels. The insert shows the upper part of the spectrum measured with optimum resolution. b) Capture from the atomic 1s-level. The insert shows the x-ray spectrum from the coincidence measurement.

nance hypothesis, Delorme[33] showed that a single axial form factor $F_A(q)$ is needed in the soft pion limit. The axial form factor at zero momentum transfer $F_A(0)$ is directly related to the rate for β- decay and it is assumed to behave with momentum transfer like the magnetic form factor measured in electrom scattering: $F_A(q) = F_A(0)F_M(q)/F_M(0)$. Using the latest experimental data[34,35] on $F_M(q)^2$ one arrives at a prediction which is 1.2 standard deviation larger than the experimental value. Furthermore, the Λ_γ^{1s} obtained is in very good agreement with the new measurement of (γ, π^+) by Audit et al.[36]. Apart from kinematic factors, the two processes differ only in the center of mass photon energies k_γ^- and k_γ^+ and the pion wave function distortion factors R_s^- and R_s^+. Using the values obtained from the Krell-Ericson potential, $R_s^- = .654$ and $R_s^+ = .957$, the combined results of the two experiments yield $F_A(k_\gamma^-)/F_A k_\gamma^+) = 1.14 \pm .07$ as compared to $F_M(k_\gamma^-)/F_M(k_\gamma^+) = 1.12$ from electron scattering[34,35].

Phenomenological wave functions have been constructed for the ^{6}Li g.s. and the 3.56 MeV $J^\pi = 0^+$ T = 1 excited state,

TABLE 2. Radiative π^- Capture Rates and Branching Ratios in ^{6}Li

Final State	R %	R %[45]	R_{1s} %	R_{2p} %	$\Lambda_\gamma^{1s}(10^{15}s^{-1})$	$\Lambda_\gamma^{2p}(10^{10}s^{-1})$
^{6}He 0^+ g.s.	.34 ± .03	.31 ± .03	.47 ± .05	.26 ± .05	1.39 ± .16	5.8 ± 1.9
^{6}He 2^+ 1.8 MeV	.11 ± .01	.15 ± .03	.11 ± .04	.11 ± .04	.33 ± .12	2.5 ± 1.1
All	4.02 ± .3	4.39 ± .58	4.69 ± .33	3.57 ± .44	13.9 ± 1.3	81. ± 24.

Pionic atom data
Refs. 38, 46. $w_{1s} = .4 \pm .09$ $w_{2p} = .6 \pm .09$ $\Gamma_a^{1s} = 195 \pm 12$ eV $\Gamma_a^{2p} = .015 \pm .004$ eV

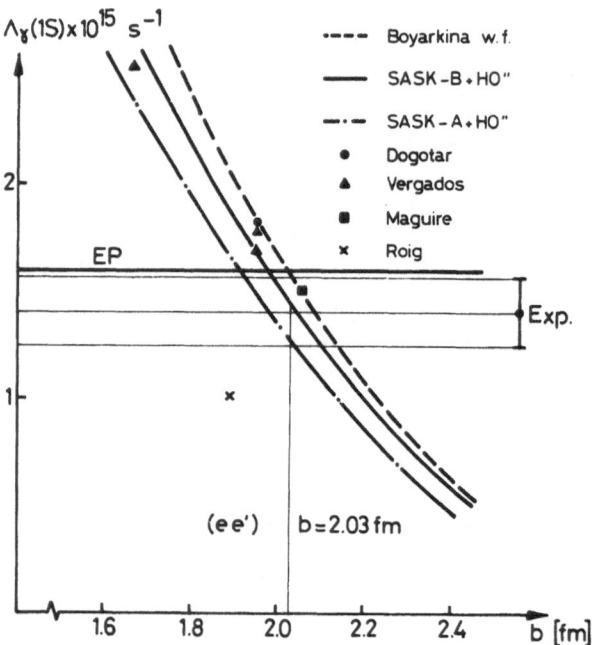

Figure 7.   Calculated rate $\Lambda_\gamma^{1s}(1^+ \to 0^+)$ for the phenomenological wave function[34] and structure type wave functions[37] versus H.O. parameter b.   The discrepancies between previous calculations[6,40,41,39,11] seem to be associated with the H.O. parameter chosen.

using the static properties of ^{6}Li and the excellent electron scattering data of Bergstrom[34]. The remaining ambiguity appears to be in the relative amount of L = 1 configuration in the 0$^+$ state. The M1 form factor is equally well fitted with two wave functions: SASK −A 26% L = 1 and SASK −B ∼ 0 % L = 1

Using harmonic oscillator (H.O.) radial wave functions, the

calculated Λ_γ^{1s} rate is represented in Fig. 7. One notes a
strong dependence on the H.O. parameter b; using the value
determined from the best fit to the M1 form factor[35], a light
preference for the SASK -B wave function is found. The discrepancies between calculations using structural type wave
functions seems to be associated with the different values
used for the H.O. parameter. The $\Lambda_\gamma^{2p}(1^+ \to 0^+)$ measured rate
can be used to test the momentum dependent terms contribution
to (1). In this case A, B and C terms contribute. Using the
π^- hydrogenic wave function with a distortion coefficient
$R_{2p}^- = 1.1$ and the set of Ref. 4 for the values of A, B, C,
and D, one obtains the curves presented in Fig. 8 for the
different nuclear wave functions. The B and C terms contri-
bute to about 40% of the Λ_γ^{2p} rate and the discrepancy between
different calculations is mostly associated with the differ-
ent values used for these coefficients.

It is interesting to note the similarity between the
(π^-,γ) operator multipole expansion and the corresponding
transverse form factor's multipoles. For 1s capture, if a
constant π^- wave function inside the nucleus is assumed, the
(π^-,γ) multipole of order J is identical to the isovector
part of the transverse form factor of the same order with
opposite parity, provided the small current density contribution to the latter is neglected. For 2p capture this similarity is lost due to the important momentum dependent terms
contribution and the appearance of the π^- wave function
$\phi_\pi \sim r$ which favors surface capture. The M1, E2 or M3 components contribute to the transverse form factor associated
with the transition to the 5.36 MeV 2^+ T = 1 level, and the
analog 1.8 MeV in ^{6}He can be excited by emission of a E1, M2
or E3 hard photon. In Fig. 9, the ratio $F_T(2^+)/F_{M1}(0^+)$ is
plotted versus momentum transfer q, the corresponding
$\Lambda_\gamma^{1s}(2^+)/\Lambda_\gamma^{1s}(0^+)$ agree perfectly with (e,e') data, whereas
the same ratio for 2p capture is about 2 times bigger. This
ratio is overestimated by Dogotar et al.[11], using the Barker
wave function. It is not absolutely clear whether the higher
value of the equivalent ratio $a(2^+)/a(0^+)$ in (γ,π^+)[36] should
contribute to the important ^{4}He + n + n continuum.

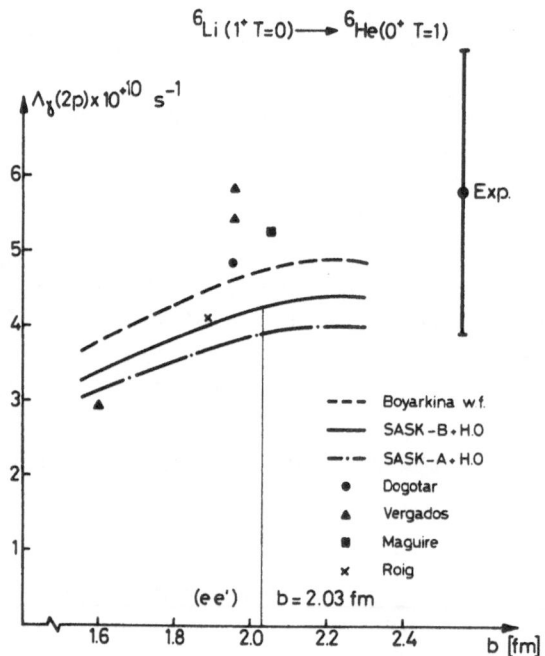

Figure 8. Calculated rate $\Lambda_\gamma^{2p}(1^+ \to 0^+)$. The references are
the same as in Fig. 7. Here the discrepancy is mostly associ-
ated with the different values used for the momentum dependent
terms B and C.

4.2 Branching Ratios to Low Lying States

        For the selfconjugate nuclei ^{10}B, ^{12}C, ^{14}N, the level
spacing in the bound states region of the final nuclei and
the concentration of strength into a few levels allowed the
Berkeley group[2,47] to isolate certain transitions even with
a 2 MeV resolution. Whereas their analysis was strongly
guided by theoretical considerations, the observed transitions
appear as clear peaks in the spectra with the improved resolution (Fig. 10). New energy photon spectra for ^7Li, ^9Be and
^{11}B were also measured by the LMZ collaboration completing,
with the exception of ^{13}C, the study of p-shell stable nuclei.

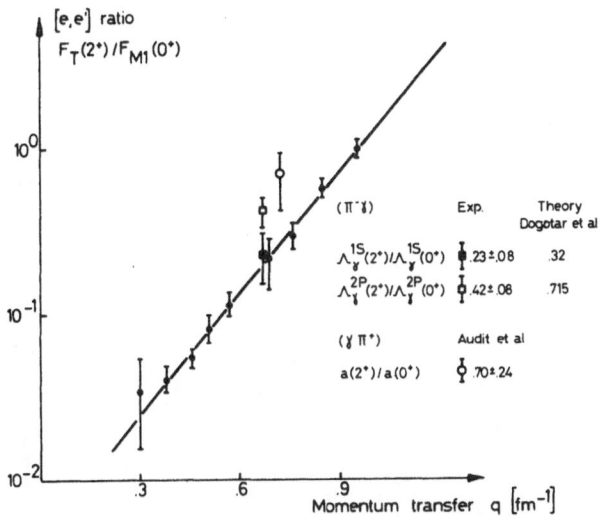

Figure 9.   The ratio of the transverse form factors[34,43] for the transition to the 2^+ and 0^+ T=1 states in ^{6}Li is plotted versus momentum transfer q.   The corresponding ratio for 1s capture agrees well with (e,e') data whereas for 2p capture this ratio is about 2 times larger.   The same ratios for $(\gamma, \overset{\bullet}{\pi}{}^+)$[36] and from theory[11] are also given.

When the states are reasonably spaced and their exact energy positions known as is the case for the four first excited states in ^{12}B (Fig. 10b) a good separation is obtained from the well known line shape of the instrument.   For obvious structure slightly above the neutron break-up threshold the branching ratio is extracted after substraction of a continuum contribution due to the three body final state n + $^{Z-1}_{N}$A+γ (Fig. 10a and 10c).   In some cases the .9 MeV resolution is still not enough to separate states, for example in ^{14}N (Fig. 10c), the peak at 6.8 MeV in ^{14}C is not resolved.   In this region Dogotar et al. indicate that 2 states are strongly excited, a 2^+ and a 3^-.   High energy resolution photon spectra for ^{16}O, ^{18}O, ^{28}Si, ^{32}S and ^{40}Ca have also been measured by the LMZ collaboration.   The high density of states in the s-d shell requires further hypotheses.   For the ^{28}Si case

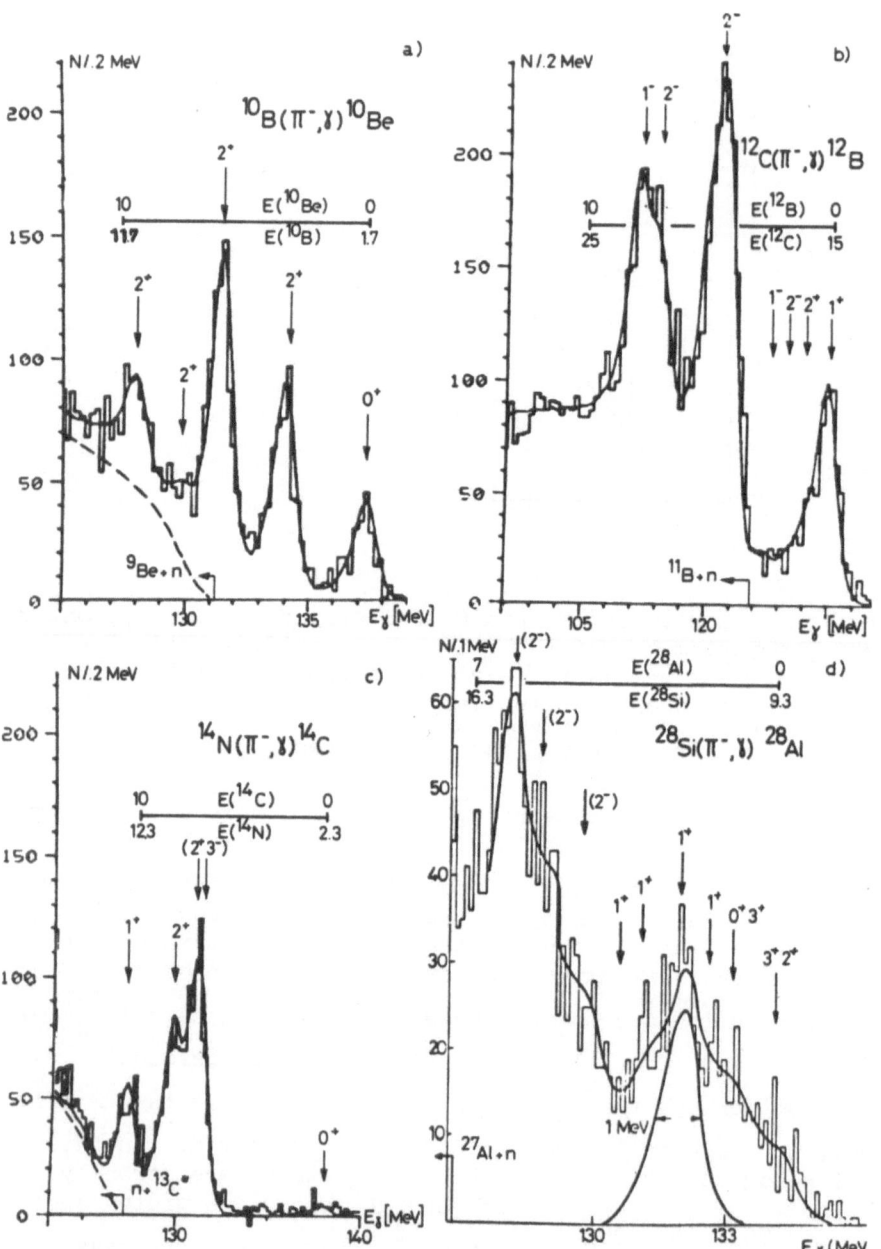

Figure 10. Upper part of the photon spectra from radiative
π^- capture in ^{10}B (a), ^{12}C (b), ^{14}N (c) and ^{28}Si (d).

(Fig. 10d), it was assumed that besides the g.s. doublet $(2^+, 3^+)$ and the doublet at about 1 MeV in ^{28}Al, there would be only transition strength to the analog of the M1 states excited at 180° electron inelastic scattering up to 3.5 MeV excitation. Above 3.5 MeV a minimum of 3 lines at 4.1, 5.2 and 6.1 MeV in ^{28}Al is needed to account for the observed spectrum.

A summary of the measured branching ratios is given in Fig. 11 where the available transverse form factors $F_T^2 (k_\gamma)$ at momentum transfer $q = k_\gamma$ involving the analog state in the target nucleus are also represented. The scale to the right is shifted upwards to match the two representative point for the ^{6}Li$\rightarrow$^{6}He g.s. transition. An evident correlation is seen between these quantities, furthermore a constant ratio $R_\gamma/F_T^2 (k_\gamma)$ is observed for many states. This is a little surprising since the branching ratios involve the total absorption widths Γ_a^{1s} and Γ_a^{2p} and the 1s and 2p contributions vary with A. A close examination of these relations would certainly be very interesting. The excitation of the analogs of the M1 isovector states seen in electron scattering arises mainly through terms of the transition operator of the form $[\vec{\sigma} \times Yo]^1 t^+$; a proton is changed into a neutron without changing its spacial quantum numbers. The Dubna group performed extensive calculations for p-shell nuclei using Cohen-Kurath type wave functions[11], their results are also given in Fig. 11. Qualitatively the distribution of the T=1 M1 strength is well understood in the 1p shell. In the pure SU(4) limit (no spin orbit coupling) one expects some strength only in 4N + 2 nuclei[9], i.e., ^{6}Li, ^{10}B and ^{14}N. Since in the intermediate coupling model of Kurath there is a large mixing of configurations of different permutation symmetries in the middle of the shell. (A = 11 - 13), but only marginal impurities at the beginning and the end of the shell, one expects low strength in ^{7}Li and ^{9}Be and some strength through impurities in ^{12}C and ^{13}C. In ^{10}B the strength is distributed over many states since two doorways are available in this case. The strength in ^{14}N is

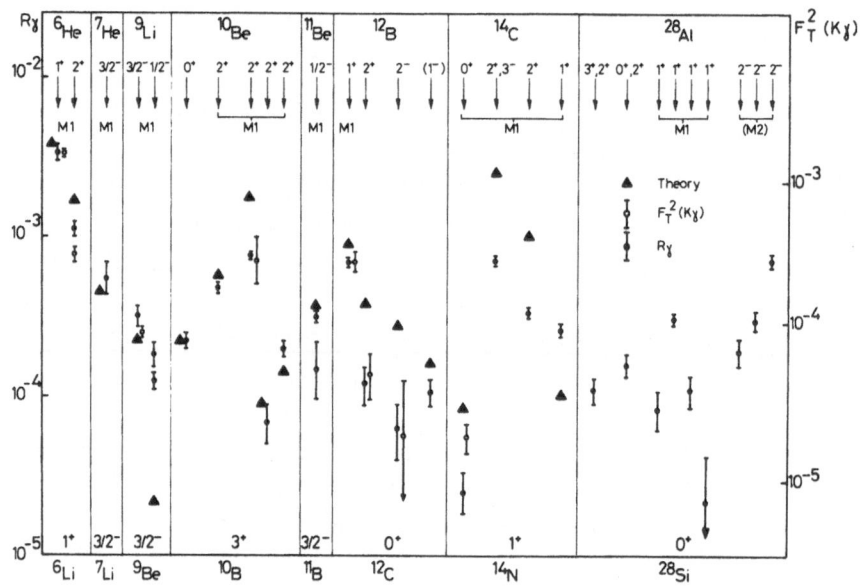

Figure 11. Summary of the measured branching ratios. The available transverse form factors at the same momentum are also given. The scale to the right is shifted upwards to match the respresentative points for the ^{6}Li→^{6}He gs transition. The calculated branching ratios[11] with Cohen-Kurath type wave functions are also shown.

split due to admixtures of higher configurations. As seen
in Fig. 11, the agreement between theory and experiment is
not always satisfactory, especially in ^{14}N where it is known
from electron scattering that more complicated structure or
tensor force is necessary to explain the measured form factor to the T=1 0^+ state. The excitation of M1 analogs in
the s-d shell is clearly established. The Berkeley group
observed recently M1 strength in ^{20}Ne[48]. In ^{28}Si the distribution of strength among the observed states is about the
same as the distribution of B(M1), and M1 strength is also
found in ^{32}S.[49]

Very recently, giant M2 resonances in ^{28}Si, ^{90}Zr and
^{208}Pb, prediced within the framework the MSI-RPA particle-
hole model have been confirmed by high resolution inelastic
electron scattering at Darmstadt[50]. The multipolarities of
these isovector excitations were identified without doubt in
^{28}Si. The excitation energies roughly follow an Ex $\approx$ 44 A$^{-1/3}$
law (14.5 MeV in ^{28}Si) and the fragmentation of strength is
maximum in this nucleus. Marked structure is found at about
13.3, 14, 14.3, 14.6 and 15.3 MeV in ^{28}Si. The strong transitions observed in (π^-,γ) (Fig. 10d), correspond to analog
states in ^{28}Si at about 13.3, 14.5 and 15.4 MeV and could be
tentatively identified as analogs to the M2 giant resonance
states seen in electron scattering. It happens that these
states are bound in ^{28}Al, a fact which would make their study
easy by excitation through radiative capture. Starting with a
0^+T = 0 target these states are excited through terms of the
form $[\vec{\sigma} \times Y_1]^{J=2}t^+$. In ^{12}C, ^{16}O, and ^{18}O strong 2$^-$ components
have been identified at excitation energies corresponding to
the giant dipole resonances but also, in oxygen isotopes,
at lower excitation energies, i.e., transition to the first
excited state in ^{16}N and to a low level in ^{18}N.

5.   TRANSITIONS TO THE GIANT RESONANCE REGION

Going to higher excitation energies in the residual nucleus, above the neutron breakup threshold, analogs to the

well known giant dipole resonance (GDR) states seen in photoabsorption or electron scattering are also excited in (π^-,γ),
mainly through terms of the transition operator of the form
$[\vec{\sigma} \times Y_1]^{J=0,1,2}t^+$. In the shell model picture they correspond to the formation of 1p-1h type excitation leading to a
coherent nuclear state through the residual particle-hole
interaction. These states appear as well marked structure
in the hard photon spectra, superimposed on an apparent
structureless continuum background. Certainly the formation
of these resonant states compete with direct emission of one
or more neutrons and to compare the experimental data to the
different theories one is faced with the problem of how to
substract this background. A further difficulty is that the
multipolarities of these excited states cannot be inferred
from the hard photon spectrum alone, only n-γ correlation
could provide this information, except in particular cases
where multipolarities can be determined by comparison with
other experiments. Furthermore, theoretical calculations are
not very predictive concerning the exact position and width
of these resonances.

5.1 General Description of the GDR in 1p Shell

As already pointed out[51], at the beginning of the p shell,
one expects that the dipole type excitation 1s → 1p gives the
main contribution. With increasing occupation of the 1p
valence shell, the role of valence excitation 1p → 2s-1d will
increase, whereas the 1s - 1p excitations should be gradually
blocked. Since typical single particles energy spacings are
1s1/2 → 1p3/2 = 28 MeV and 1p 3/2 →1d5/2 = 16 MeV the E1
strength will be concentrated in regions separated by about 12
MeV. We have a configuration splitting of the GDR at the beginning of the shell. This picture is well supported by the
extensive calculations of the Dubna group[61] and by the LMZ
data. In ^{6}Li (Fig. 6) one clearly sees a double peaked GDR
at 110 MeV and around 120 MeV. In fact, the lower part of the
photon spectrum looks like the free ^{4}He spectrum measured by
the Berkeley group[4]. The same GDR splitting is observed in
^{7}Li and ^{9}Be (Fig. 12). In the middle of the shell a broad
structure appears in ^{10}B whereas in the second half of the

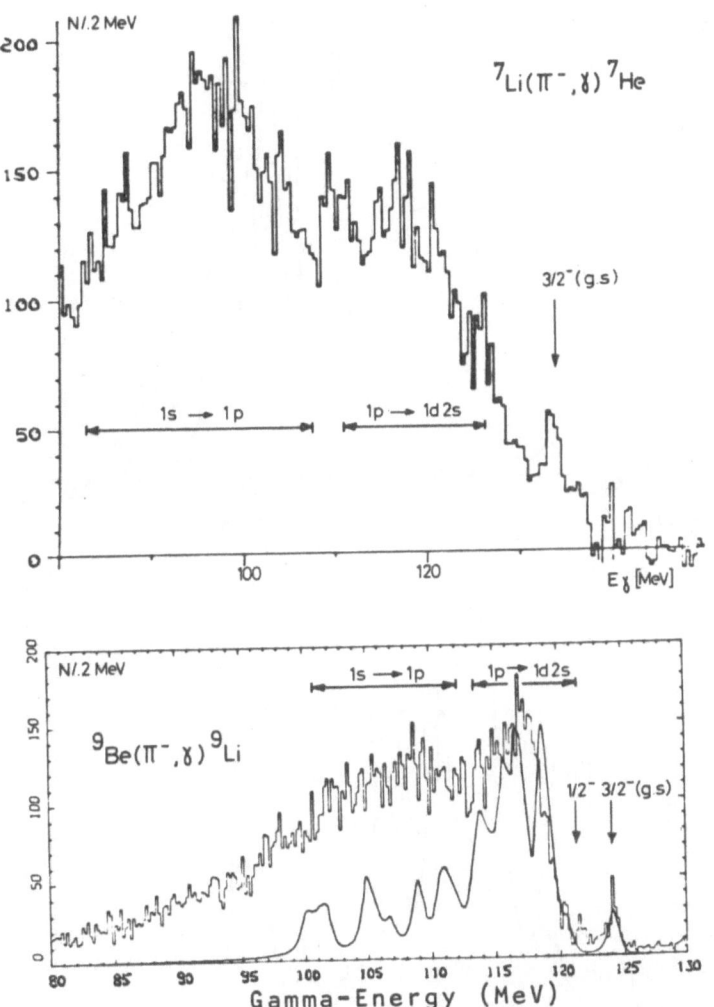

Figure 12.   Photon spectra for ^{7}Li and ^{9}Be measured with the SIN pair spectrometer. The theoretical calculation of the Dubna group[61] folded with the response curve and acceptance of the spectrometer is shown for ^{9}Be.   One can see two broad peaks in the continuum corresponding to the configurational splitting of the GDR.

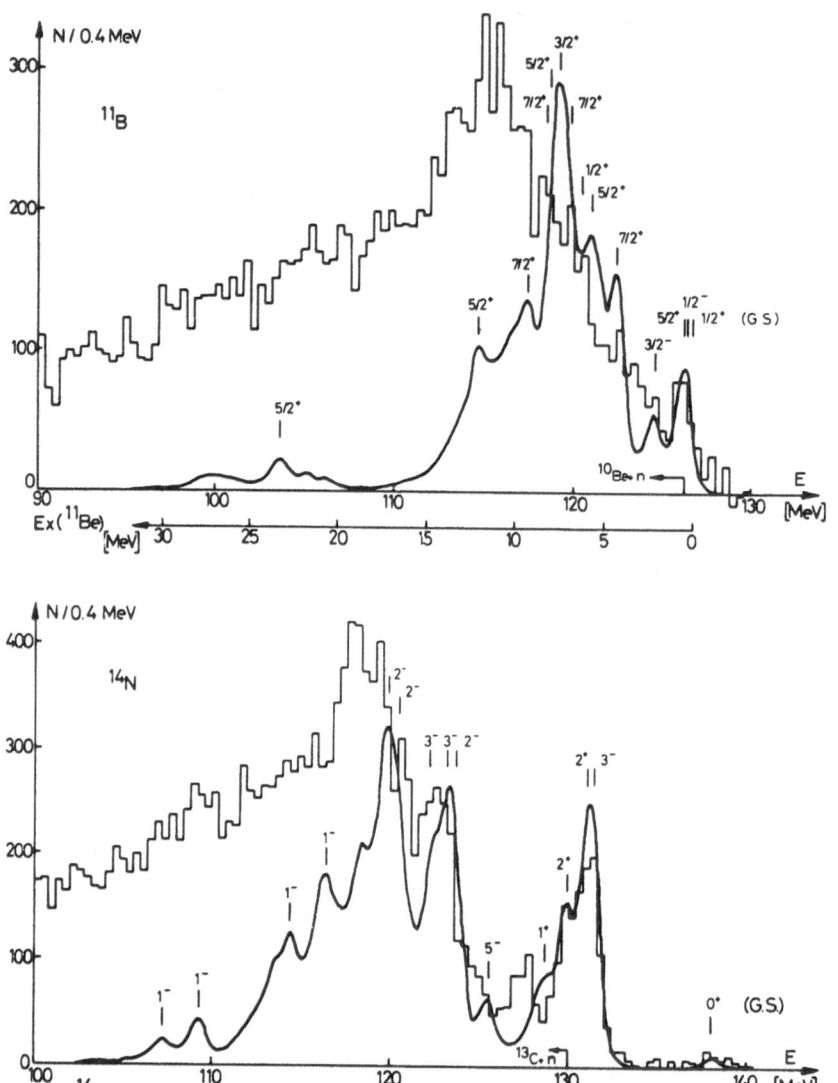

Figure 13.   Photon spectra for ^{11}B and ^{14}N measured with the
SIN pair spectrometer.   A single well formed GDR is seen in
the second half of the p-shell.   The solid curves result from
a theoretical calculation of the Dubna group[61] folded with
the response curve and acceptance of the spectrometer and
normalized to the gs transition.

shell a single, well formed GDR resonance is seen in ^{11}B, ^{12}C, ^{14}N and ^{16}O. In Fig. 13 the theoretical preductions of the Dubna group[61], folded with the resolution and acceptance curve of the spectrometer, are compared with the measured spectra for ^{11}B and ^{14}N. The positions of resonances are better reproduced in the latter nucleus but in both cases the theoretical strength normalized to the g.s. transition was multiplied by a factor ~ .3. The same type of comparison is also shown for ^{9}Be in Fig. 12. To account for the missing strength at lower photon energies, $2\hbar\omega$ type excitation is invoked. In the s-d shell, a well marked structure at excitation energies corresponding to the giant dipole resonance is observed only in ^{18}O and ^{28}Si.

5.2 Radiative Capture in ^{16}O and ^{18}O

In the spectra of Fig. 14, it is seen that the GDR in ^{16}O and ^{18}O is well marked. For ^{16}O, seven components, four 1^- states and three 2^- states were identified[13] by comparison with other experiments such as photoabsorption, (p,γ) and (e,e'). To account for the direct process ^{15}O (π^-,nγ) ^{15}N, two pole model contributions were used assuming a residual nucleus ^{15}N left in its $1/2^-$ gs or in the first negative parity excited state $3/2^-$ at 6.3 MeV. The total branching ratio of 2.27 $\pm$ .24% is in good agreement with the earlier Berkeley value[52] of 2.24 $\pm$ .43%.

Many theoretical calculations exist for radiative pion capture in ^{16}O. In the simplest case one considers a closed shell ^{16}O gs and 1p-1h type excitation for the dipole states. The comparison with theory is further complicated since two values are quoted for total Γ_a^{2p} width 12 $\pm$ 4[53] or 4.7 ev[54], but no matter which one is chosen the ground state quartet branching ratio is overestimated by this type of calculation and requires at least ground state correlation in the form of 2p - 2h admixtures to the ^{16}O gs. Eramzhyan[55] also considers 2p - 2h admixtures for the dipole states. If one looks at

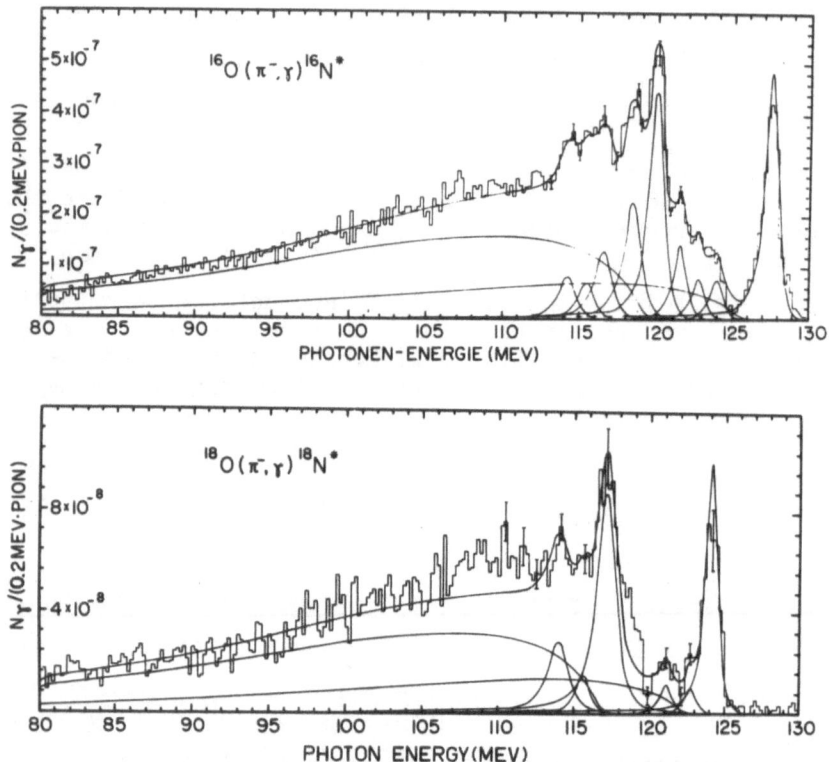

Figure 14.   Photon spectra from π^- capture in ^{16}O and ^{18}O.
The solid curves are derived by folding the contributions of
two pole-models, Breit-Wigner-resonances and lines with the
response function of the SIN pair spectrometer.

the relative distribution of the 1^- and 2^- strength over the
different dipole transition (Fig. 14) one immediately sees
that the theoretical 1^- strength is concentrated around 26 MeV
whereas experimental 1^- states are spread between 17 and 25
MeV with a main component at 22 MeV. This supports the observation from photoabsorption studies[56] that more complicated
structure is necessary to describe the 1^- states in ^{16}O. For
the 2^- states another picture emerges, the predicted positions
and the relative strengths agree fairly well with experiment,
telling us that the 2^- states have rather pure 1p-1h configurations; the main component at 20.4 MeV being of rather
pure spin-isospin type.

New theoretical developments are under way on radiative π^- capture in ^{16}O. The DUBNA group[57] has extended their earlier calculations using R matrix theory to the particle decay of the dipole states, and this will be particularly useful for the interpretation of future γ-n coincidence experiments. Balaskov and Wünsch[58] following Ohtsuka and Ohtsubo[59] developed a unified description of direct and resonance mechanisms using the continuum shell model and 1p-1h excitation. The prediction for the latter folded with the response function of the spectrometer and norlaized to the gs quarted is presented in Fig. 16. It is evident that the energy positions are well reproduced but the relative strength to the 2^- state at 20.4 MeV is overestimated. Roughly 70% of the available strength is concentrated to resonance-excitation. An even better agreement is found in the case of ^{12}C. This is no doubt a nice step forward towards the solution of the background problem.

The ^{18}O spectrum is very similar to the ^{16}O one. The only effect of the two extra neutrons seems to be in a change of the relative positions of the 1^- and 2^- states and a somewhat more complicated structure at high excitation energies in ^{18}N.

6. HEAVY NUCLEI

For nuclei with A > 40, the high resolution (π^-,γ) data are limited to ^{165}Ho, ^{208}Pb and ^{209}Bi. This last nucleus was also measured by the Berkeley group[41] and the photon spectrum suggested a peak corresponding to Ex(^{209}Pb) = 7.9 MeV which was tentatively identified as the analog of an isovector E2 resonance at Ex = 26.5 MeV in ^{209}Bi. The new higher statistic LMZ data do not confirm this peak, the spectrum is structureless as for ^{165}Ho and ^{208}Pb. An explanation for this was put forth by Ebert and Meyer-ter-Vehn[60] who made a calculation for both (π^-,γ) and (μ^-,ν) capture on ^{208}Pb using realistic RPA nuclear wave functions. The initial state

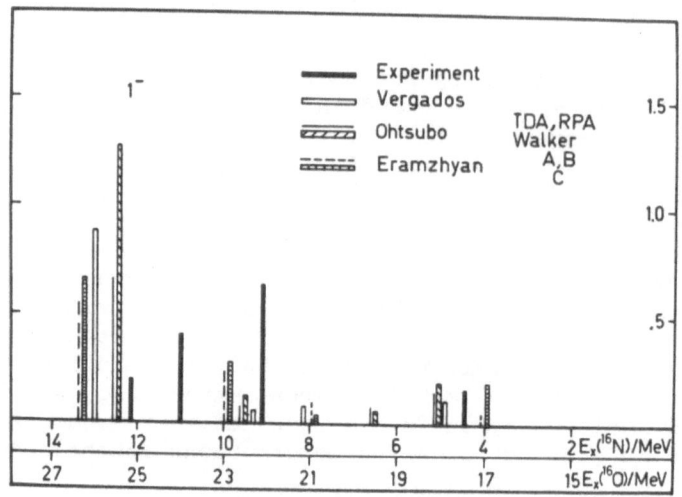

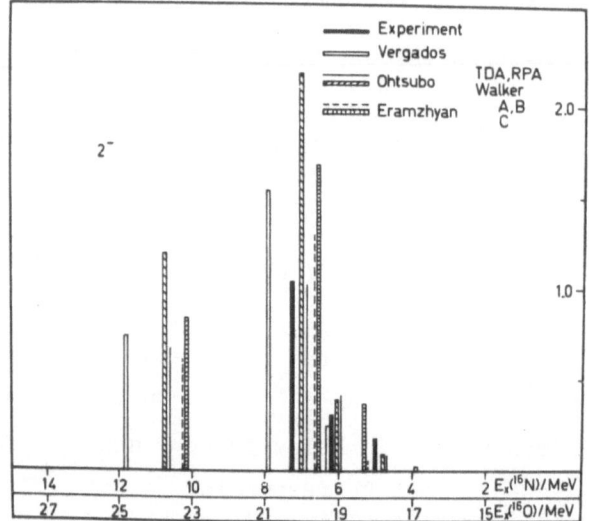

Figure 15. The relative distributions of experimental strengths $R_\gamma(1^-)/R_\gamma(g.s.)$ for 1^- states (top) and $R_\gamma(2^-)/R_\gamma(g.s.)$ for 2^- states (bottom) are compared with the results of the theoretical calculations by Vergados[10], Ohtsubo[8] and Eramzhyan[55]. The experimental distribution of strength agrees with theory only for the 2^- states.

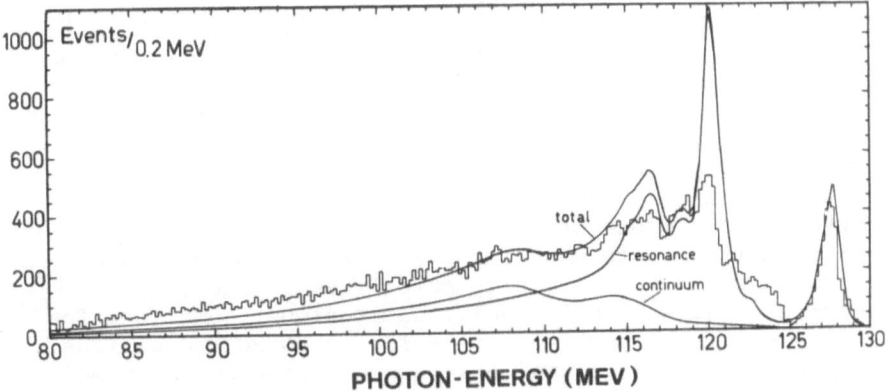

Figure 16. Comparison of the photon-spectrum for ^{16}O with the results of the continuum-shell model calculation[59]. The theoretical strength folded with the response curve and acceptance of the pair spectrometer is normalized to the bound states experimental strength.

$\phi_\pi \sim r^3$ $\ell_\pi = 3$ favors surface capture with $\ell_\gamma = 3$ (higher photon partial waves are cut off since $k_\gamma R = 4\hbar$- for $R(Pb) = 6.7$ fm). The transition operator therefore contains terms $[\vec{\sigma} \times Y_L]^J$ with $L = 0$ to $\ell_\gamma + \ell_\pi = 6$ and starting with an initial nuclear state 0^+ will excite states ranging from 1^+ to 7^+ rather uniformly. On the contrary, μ capture with $\ell_\mu = 0$ occur and appear more selective for the excitation of isovector resonances with low multipolarities ($J = 0$ to 2).

7. SUMMARY AND CONCLUSION

For very light nuclei there are several important results, i.e., a new value for a_{nn} and an improved measurement of $|a_1 - a_3|$, which are of fundamental significance for nuclear physics.[3] Some interesting properties of the three nucleons wave function have also been established.

An agreement within 10% was found between 1s radiative

capture, (γ, π^+), electron scattering and weak processes involving the ^{6}Li gs and the A=6 T=1 0^+ isobaric states using
the EP treatment or shell model and the impulse approximation.
This gives confidence in the latter treatment for extraction
of valuable information on nuclear wave functions. The test
of momentum dependent terms is limited by the present accuracy in π^- atom data and better measurements in this field
are needed to sharpen the comparison with theory. The selective excitation through **radiative capture of** Ml and M2
analogs in p and s-d shells is established. The (π^-, γ)
spectroscopy with .9 MeV resolution is almost completed for
p shell nuclei. Branching ratios for about 20 transitions
to isolated states are available and a strong correlation
with the isovector part of the transverse electromagnetic
form factor is observed and needs further investigation.
Cohen-Kurath type wave functions allow a qualitative understanding of the isovector Ml strength distribution in the p
shell, but are insufficient for a quantitative description
of all the transitions observed. With the present resolution
the systematic study of s-d shell nuclei is limited by the
high density of states.

At higher excitation energies a configuration splitting
of the GDR is observed at the beginning of the p shell, while
in the s-d shell, the GDR is well marked only in ^{18}O and ^{28}Si.
The structure of 1^- and 2^- states is different in ^{16}O; the
2^- states should perhaps be interpreted as belonging to the
giant M2 resonance with a different excitation energy law
since they are probably of the same nature as the strongly
excited bound states observed in ^{28}Al. A promising approach
for the quantitative description of the photon spectra for
high excitation energies seems to be the continuum shell model.

For nuclei with A > 40, the spectroscopy with the present
resolution does not look promising and a new type of experiment selecting specific transitions is needed here.

ACKNOWLEDGEMENTS

The author wishes to thank all his colleagues of the LMZ
collaboration and particularly Professor C. Joseph and P.
Trüol for many enlightening discussions.

REFERENCES AND NOTES

*Work supported in part by Swiss National Science Foundation (Fond. National Suisse pour la Recherche Scientifique).

1. W.K.H. Panofsky et al., Phys. Rev. 81 (1951) 565.
2. J.A. Bistirlich et al., Phys. Rev. Lett. 25 (1970) 950.
3. J.C. Alder et al., submitted to NIM (1978).
4. H.W. Baer, K.M. Crowe and P. Trüol, Advances in Nuclear Physics 9 (1977) 177, Plenum (New York),
5. M. Ericson and M. Rho, Phys. Reports 5C (1972) 57.
6. W. Maguire and C. Werntz, NP A205 (1973) 211.
7. C. Werntz contribution to this conference.
8. H. Ohtsubo et al., NP A224 (1974) 1964.
9. N.C. Mukhopadhyay and F. Cannata, Phys. Lett. 51B (1974) 225.
10. J.D. Vergados, Phys. Rev. C12 (1975) 1278.
11. G.E. Dogotar et al., NP A282 (1974) 474.
12. G. Backenstoss Annual Review of Nuclear Science 20 (1970) 467.
13. G. Strassner et al., submitted to Physical Review.
14. V.T. Cocconi et al., Nuovo Cimento 22 (1961) 494.
15. P. Trüol et al., Phys. Rev. Lett. 32 (1974) 1268.
16. A.C. Phillips and F. Roig, NP A234 (1974) 378.
17. J. Spuller et al., Phys. Lett. 67B (1977) 479.
18. R. MacDonald et al., Phys. Rev. Lett. 38 (1977) 746.
19. M.D. Hasinoff et al., Proc. VII Int. Conf. on High Energy Physics and Nuclear Structure, cont. C-19, p. 48, Zürich 1977.
20. B. Kühn, Sov. J. Part. Nucl. 6 (1976) 139.
21. E.M. Henley and D.H. Wilkinson, in Few Particle Problems in the Nuclear Interaction (North-Holland, Amsterdam (1972) 229.
22. R.M. Salter et al., NP A254 (1975) 241.
23. M. Bander, Phys. Rev. 134 (1964) B1052.
24. W.R. Gibbs et al., Phys. Rev. C11 (1975) 90.
25. G.F. de Teramond, Phys. Rev. C16 (1977) 1976 and private communications.
26. R. Omnes, Nuovo Cimento 8 (1958) 316.
27. P.U. Sauer, Phys. Rev. C11 (1975) 1786.
28. A. Coon et al., NP A287 (1977) 381.
29. J.L. Friar and B.F. Gibson, Phys. Rev. C17 (1978) 1752.
30. J.A. Bistirlich et al., Phys. Rev. Lett. 36 (1976) 942.
31. D. Renker et al., to be published in Phys. Rev. Lett.
32. A.C. Phillips and F. Roig, contribution to the Sixth Int.

Conf. on High Energy Physics and Nuclear Structure (Santa Fe, 1975) Los Alamos Report LA-6030-6, p. 170.

33. J. Delorme, NP $\underline{B19}$ (1970) 573.
34. J.C. Bergström et al., NP $\underline{A251}$ (1975) 401.
35. J.B. Cammarata, T.W. Donnelley, $\underline{NP}$ A267 (1976) 365.
36. Audit et al., Phys. Rev. $\underline{C15}$ (1977) 1415.
37. A.N. Boyarkina, Bull, Aca. Sci. USSR (Phys.) 28 (1964) 255.
38. G. Backenstoss et al., NP $\underline{B66}$ (1973) 125.
39. F. Roig and P. Pascual, NP $\underline{B66}$ (1973) 173.
40. J.D. Vergados and H.W. Baer, Phys. Lett. $\underline{41B}$ (1972) 560.
41. J.D. Vergados, NP $\underline{A220}$ (1974) 259.
42. R. Neuhausen et al., NP $\underline{A164}$ (1971) 497.
43. O.A. Zaimidoroga et al., Sov. Phys. JETP $\underline{21}$ (1965) 848.
44. H.W. Baer et al., Phys. Rev. $\underline{C8}$ (1973) 2029.
45. W.W. Sapp et al., Phys. Rev. $\underline{C5}$ (1972) 690.
46. H.W. Baer et al., Phys. Rev. $\underline{C10}$ (1974) 1140.
47. J.K. Novak et al., Proc. VII Int. Conf. on High Energy
48. Physics and Nuclear Structure, cont. C-10, p. 39, Zürich 1977.
     J.C. Alder et al., contribution to this conference.
49. W. Knüpfer et al., Nuclear Physics Institute preprint,
50. Darmstadt 1978.
     H.W. Baer in Seventh International Conference on High
51. Energy Physics and Nuclear Structure, Zürich 1977, Birkhäuser Verlag, Basel.
     J.A. Bistirlich et al., Phys. Rev. $\underline{C5}$ (1972) 1867.
52. W. Sapp et al., Phys. Rev. $\underline{C5}$ (1972) 690.
53. H. Koch et al., Phys. Lett. $\underline{28B}$ (1968) 279.
54. R.A. Eramzhyan et al., Nucl. Phys. $\underline{A290}$ (1977) 294.
55. W.L. Wang and C.M. Shakin, Phys. Rev. $\underline{C5}$ (1972) 1898.
56. R.A. Eramzhyan et al., contribution to this conference.
57. V.V. Balashov and R. Wünsch, contribution to this confer-
58. ence.
     N. Ohtsoka and H . Ohtsubo, Proc. Int. Conf. on Nucl.
59. Structure, Tokyo (1977) and private communication.
     K. Ebert and J. Mayer-Ter-Vehn, Phys. Lett. $\underline{77B}$ (1978) 24.
60. G.E. Dogotar et al., Joint Institute for Nuclear Research preprint E2-11275 Dubna 1978 and private communication.

INVESTIGATION OF ANALOGS OF M1, E1 AND M2 TRANSITIONS IN ^{10}B, ^{12}C AND ^{14}N THROUGH RADIATIVE PION CAPTURE[+]

J.C. Alder, W. Dahme, B. Gabioud, C. Joseph, J.F. Loude, N. Morel, H. Panke, A. Perrenoud, J.P. Perroud, D. Renker, G. Strassner, M.T. Tran, P. Truöl and E. Winkelmann

Lausanne-München-Zürich Collaboration, SIN, Villigen, Switzerland

For the selfconjugate nuclei ^{10}B, ^{12}C and ^{14}N some of the most convincing evidence was compiled, which demonstrated the potential of the radiative pion capture reaction[1,2]. The reasons for this are twofold. Firstly the level spacing in the bound state region in the final nuclei ^{10}Be, ^{12}B and ^{14}C is sufficiently large, that even with a resolution of 2 MeV certain transitions can be isolated experimentally. Secondly the observed transition strength is concentrated into a few levels, mainly analogs to strong isovector M1-transitions in the target nucleus. In addition a considerable fraction of the reaction proceeds through spin-isospin dipole states with prominent $J^{\pi} = 1^-$ and 2^- structures observed in ^{12}C. In view of the impact of these findings for future radiative pion capture experiments, it seemed appropriate to remeasure these photon spectra with improved resolution (900 keV FWHM at 129.4 MeV) and statistics in our pairspectrometer at SIN[3]. Figure 1 shows the result, where only the important part of the spectrum above 100 MeV is displayed. In Table I we present the photon energies, excitation energies and branching ratios for the different transitions. The agreement with the Berkeley results[1,2] is quite satisfactory. However, whereas in the Ber-

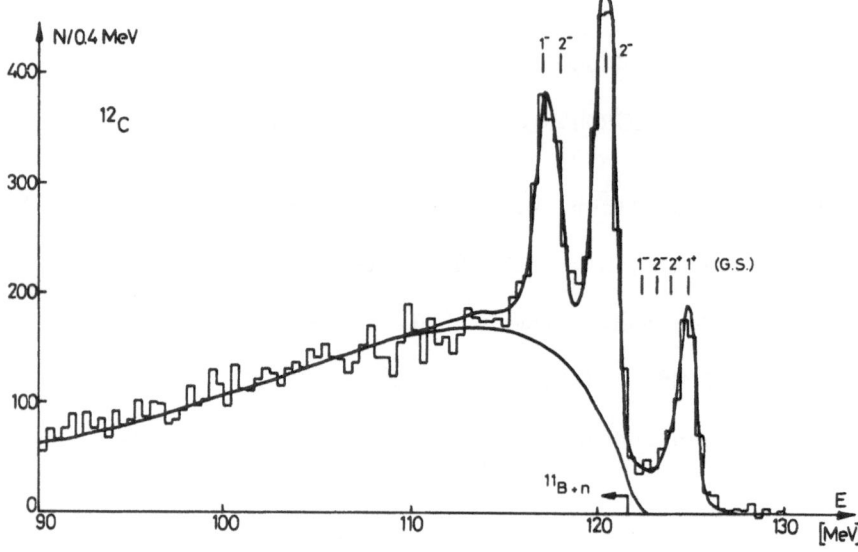

FIGURE 1a

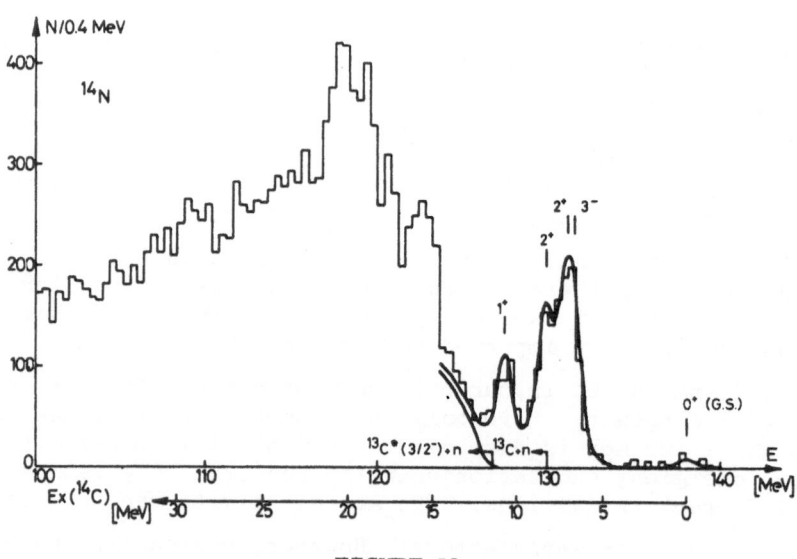

FIGURE 1b

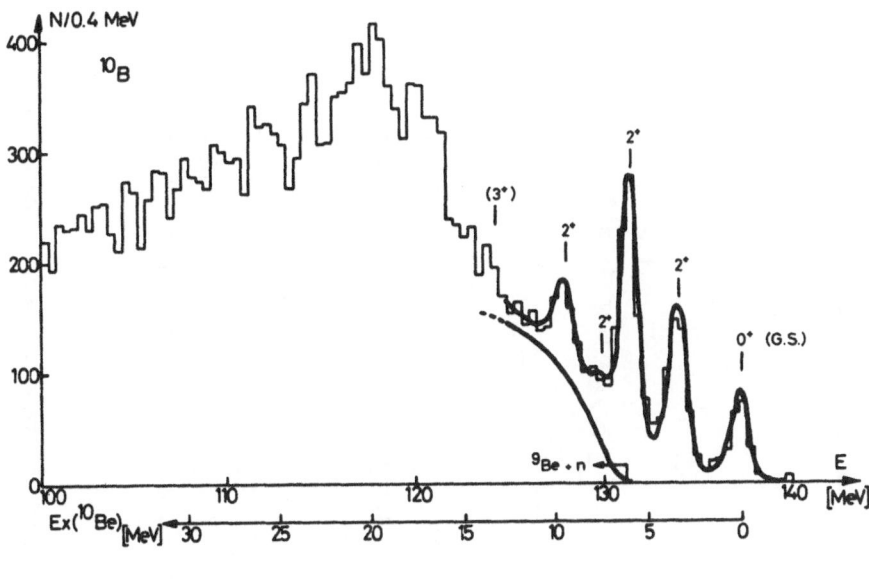

FIGURE 1c

keley analysis the fit to the experimental spectrum for ^{10}B
and ^{14}N was strongly guided by theoretical considerations, the
observed transitions here manifest themselves as clear and
separated peaks in the spectrum, such that the photon energy
and consequently the level position can be determined to an
accuracy below 100 keV. Furthermore, with the transition energies as well as the resolution function known, the separate
contributions for the first four bound states in ^{12}B within
2.6 MeV could be determined. Although some of the results
presented here are still preliminary as far as the absolute
normalization is concerned, the following conclusions are already apparent:

^{10}B(^{10}Be): The first three peaks in the spectrum correspond to the 0^+-^{10}Be ground state and two 2^+-states. The latter two are identified as the analogs of M1-states in ^{10}B. The
shell-model calculations of Dogotar et al.[4] using Cohen-Kurath
wave-functions[5] are in good agreement with the data for the

E_γ [MeV]	J^π	E^x [MeV]	$E^{x,a)}$ [MeV]	R [*10^{-4}]	$R^{,b)}$ [*10^{-4}]	Type of Transition	R(theory) [*10^{-4}]
^{12}C, ^{12}B 0^+							
124.96	1^+	0.	15.11	6.9 ± .35		M1	8.4[7] 6.8[4] 8.4[8]
124.02	2^+	0.95	16.11	1.2 ± .3		E2	3.5 3.5 4.9
123.31	2^-	1.67	16.58	.65 ± .25	9.1 ± .4	M2	2.5 .7
122.37	1^-	2.62	17.23	1.06 ± .20		E1	1.2 2.2
120.64	2^-	4.37	19.5	18.73 ± .5	18.5 ± 1.9	M2-SI	61.9[10]
118.0		7.2	22.3	5.4 ± .5	15.9 ± 1.6		10.9
117.2	1^-	8.1	23.2	6.8 ± .5		E1-SI	
^{10}B, ^{10}Be 3^+							
137.47	0^+	0.	1.74	2.31 ± .15	2.5 ± .4		2.3[4] 3.6[2]
134.14	2^+	3.37	5.11	4.88 ± .2	4.4 ± .7	M1	5.7 8.5
131.60	2^+	5.95	7.48	7.64 ± .3	10.5 ± 1.3	M1	17.9 16.9
^{14}N, ^{14}C 1^+							
138.11	0^+	0.	2.30	.25 ± .07	0.3 ± .2	M1	.9[4] 1.0[2]
131.17	2^+	7.01	9.17	7.06 ± .2	7.7 ± .9	M1	10.3 12.2
129.88	2^+	8.31	10.44	3.41 ± .25	4.0 ± .6	M1	10.3 12.2

TABLE I

a) Excitation energy of analog state in target nucleus.

b) Ref. 1,2.

first two transitions, overestimate however the rate to the
third level. A similar disagreement exists for the B(M1)-
value of this level. An alternative study, discussed in detail in Ref. 2, yields the correct relative populations, but
overestimates the strength by a factor of 1.5. The Berkeley
data[2] in the region between 8 and 12 MeV excitation energy
were fitted with two levels at 8.6 and 10.1 MeV. Our data
seem to indicate a single transition at 9.5 MeV. This level
is also seen in ^{9}Be(p,π^+) and ^{9}Be(n,n')[6] and is presumably
also 2$^+$, the predicted branching ratio amounts to 25% of the
first 2$^+$ state[4]. For the giant resonance region no calculations are available. Experimentally evidence for a doubly
peaked structure at 17 and 20 MeV is visible. In the equivalent energy range in ^{10}B 1$^-$ and 2$^-$ (T=1) states are observed
in photo-nuclear reactions.

^{12}C(^{12}B): A phenomenological determination of the radiative pion capture matrix elements using β-decay and inelastic
electron scattering data as input yields good agreement with
our data[7]. The same is true for Cohen-Kurath wavefunctions,
since they are known to represent well the weak and electromagnetic processes connecting the ^{12}B- ground state or its
analog at 15.1 MeV in ^{12}C with the ^{12}C ground state[4,8]. For
the other bound states the input data as well as the wave-
functions are not nearly as well determined and consequently
only poor agreement with our data results. The two prominent
peaks above the nonresonant continuum at 120.6 and 117.5 MeV
were identified in the first high-resolution π^-, γ-experiment[1]
as the 2$^-$ and 1$^-$ member of the spin isospin modes of the generalized Goldhaber-Teller model[9]. Our experiment indicates
that the 1$^-$ (upper peak) is split into two states. Skupsky[10]
predicts two 1$^-$ levels within 0.5 MeV at E = 7.5 MeV, but
their combined strength falls below the experimental one.
Furthermore 50% of the 1$^-$ strength in his calculation is expected at E* = 11 MeV, where experimentally no peak is observed, and the 2$^-$ strength for the 4.4 MeV level is too high
by a factor of four. In a continuum shell model calculation
by Ohtsubo[11], where the quasifree background is treated on
equal footing with resonance, the 7.2 MeV state would be

another 2^- state, the 8.1 MeV a 1^- level. This calculation gives an excellent description of the spectrum shape down to the lowest photon energies.

$\underline{^{14}N(^{14}C)}$: Since our data are in good agreement with the Berkeley results[2] and newer theoretical predictions[12] do not differ significantly from previous calculations, the discrepancy between experiment and theory still persists. For the 1^+ level, previously found at 11.3 MeV we could determine the level position more accurately to 10.7 MeV. Its analog has been observed recently in backward inelastic electron scattering for the first time[13]. In the giant resonance domain a large number of 1^- to 3^- states is expected, such that a spectroscopy of individual levels is impossible. The clustering into two groups around 15 and 20 MeV observed in our experiment however is supported by recent shell model calculations[12].

REFERENCES AND NOTES

+Work supported by BMFT, Bonn, W. Germany and Swiss National Science Foundation.

1.   J.A. Bistirlich et al., Phys. Rev. Letters 25, 689 (1970).
2.   H.W. Baer et al., Phys. Rev. C10, 1140 (1974).
3.   J.C. Alder et al., submitted to Nucl. Instr. and Methods.
4.   G.E. Dogotar et al., Nucl. Phys. A282, 474 (1977).
5.   S. Cohen, D. Kurath, Nucl. Phys. 73, 1 (1965).
6.   F. Ajzenberg-Selov, T. Lauritsen, Nucl. Phys. A227, 1 (1974).
7.   C.F. Maguire, C. Werntz, Nucl. Phys. A205, 211 (1973).
8.   H. Ohtsubo et al., Nucl. Phys. A224, 164 (1974).
9.   F.J. Kelly, H. Überall, Nucl. Phys. A118, 302 (1968).
10. S. Skupsky, Nucl. Phys. A178, 289 (1971).
11. H. Ohtsubo, private communication.
12. G.E. Dogotar et al., Dubna-Preprint E2-11275 (1978).
13. A. Richter, Int. School on Electro- and Photonuclear Reactions, Erice (Sicily) (1976).

RADIATIVE PION CAPTURE IN MEDIUM MASS NUCLEI: ^{32}S, ^{40}Ca +)

J.C. Alder, W. Dahme, B. Gabioud, C. Joseph, J.F.

Loude, H. Medicus, N. Morel, H. Panke,

A. Perrenoud, J. P. Perroud, D. Renker,

G. Strassner, M.T. Tran, P. Truöl,

and E. Winkelmann

Lausanne - Munchen - Zürich Collaboration, SIN,

Villigen, Switzerland

The comparison between enelastic electron scattering, photo-absorption and radiative pion capture for 1p - shell nuclei has shed some light on the structure of the isovector M1 and spin-isospin transitions[1]. The principle reason for this is, that in the π^-, γ -reaction the $T_z = +1$ analog states to the $\Delta T = 1$ excitations in the target nucleus are observed, whereas with conventional electromagnetic probes $\Delta T = 0$ and $\Delta T = 1$ can be excited. It is found thatin the impulse approximation matrix-elements of the type $<J_f || \tau^+ j_L(kr) \{\sigma x Y_L\}^J || J_i>$ contribute to the radiative pion capture rate, and are thus identical to the spin-density part of the electromagnetic M1, the spin-isospin dipole or quadrupole operator for different JL, respectively. To test whether these ideas are still meaningful for nuclei in the s-d shell, we selected target nuclei with known M1-levels. Good cases are ^{28}Si, already discussed previously[2] and ^{32}S, for which we present preliminary results here. We also include some results on ^{40}Ca, which had been measured previously with inferior resolution[3] and which were obtained in the process of background studies for a radiative muon capture experiment[4]. Figure 1 shows the results.

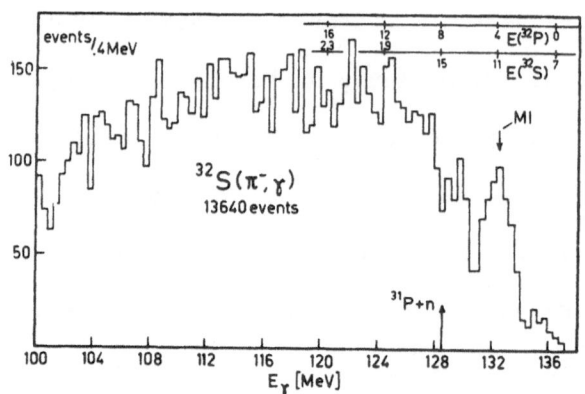

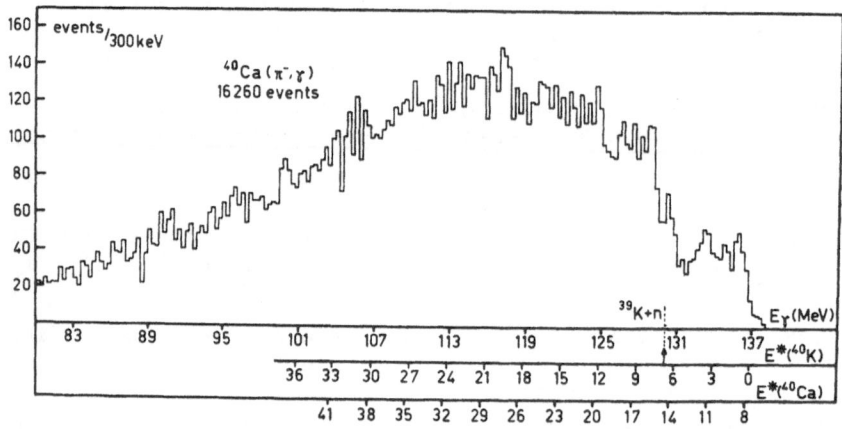

FIGURE 1

Although the data have not been completely analyzed, the total
branching ratios are found to agree with previous values, namely (1.9 ± 0.4) % for ^{40}Ca[3] and (1.8 ± 0.1) % for ^{32}S.[5] Both
spectra indicate that even a resolution of 900 keV FWHM at 130
MeV does not suffice to isolate individual transitions clearly

For ^{32}S, however, a 2.4 MeV wide peak centered at 132.6
MeV dominates the bound state region of ^{32}P. Here we observe
probably the analogs to a group of M1-states in ^{32}S with J = 1
which have been isolated in backward electron scattering[6,7].

Transitions to the ^{32}P ground state are comparatively weak. A theoretical calculation by Eramzhyan et al.[8] allocates in a simple particle-hole model 14% of the total transition strength to these $0\,\hbar\omega$ - $1d_{5/2} \to 1d_{3/2}$ transitions half of it to 1^+-levels. The major part of the branching ratio is expected to go to $1\hbar\omega$ - (0^- to 4^-, 33%) and $2\hbar\omega$ - (1^+ to 5^+, 37%) transitions, the remainder to $3\hbar\omega$ - excitations. Since the pion gets captured preferentially from the 2-p level, 1^+ to 3^+ and 1^-, 2^- levels dominate. The predicted configurational splitting of the $1\hbar\omega$ - excitations into those with a $1d_{5/2}$ hole around 120 to 125 MeV and those with a $1p_{3/2}$ hole around 110 to 115 MeV does not manifest itself as clearly as a similar splitting in the 1p shell.

For ^{40}Ca, despite the importance for shell model studies, no complete calculations have been published. Guy and Eisenberg[9], on the basis of a simple particle-hole picture find a considerable fragmentation of strength, e.g. for $J^{\pi} = 1^-$ at least eight states with partial strength between 5 and 27% contribute. Preliminary results of Ebert[10], using Migdal's quasiparticle approach similar to calculations for ^{208}Pb (π^-,γ)[11] indicate, however, that most of the strength for E^* (^{40}K) < 12 MeV goes to 2^- and 3^- levels. This is consistent with the experiment, where we find:

a) almost no strength to the ^{40}K - ground state and first excited state ($4^-,5^-$)

b) transitions to groups at $E^*(^{40}$K) = 0.8 (2^-), 2.1 (2^-:2.05; 3^-:2.07) and 3.3 [(2,3)$^-$: 3.23] MeV with relative strength 2.5:1:2.5

c) additional stronger transitions to levels near 6.5, 7,9 and 10.1 MeV. The analog of the 0.8 (2^-) level is also seen in backward inelastic electron scattering[6] at 8.43 MeV, which besides another peak at 10.34 (1^+, 2.7 MeV in ^{40}K) shows no other structure at higher energies. Since the giant dipole resonance in ^{40}Ca lies at 19.5 MeV ($\sim$ 12 MeV in ^{40}K), considerable strength in our reaction is observed below and inelastic electron scattering is inconclusive in this energy region, it appears worthwhile to perform a combined theoretical analysis of the electromagnetic and pion processes to disentangle the different collective excitation modes.

+Work supported by BMFT, Bonn, West Germany, and Swiss National Science Foundation.

REFERENCES

1. H.W. Baer, K.M. Crowe, P. Truöl, Adv. Nucl. Phys. 9, 177 (1977).
2. J.C. Alder et al., contribution to VIIth Int. Conf. on High-Energy Physics and Nuclear Structure, Zürich (1977).
3. J.A. Bistirlich et al., Phys. Rev. C5, 1867 (1972).
4. A. Frischknecht et al., Helv. Phys. Acta (1978).
5. H. Davies, H. Miurhead, J.N. Woulds, Nucl. Phys. 78, 673 (1966).
6. L.W. Fagg et al., Phys. Rev. C4, 2089 (1971).
7. D. Sober, CUW, private communication.
8. R.A. Eramzhyan et al., Dubna Preprint E2 - 11234 (1978).
9. R. Guy, J.M. Eisenberg, Phys. Letters 33B, 137 (1970).
10. K. Ebert, private communication.
11. K. Ebert, J. Meyer-ter Vehn, Nucl. Phys., to be published.

BRANCHING RATIO FOR RADIATIVE PION CAPTURE WITH

RESPECT TO ABSORPTION OF STOPPED NEGATIVE PIONS

Carl Werntz

The Catholic University of America

Department of Physics

Washington, D. C.

An initial effort to predict the branching ratio of radiative pion capture with respect to absorption of stopped negative pions is presented. The theory is developed by analogy to Primakoff's theory for the total $\mu-$ capture rate[1,2]. Unlike muons negative pions are captured or absorbed from a mixture of atomic states, the ℓ values increasing with Z. The branching ratio can be written as a weighted sum of the ratio for each contributing atomic state

$$R = \sum_{n,\ell} \omega_{n\ell} \frac{\Lambda_{n\ell(\pi\gamma)}}{\Lambda_{n\ell(\pi)}} . \qquad (1)$$

Because the optical potential is much larger than the atomic binding energies the wave functions for given ℓ but different n should have almost identical shapes within the nucleus. The ratio of the radiative capture rate $\Lambda_{n\ell(\pi\gamma)}$ to the absorption rate $\Lambda_{n\ell(\pi)}$ should be independent of n.

The gamma spectrum following radiative pion capture is well known for a representative sampling of nuclei[3]; the spectrum has a somewhat asymmetric shape, very similar throughout the periodic table, with a peak at ~ 115 MeV gamma energy and a width at half maximum of ~ 30 MeV. Studies of transitions to individual states show that the impulse approximation is an accurate method to use for evaluating the radiative capture rate[3]. The spectral shape suggests applying closure to the sum over the final states of the square of the single body matrix elements. Representing the ground state by a single Slater determinant with no unpaired nuclear spins the expectation value of the quadratic products of single nucleon operators can be evaluated following methods outlined in Ref. 4. An equation analogous to the Primakoff equation is obtained:

$$\Lambda_\ell(\pi\gamma) = 4\pi \, \vec{k} \, C \, (\bar{k}) \, b\pi\gamma \, Z \, \rho \, (o) \, / \, m\pi A$$

$$(2)$$

$$x \left[1 - \frac{A-Z}{ZA} \gamma\ell\delta + \gamma\ell\delta/A - {}^\beta/A \right]\left[1 + X_\ell(\pi\gamma) \right] \int dr \, r^2 f(r) |Q_\ell(r)|^2$$

The parameter δ is the Fourier transform of the product of $\delta(o)$ and the np correlation function, evaluated for the average $\bar{k}$, and γ_ℓ is defined by

$$\gamma_\ell = \int r^2 \, dr \left| Q_\ell(r) \right|^2 f^2(r) \Big/ \int r^2 \, dr \left| Q_\ell(r) \right|^2 f(r). \quad (3)$$

In the above two equations the nuclear matter density $\rho(r) = \rho(0) \, f(r)$ where $f(r)$ is taken to be a two parameter Fermi shape[5], $X_\ell(\pi\gamma)$ is the ratio of pion momentum dependent terms to the S-wave term in the photo-capture amplitude, the latter being $\sqrt{b\pi\gamma} \, \hat{\epsilon}\lambda \cdot \sigma\tau^{(-)}/\sqrt{2}$, and $C(k)$ is the kinematic factor $(1 + m_\pi /M) (1 + k/M)/(1 + k/M_A)$.

The term $\gamma_\ell\delta/A - \beta/A$ is a correction to the correlation term because of neutron-proton pairs in identical space states. We have estimated its value to be $-.05$ independent of A so it has been dropped at the present level of approximation.

The absorption rate for a pion in an atomic state can be calculated from the absorptive part of the pion optical potential. Separating out the iso-vector from the iso-scalar terms, the absorptive part of the optical potential can be written as

$$V_{ab} = -\frac{4\pi}{2} \frac{2Z}{\pi} \{[1 + \frac{a_1^2}{a_1^2 + a_0^2}(\frac{N-Z}{A})] \text{ Im } \tilde{B}_0 \delta^2(r)$$

$$-[1 + \frac{b_1^2 - \frac{1}{2}bo^2}{b_1^2 + \frac{1}{2}bo^2}(\frac{N-Z}{A})] \text{ Im } \tilde{C}_0 \vec{\nabla}\cdot\delta^2(r)\vec{\nabla}\}$$

(4)

a_o and a_1 are the strengths of the interaction responsible for the absorption of s-wave pions by spin singlet, iso-spin triplet or by spin triplet, iso-spin singlet nucleon pairs, respectively, while b_o and b_1 are the strengths for the absorption of p-wave pions by pp singlet or np triplet pairs, respectively. Numerically, the coefficients of $(N - Z)/A$ are predicted to be very close in value[6,7] so that the overall branching ratio of radiative pion capture to absorption can be written (for a particular ℓ value)

$$\frac{\Lambda_\ell(\pi\gamma)}{\Lambda_\ell(\pi)} = \frac{\bar{k}b\pi\gamma C(\bar{k})}{\text{Im}\tilde{B}_0\delta(o)} [\frac{1 - \frac{1}{4}\gamma_\ell\delta(1 + \frac{N-Z}{A})}{1 + \gamma(\frac{N-Z}{A})}]\frac{\Phi_\ell(A,Z)}{2},$$

$$\Phi_\ell(A,Z) = [1 + \chi_\ell(\pi\gamma)]/[(1 + \chi_\ell(\pi)\gamma_\ell]$$

(5)

The constant $\gamma = a_o^2/(a_o^2 + a_1^2) = 0.80$ and the parameter $\chi_\ell(\pi)$ measures the ratio to momentum dependent terms to S-wave terms in the absorption process.

The values adopted for the constants at the beginning of Eq. (5) are[3,8,5]: $B_{\pi\gamma} = 20.6 \times 10^{-4} m_\pi^{-2}$, $\text{Im}\tilde{B}_0 = .0510_{m\pi}^{-4}$, and $\delta(0) = .479_{m\pi}$. Using Primakoff's method, but evaluating the constant for half nuclear density and $\bar{k} = 115$ MeV, $\delta = 2.68$. (For muon capture $\delta = 3.15$, Ref. 9).

The weighted sums $\varepsilon_\ell\omega_\ell\Phi_\ell$ and $\varepsilon_\ell\gamma_\ell\omega_\ell\Phi_\ell$, implicit in Eq. (5), were evaluated by (1) using values for selected

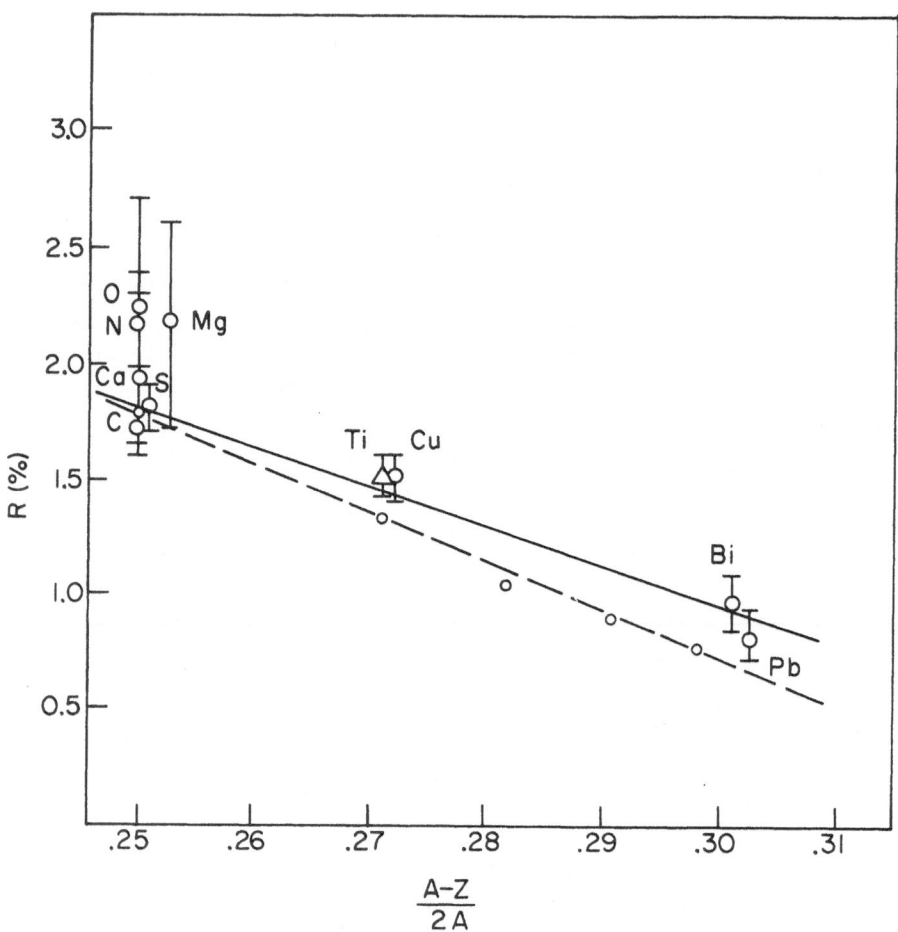

Figure 1.   Ratio of radiative pion capture to absorption.

nuclei sent to us by Dr. Melvin Leon of LASL and (2) evalu-
ating the ℓ dependent coefficients γ_ℓ . $\chi_\ell(\pi\gamma)$, and $\chi_\ell(\pi)$
(all ratios) using the simplifying assumption $Q_\ell(r) \propto r^\ell$.
Both $\Sigma_\ell \omega_\ell \Phi_\ell$ and $\Sigma_\ell \gamma_\ell \omega_\ell \Phi_\ell$ were found to be essentially
constant with average values of 0.82 and 0.66, respectively.
The theoretical points are shown by black dots in Figure 1
where they are compared to the data compiled in Ref. 3.
The solid line is simply an "eyeball" fit.

REFERENCES

1. H. Primakoff, Rev. Mod. Phys. 31, 802 (1959).
2. B. Goulard and H. Primakoff, Phys. Rev. C10, 2034 (1974);
    Phys. Rev. C11, 1894 (1975).
3. H. W. Baer, K. M. Crowe, and P. Truoel, Adv. in Nucl.
    Phys. 9, 177 (1977).
4. M. L. Goldberger and K. M. Watson, Collison Theory,
    John Wiley & Sons, Inc., New York, London, Sydney,
    1964. Section 11.2.
5. H. R. Collard, L.R.B. Elton, and R. Hofstadter,
    Landolt-Bornstein, New Series, Group I, Vol. 2,
    Nuclear Radii, Springer, Berlin, Heidelberg, New York,
    1967; edited by H. Schopper. Section 1.1.
6. P. P. Divakaran, Phys. Rev. 139B, 387 (1965).
7. A. Figureau and M. Ericson, Nucl. Phys. B10, 349 (1969).
8. C. J. Batty, S. F. Biagi, E. Friedman, S. D. Hoath,
    J. D. Davies, G. J. Pyle, and G.T.A. Squire, Phys. Rev.
    Letts. 40, 931 (1978).
9. N. C. Mukhopadhyay, Phys. Letts. 30C, 1 (1977).

GIANT RESONANCE EXCITATION IN VARIOUS PROCESSES. EXAMPLE: ^{13}C

H.R. Kissener
Zentralinstitut für Kernforschung
Rossendorf, DDR

R.A. Eramzhyan
Joint Institute for Nuclear Research
Dubna, USSR

Excitation functions for radiative capture of stopped pions, electro- and photoexcitation through the giant resonance region were calculated for 1p shell nuclei, using shell model functions[1] which span the full space of 0 $\hbar\omega$ and 1$\hbar\omega$ excitations. Kinematical conditions for the ee' process (E_o = 70 MeV, θ = 180°) were chosen such as to have the same momentum transfer as in the (π^-,γ) process and to exhibit magnetic multipole contributions.

The calculation generally yields close analogy between M1 type resonances[2] and also correspondence between spin-dipole resonances in (π^-,γ) and M2, $T_>$ strength in (ee'). Theoretical M2 strength distributions in backward electron scattering peak at lower energy than transverse E1 strengths for all 1p shell nuclei considered, in accordance with the sequence of 2$^-$ and 1$^-$ levels seen in the electroexcitation of ^{12}C and ^{16}O. The M2, $T_>$ strength is obtained as a doubly-humped structure with peaks due to $1p_{3/2} \rightarrow 1d_{5/2}$ and $1p_{3/2} \rightarrow$ 2s transitions; it is even more concentrated to individual levels than the transverse E1, $T_>$ strength is. This feature is illustrated in Fig. 1 for the ^{13}C case. The transverse E1, $T_>$ strength is essentialy due to $1p_{3/2} \rightarrow 1d_{3/2}$ and 1s $\rightarrow$ $1p_{1/2}$ excitation. The photo strength distribution (vector

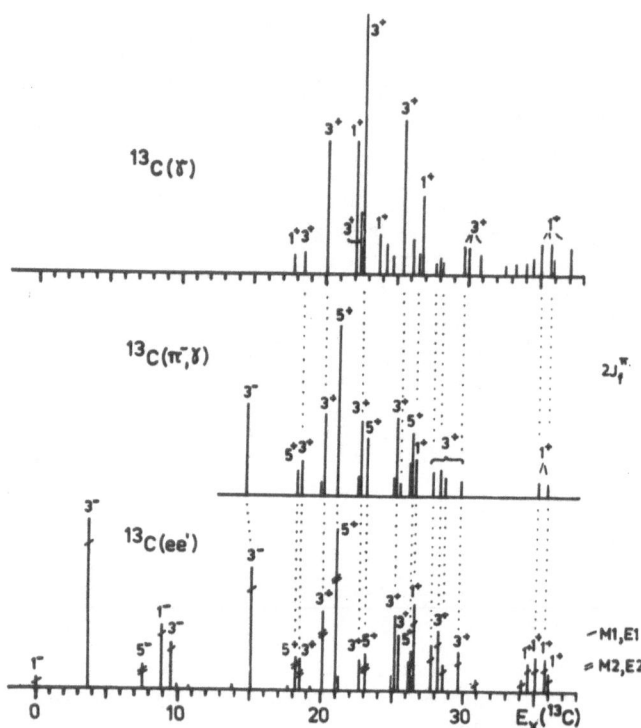

Figure 1. Calculated transition rates for photo absorption
($\int \sigma$) radiative pion capture (R), and 180°, E_o =· 70 MeV electroexcitation (dσ/dΩ) on ^{13}C. Spins (2J) and parities of
strong resonances and dominant multipole components in the
ee' cross section are indicated.

excitation) shows less similarity to the (π^-,γ) distribution
(axial vector excitation) than the magnetic scattering does;
it is mainly due to non-spinflip $1p_{3/2} \rightarrow 1d_{5/2}$ transitions, as
for longitudinal Cl excitation.

   Our calculated $\Delta T = 0$ strengths in both photo- and electroexcitation are less concentrated to individual resonances
and are spread over a larger energy region than the $\Delta T = 1$
strengths. Hence, they do not obscure the analogy between the
different excitation types. The M2, $T_<$ strength peaks close
to the region of the M1, $T_>$ resonance. The transversal E1,
$T_<$ strength again shows two peaks between 15 and 25 MeV with

a long high-energy tail. On the contrary, 2p-1h calculations[3] using Tabakin residual interactions give also strong
M2, $T_<$ excitation at higher energy.

The electroexcitation data for ^{13}C through the giant
resonance region[4] are qualitatively described by our model
(Fig. 2), except for some shift of maxima around 20 MeV and

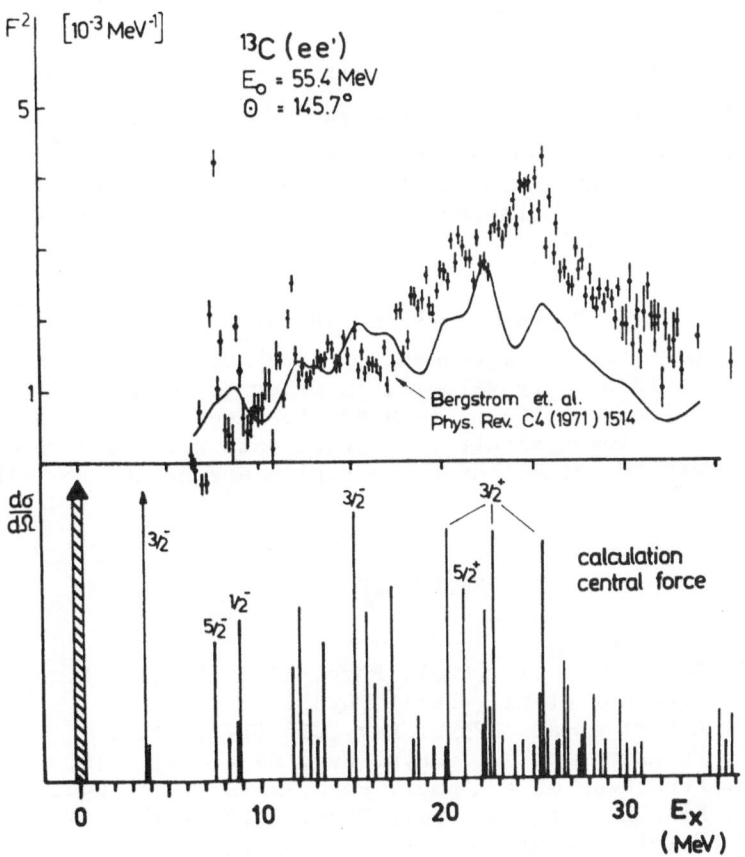

Figure 2. Experimental[4] and calculated form factors for the
^{13}C(ee') reaction with E_o = 55.4 MeV, θ = 145.7°. The smooth
curve corresponds to spreading of lines with assumed constant
width (1.5 MeV).

for the lack of transition strength above the main peak. In
this region $2\hbar\omega$ quadrupole excitations (which are not included
in our model) should also contribute, as in the (π^-,γ)
reaction[5]. The narrow resonances in the form factor below
15 MeV are also reproduced (the smooth curve in Fig. 2a
corresponds to spreading intensities in Fig. 2b with an
arbitrary width of 1.5 MeV). More restricted calculations[3]
provide roughly the same imtegrated strength as our model;
however, they underestimate the low-energy region and shift
a large amount of strength to energies above 30 MeV where
the simple model should no longer work.

Our model also provides a fair description of the gross
structure of photo reaction data on ^{13}C[6,7,8] apart from
absolute cross sections and the decay branch to the ^{12}C
ground state. Continuum shell model calculations can improve
on details of decay branches but not of the gross structure
of the resonance[9,10].

In view on the agreement for the photo- and electroexcitation functions, a measurement of the $^{13}C(\pi^-,\gamma)$ excitation
function would be very welcome, in order to compare with
other reactions on this nucleus and to check model predictions. Also, low-q electroexcitation data with good resolution through the giant resonance region are needed for other
nuclei.

REFERENCES

1.   G.E. Dogotar et al., Nucl. Phys. A282 (1977) 474.
2.   H.R. Kissener et al., Nucl. Phys., in print.
3.   D.J. Albert et al., Phys. Rev. C16 (1977) 503.
4.   J.C. Bergstrom et al., Phys. Rev. C4 (1971) 1514.
5.   R.A. Eramzhyan et al., Nucl. Phys. A290 (1977) 294.
6.   H.K. Kissener et al., Nucl. Phys. A219 (1974) 601.
7.   B.H. Patrick et al., J. of Phys. G1 (1975) 874.
8.   R. Koch and H.H. Thies, Nucl.Phys. A272 (1976) 296.
9.   M. Maragoni et al., Nucl. Phys. A277 (1977) 239.
10. J. Höhn et al., submitted to Nucl. Phys.

NUCLEON SPECTRA FROM THE $^{16}O(\pi^-,\gamma)$ REACTION

R.A. Eramzhyan, M. Gmitro, and L.A. Tosunjan

JINR Dubna, Laboratory for Theoretical Physics

101000 Moscow, P.O Box 79, U.S.S.R.

The main features of the hard ($\leq$ 130 MeV) gamma ray spectra of the (π^-,γ) reaction on light nuclei are apparently understood[1] in terms of the nuclear shell model and impulse approximation. To make a step further an analysis of the distributions of outgoing nucleons and in particular the coincidence type observations ar needed. Presented here is an extension of our earlier[2] n hole – n particle (n = 0, 1, 2) shell model description of $^{16}O(\pi^-,\gamma)^{16}N$ reaction. We have calculated the decay characteristics of the ^{16}N excitation. The formalism of reduced widths of the R-matrix theory was employed.

In Table I we present the population probabilities of the ^{15}N nuclear states. The values R_+ and R_- correspond to the decay of positive and negative parity states of ^{16}N respectively. We would like to underline the strong population of the ^{15}N levels which lie above the particle emission threshold. The calculation shows that the $(\pi^-,\gamma nn)$ and $(\pi^-,\gamma np)$ decay is about 20% of the total radiative capture yield.

It can be seen from the table that two levels of ^{15}N, namely the $1/2^-$ (g.s.) and the $3/2^-$ (6.3 MeV) states are most populated via decay of the ^{16}N resonance states excited in the radiative pion capture reaction. Our detailed calcula-

TABLE I

^{15}N nuclear states		Level population (10^{-4})		
J^{π}	E_x (MeV)	R_+	R_-	$R=R_+ + R_-$
$1/2^-$	0.0	12	63	75
$1/2^-$, $5/2^+$	5.3	7	1	8
$3/2$	6.3	23	31	54
$5/2^+$	7.2	5	0.2	5
$3/2^+$	7.3	1	< 0.1	1
Remaining Bound States		5	< 0.1	5
E_x above Threshold		42	2	44
Total		95	97	192

tions[4] show that observation of the partial spectra of neutrons connected with the population of the particular ^{15}N levels may disclose some of the needed details of nuclear structure.

The observed total neutron spectrum of the ^{16}O$(\pi^-,\gamma n)$ reaction shows up a well developed high-energy tail[3]. In the calculations we were able to understand this interesting feature. We have obtained the yield of 0.28 neutrons per one emitted gamma-ray in the energy interval E_n 10 MeV (0.20 neutrons per gamma-ray if E_n > 15 MeV). Such an enhanced yield of "hard" neutrons comes out due to the decay of highly excited 1^+, 2^+ and 3^+ levels of ^{16}N. These "quadrupole" excitations (operator $[\sigma\cdot Y_2]_{J+}$) contribute 44% to the branching ratio of the radiative pion capture reaction on ^{16}O (see Ref. 2). Simultaneously, and in an agreement with observation, our calculation of the neutron spectrum shows up the typical peak at $E_n = 4 \div 5$ MeV. This part of the neutron spectrum is due to the decay of the dipole (operator $[\sigma\cdot Y_1]_{J-}$) excitations in ^{16}N.

The characteristics which we have considered are all sensitive to the detailed wave functions of the initial, intermediate and final nuclear states involved in the reaction. A critical experimental and theoretical study of these characteristics may constitute a valuable step towards a deeper understanding of the ractions which involve simultaneously bound states and continuum nuclear excitations.

REFERENCES

1. H.W. Baer in Proc. 7th Int. Conf., Zurich, 1977.
2. R.A. Eramzhyan, M. Gmitro, R.A. Sakaev and L.A. Tosunjan, Nucl. Phys. A290 (1977) 294.
3. W.C. Lam, K. Gotow, B. MacDonald and W.P. Trower, Phys. Rev. C10 (1974) 72.
4. M. Gmitro, L.A. Tosunjan and R.A. Eramzhyan, to be published.

THE UNIFIED DESCRIPTION OF DIRECT AND RESONANCE PROCESSES IN THE RADIATIVE PION CAPTURE BY LIGHT NUCLEI

V.V. Balashov, Nuclear Research Institute, Moscow State University, Moscow, U.S.S.R.

R. Wünsch*, Joint Institute for Nuclear Research Dubna, U.S.S.R.

Developing the idea of the unified description of direct and resonance mechanisms of nuclear disintigration by high-energy particles[1], we have calculted the spectrum dR/dE of hard γ-rays in the $^{16}O(\pi^-,\gamma n)^{15}N$ reaction in a 1p-1h continuum shell-model version[2]. The pion capture from the mesoatomic s- and p-orbits has been taken into account. In Fig. 1 the calculated spectrum folded with a Breit-Wigner shape is compared with the available experimental data[3].

The unified description of direct and resonance processes gives a proper way for the analysis of the $(\pi^-,\gamma n)$ coincidence experiments[4] which are much more informative than measurements of the hard γ-ray spectra only. The γ-n angular correlation functions measured in the $(\pi^-,\gamma n)$ experiments for different channels of nuclear disintigration are not only sensitive to the absolute values of the different multi-polarity contributions but also to their relative phases. The proposed approach allows one, without changing the model, to trace the transformation of the angular correlation functions as well as the variation of the ratio between different disintegration channels from the region of well-isolated resonances to that of dominating direct processes. An example of our calculations is shown in Fig. 2.

The results received come against the naive assumption

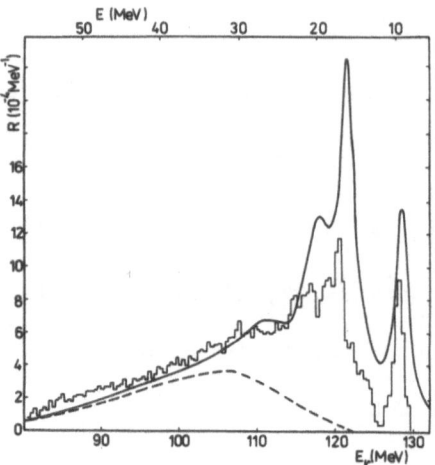

Figure 1.   The spectrum of hard γ-rays in the radiative pion capture by ^{16}O.   Heavy line: our calculation.   Dotted line: contribution of the single-particle continuum without the $1d_{3/2}$ and $1f_{7/2}$ single-particle resonances.   Histogram: experiment.

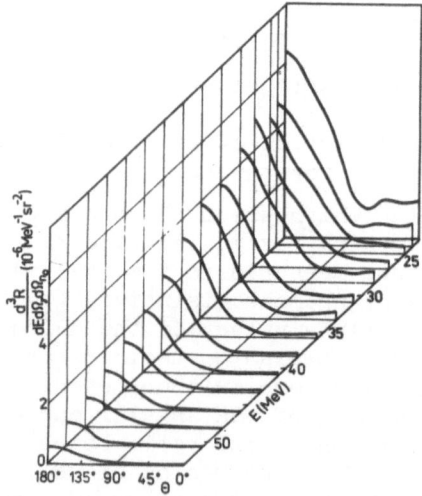

Figure 2.   γ-n correlation function $dR/dE\,d\Omega_\gamma\,d\Omega_n$ of the reaction ^{16}O$(\pi^-,\gamma n_o)^{15}$N$_{g.s.}$ as a function of the excitation energy E of the intermediate nucleus ^{16}N.

of the additivity of direct and resonance (giant resonance) contributions to the probability of the radiative pion capture which is often accepted in the analysis of the $(\pi^-,\gamma n)$ data.

REFERENCES AND NOTES

*On leave from Zentralinstitut für Kernforschung Rossendorf, GDR
1. V.V. Balashov, Proc. of the IV Int. Conf. on High-Energy Physics and Nuclear Structure, Dubna 1971, p. 167.
2. I. Rotter et al., Particles and Nucleus (USSR) 6 (1975) 435; H.W. Barz et al., Nucl. Phys. A275 (1977) 111.
3. J.C. Alder et al., 7th Int. Conf. on High-Energy Physics and Nuclear Structure, Zürich 1977, Abstract Volume of Conf. Proc. C16.
4. W.C. Lam et al., Phys. Rev. Lett. 28 (1972) 108; Phys. Rev. C10 (1974) 72.

TOTAL CROSS SECTION MEASUREMENTS FOR THE PHOTOPRODUCTION OF
CHARGED PIONS FROM THRESHOLD TO 350 MeV

Edward C. Booth

Boston University

I. INTRODUCTION

Total cross-section measurements for charged photo pions have been made by detecting the e^+ decay positron from a μ^+ meson stopped in the target and by measuring the radiactivity of the daughter nucleus.

The results separate into those obtained near the pion threshold and those which are taken over the Δ (1232) resonance region up to about $E_\gamma = 350$ MeV. Total cross-section measurements done by μ^+ decay are restricted to pion energies of less than 20 or 30 MeV because of pion loss from the target, and to light nuclei because of the Coulomb barrier.

Most of the high energy experiments have been done using the radioactivity of heavy nuclei, requiring a sum over many bound states of the daughter nucleus, while a goal of threshold measurements has been to measure the cross-sections to successive individual levels of the daughter nucleus.

II. MOTIVATION

The nuclear pion production process is interesting both in itself and as a tool for nuclear structure studies. The photo pion measurements are part of a system of complementary or inverse reactions as shown in Fig. 1. Charged photo-pion

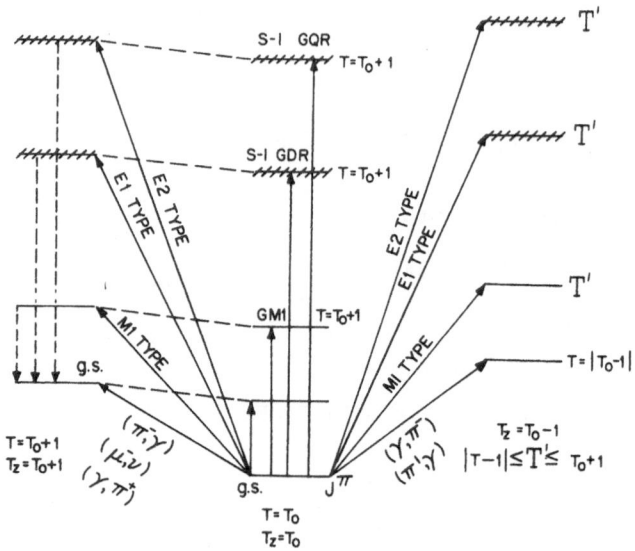

Figure 1. $\gamma\pi^{\pm}$ and other reactions populating similar nuclear levels. The analogue states are connected with horizontal dashed lines. Near threshold the $\pi^{\pm}$ reactions are primarily spin-flip to $T_0 \pm 1$ states[29].

production near threshold is generally from a T_0 to a $T_0 + 1$ level in an adjacent nucleus. It is a good test of the theory for these reactions to require agreement among all transitions to the same or analogue states. The $\gamma\pi^{\pm}$ reaction changes T_3 while providing variable momentum transfer, and is free of some of the problems involved in interpreting pion capture from various atomic orbits. Photo-pion production, especially at high energies, involves more of the nuclear volume than do pion scattering and charge exchange reactions. There is interest in all three aspects of nuclear pion production, the production mechanism itself, the distortion of the outgoing pion wave by the nuclear potential, and in the structure of the daughter nucleus. Nuclei such as ^{2}H, ^{3}He, ^{6}Li, ^{12}C and ^{16}O offer targets where the nuclear information is known, emphasizing the production mechanism and the final state interaction. The study of $\gamma\pi^-$ and $\gamma\pi^+$ reactions offers alternative cases where the cross-sections are more and less sensitive to the pion-nuclear potential because of the attraction or repulsion of pion wave function by the nucleus[1]. Very near threshold $(E_\pi < 4$ MeV) the long mean free path of the pion minimizes the effect of the pion-nuclear interaction, calculated using an optical potential, but even there the absorption is large as shown in Fig. 2[1]. At higher energies the final state corrections become much more important and there we find disagreement between calculations and experiment.

Some questions may be raised about models for the production mechanism on the following topics[2]: the validity of the impulse approximation, charge exchange contributions, the accuracy of the fundamental amplitudes for pion production from the nucleon, the importance of meson current effects, the need for Lorentz-Lorenz corrections, the effect of the Fermi motion of the nucleons, and the need for relativistic calculations of the fundamental amplitudes. There are also problems with transformation to the nuclear cm system, the question of extrapolation off the mass-shell, and the proper number of partial waves to use for the pion. The final state interaction involves the proper choice of the pion-optical potential, the

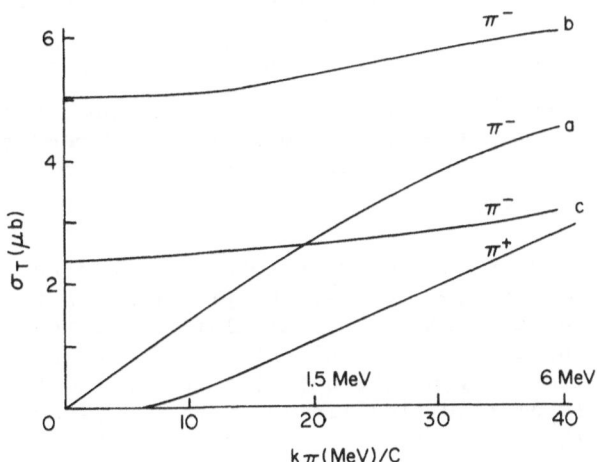

Figure 2. Threshold cross-section for $\pi^\pm$ production on ^{12}C to the daughter ground state. (a) no nuclear potential, (b) Coulomb potential only, (c) full nuclear plus Coulomb potential for π^-. The π^+ has the full nuclear plus Coulomb potential[1].

number of times the pion scatters, Fermi motion in the final
state, and the possibility of reaching the daughter state by
an indirect process through the final state interaction.

The pion nuclear optical potential is well established
for low momentum pions by π^- capture studies, and for $E_\pi > 50$
MeV by pion scattering results. One might expect agreement
between results and theory in those energy regions, with some
work to be done filling in the gap. As shown below, the potentials seem to give agreement near threshold and disagree-
ment for $E_\pi > 50$ MeV for $\gamma\pi^\pm$ reactions. More systematic data is
needed for π^+ and π^- at different energies and for a range of
nuclei in the hope of learning more about the optical potential. This problem can be expolored by differential cross-
section measurements using a pion spectrometer. However, a
few good cases remain for radioactivity work.

In using pion spectroscopy as a tool for nuclear structure studies, once nuclear interactions are understood the
targets must be chosen to exploit the advantages of the reaction. For example $\gamma\pi^\pm$ production involves transitions to
$T_3 \pm 1$ nuclei, which cannot be studied directly by electron
scattering. The low energy $\gamma\pi^\pm$ reactions, like π^- capture,
pick out transitions with large spin flip components; for example, states which are analogues to M1, M2 and E1 (spin-iso-
spin) giant resonances as seen in the $^{12}C(\gamma\pi^+)^{12}B$ and $^{16}O(\gamma\pi^+)$
^{16}N reactions described below. The $\gamma\pi^\pm$ reaction gives a separate measure of the magnetization current as opposed to a sum
of orbital and spin parts, as is seen below in the case of
$^6Li(\gamma\pi^+)^6He$.

Total $\pi^\pm$ cross-section measurements were favored near
threshold in light nuclei because the pion-nucleus final state
interaction is at a minimum, the calculations are simple, and
the nuclear structure is well understood. The goal was to establish the reliability of the calculations in the face of the
various possible difficulties listed above. This goal has been
largely accomplished. More ambitious goals such as using the
technique to measure the n-n scattering length from the $d(\gamma\pi^+)$
2n reaction, placing new restrictions on the pion optical potential from the $^{12}C(\gamma\pi^+)^{12}B$ reaction, and obtaining new nuclear
structure information from the $^6Li(\gamma\pi^+)^6He$ reaction have en-

joyed limited success because of inadequate experimental
accuracy and because of the theoretical problems listed
above.

The motivation for $\pi^{\pm}$ photo production in the resonance
region with the daughter nucleus left in the ground state has
been to test the production mechanism and to study the pion-
nuclear optical potential. In this energy region the full
momentum-dependent Hamiltonian must be used and the Δ is an
important part of the nucleon amplitude. Several pion waves
are required and some of the simplifying approximations used
in calculations near threshold may be invalid.

III. YIELD AND CROSS-SECTION

Electron bremsstrahlung are used to excite one or more
levels of the daughter nucleus with the emission of pion. The

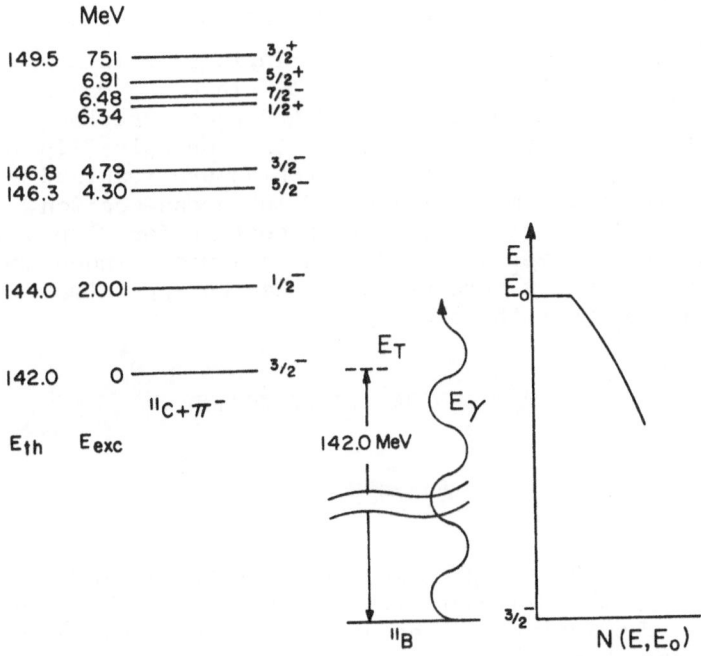

Figure 3. Particle-stable states of ^{11}C accessible in the ^{11}B $(\gamma\pi^-)$ ^{11}C reactions[40].

situation is shown in Fig. 3 for $^{11}B(\gamma\pi^-)^{11}C$, where E_γ is not a monoenergetic photon but is the upper end of the brems-strahlung spectrum with electron energy E_o. The yield as a function of E_o is $Y \propto \sum_i \int_{E_T}^{E_o} N(E,E_o)\,\sigma_i(E)dE$. In the $\pi^+ \to \mu^+ \to e^+$ method, the sum runs over all states between E_o and E_T but it only runs over all bound states when the radioactivity of the daughter is detected. The contribution from individual states can be obtained by measuring the excitation function $Y(E_o)$ and fitting to appropriately parameterized expressions for $\sigma_i(E)$. The shapes of the $\gamma\pi^+$ and $\gamma\pi^-$ cross-sections[1] are shown in Fig. 2. The step function in the π^- production cross-section is due to the Coulomb attraction term, which correspondingly depresses the π^+ cross-section near threshold. The approximation usually used for π^+ production is $\sigma = a(q/k)\,S$ where q and k are the pion and photon momenta in the cm system and S is the Coulomb barrier penetration factor. The factor a is expanded to a constant times a_p, the value for the proton. This simple relationship is appropriate within a 5 or 10 MeV of threshold, since the momentum independent $\bar{\sigma}\cdot\hat{\varepsilon}$ term produces most of the cross-section. For mixed or forbidden transitions where the non-spin flip momentum dependent terms of the Hamiltonian are required a complete calculation must be made of the cross-section. It is also possible to expand the cross-section for π^+ production in a power series in E and produce a model independent cross-section. The π^- cross-section can be approximated by a step function near threshold.

IV. THE RADIOACTIVITY METHOD

1. Advantages

Compared to using a pion spectrometer the advantages of measuring σ by the radioactivity of the daughter are the following: The activation technique is cheap, requiring no spectrometer. There is a high counting efficiency and a large (total) cross-section. Count rate problems from background are greatly reduced by counting while the beam is turned off.

The reaction is very exclusive, at least when only one or a
few states are bound. The technique is not limited by pion
decay in flight, so it can be used very near threshold. The
cross-section can be measured at high energy, without the
limit normally imposed by having a finite magnet. The reaction can be monitored by comparison with some known photonuclear reaction proceeding mainly through the E1 giant
resonance at 20 MeV.

2. Disadvantages

One is only measuring a total cross-section, whereas
$d\sigma(\theta)$ is usually more informative. Closely spaced bound
states of the daughter are usually hard to resolve. Excitation functions are necessary in any case, and these are very
time consuming, not only for statistical reasons but because
linear accelerators cannot change energies quickly. Each
case requires a different counting technique and a new type
of monitor. The real curse of the method is the presence of
non-mesic interferences. These come from three sources: target isotopes, two step activites and competing activities.
Examples of each in turn are these: $^{13}C(\gamma p)^{12}B$ interferes
with $^{12}C(\gamma \pi^+)^{12}B$, $^{12}C(\gamma p)^{11}B$ followed by $^{12}C(pn)^{12}N$ interferes with $^{12}C(\gamma \pi^-)^{12}N$, and $^{12}C(\gamma,3n)^9C$ gives a 17 MeV β^+
(127 msec) interfering with $^{12}C(\gamma \pi^-)^{12}N$ which has a 16 MeV β^+
(11 msec).

3. Threshold Measurements

The interferences greatly restrict the utility of radioactivity measurements near threshold. Table 1 presents a
list of the nuclei that might be or have been used as targets
near threshold. All of the cases could be extended through
the resonance region relatively easily. A low log ft value
corresponds to a large $\overline{\sigma \cdot \varepsilon}$ value for G-T decay, indicating a
large cross-section for the ground state transition. The
ground state spins are shown in the last column. The number
of bound states in the daughter nucleus, shown in column 6,
should be small to simplify comparison with theory.

TABLE 1
Favorable Cases for Radioactivity Measurements
(a) Work in progress, (b) Threshold data exists, (c) Resonance
data exists.

Target	Abund. %	Reaction	$\tau_{1/2}$	Radiation (MeV)	Bound States	log ft	g.s. spins
^{7}Li (a)	92.5	$(\pi^-)^7$Be	53d	$\gamma(0.48)$	2	3.3	$3/2^- \to 3/2^-$
^{9}Be	100	$(\pi^+)^9$Li	0.17s	n	2	5.5	$3/2^- \to 3/2^-$
^{11}B(b,c)	81	$(\pi^-)^{11}$C	20min	$\beta^+(0.97)$	7	3.6	$3/2^- \to 3/2^-$
^{11}B	81	$(\pi^+)^{11}$Be	14s	$\beta^+(12)$	2	6.8	$3/2^- \to 1/2^+$
^{10}B	19	$(\pi^-)^{10}$C	19s	$\gamma(1.03$	2	large	$3^+ \to 0^+$
^{12}C(a,b,c)	99	$(\pi^-)^{12}$N	11ms	$\beta^+(16.4)$	–	4.1	$0^+ \to 1^+$
^{13}C	1	$(\pi^-)^{13}$N	10m	$\beta^+(1.2)$	–	3.7	$1/2^- \to 1/2^-$
^{13}C	1	$(\pi^+)^{13}$B	19ms	$\beta^+(13)$ $\gamma(4.4)$	3	4.0	$1/2^- \to 3/2^-$
^{14}N(a)	99.6	$(\pi^-)^{14}$O	71s	$\gamma(2.31)$	–	3.5	$1^+ \to 0^+$
^{16}O(c)	99.8	$(\pi^+)^{16}$N	7s	$\beta(10.4)$ $\gamma(6.1)$	4	6.7	$0^+ \to 2^-$

The ^{9}Be(π^+) cross-section has been measured[3] using the
$\pi^+ \to \mu^+ \to e^+$ method but in that case the unbound levels also
contribute. A radioactivity measurement would pick out only
the ground and first excited states, and could be carried
through the resonance region. The large log ft value of
^{10}B for the $3^+ \to 0^+$ transitions may not predict a vanishing
$\sigma(\pi^-)$ since the angular momentum is easily provided by a
150 MeV photon. The ^{13}C(π^-) case is of interest because of
the lack of excited bound states, permitting a clean experiment at threshold and through the resonance region.

4.   Radioactivity in the Threshold Region –
Discussion of Results

^{11}B(π^-). The results of the measurement of the ^{11}B
($\gamma\pi^-$)^{11}C reaction[41] are shown in Fig. 4. It is clear that the
data cannot support the separation of the contributions from
the individual levels of ^{11}C, but a best fit can be made for
assumed values of the π^- cross-section taken to be a step
function. Agreement was found with theory[1] within a factor
of two. Experimental work was made difficult by the very
large non-mesic background.

^{12}C(π^-). The experimental situation is improved over
^{11}B(π^-) since the non-mesic two-step background is a factor
of 10 lower, and there are no excited states in ^{12}N. Nevertheless, other experimental difficulties caused the data[4]
to be inaccurate for at least 3 MeV above threshold, where
it agrees with theory at the $\pm$ 15% level. The theoretical and
experimental results are compared by Epstein and by Nagl
in the Proceedings of this Conference.

5.   Resonance Region

In the resonance region the cross-sections are up by an
order of magnitude over those in the threshold region, and
the energy interval ΔE can easily be increased to 20 MeV,

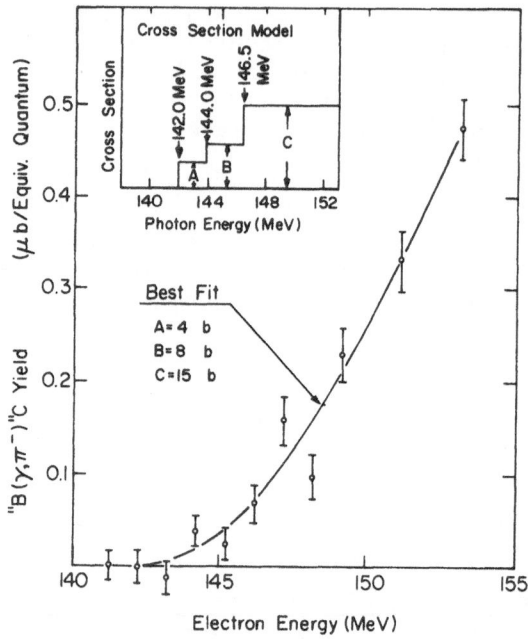

Figure 4. Low energy $^{11}B(\gamma\pi^-)^{11}C$ yield data[40]. The insert shows the three-step cross section model used to the data. The solid curve corresponds to A = 4μb, B = 8μb and C = 15μb.

giving an improvement by a factor of 100 in the ratio of mesic to non-mesic activity. In addition, the Coulomb barrier for π^+ is no longer a problem, so a large range of nuclei is open to investigation. A list of papers reporting $\pi^\pm$ production from the resonance region are given in Table 2. The number of bound states in the daughter nuclei is generally large, except in the light nuclei shown in Table 1. The usual approach in analysis is to sum over the theoretically bound single-particle states using the full interaction Hamiltonian and taking the final state interactions into account by using the potentials determined by pion scattering. An experimental difficulty is that the background must be extrapolated over a long energy range, from 140 - 350 MeV, which has apparently led to substantial errors in the older data. Since the two step n,p contribution goes as the square of the target thickness, this background can be measured through the entire energy range

TABLE 2

Results in the Resonance Region by the Radioactivity
Method

$^{11}B(\pi^-)$, Hughes[31] ; $^{60}Ni(\pi^-)$, March[32] ; $^{11}B(\pi^-)$, $^{11}B(\pi^+)$,

Dyal[33] ; $^{27}Al(\pi^+)$, Masaike[34] ; $^{209}Bi(\pi^-,\chi n)$, Yavin[35] ; $^{16}O(\pi^+)$,

Meyer[36] ; $^{51}V(\pi^-,2n)$, Meyer[36] ; $^{27}Al(\pi^+)$, Walters[37] ; $^{51}V(\pi^-,2n)$,

$^{11}B(\pi^-)$, $^{27}Al(\pi^+)$, $^{51}V(\pi^+)$, Nydahl[37] ; $^{88}Sr(\pi^+)$, $^{138}Ba(\pi^+)$,

$^{41}K(\pi^+)$, $^{65}Cu(\pi^+)$, Blomqvist[38] ; $^{27}Al(\pi^+)$, Anderson[12] ; $^{11}B(\pi^-)$,

$^{27}Al(\pi^+)$, Noga[40] ; $^{12}C(\pi^-)$, Epaneshnikov[7] ; $^{27}Al(\pi^+)$, $^{51}V(\pi^+)$,

$^{51}V(\pi^-)$, Blomqvist[9] ; $^{197}Au(\pi^-)$, Medicus[39] ; $^{197}Au(\pi^-)$, Blomqvist[9].

by using targets composed of stacks of foils[5]. The p,n contribution is linear in target thickness so that the stacked foil trick will not work but improved data on γp and p,n reactions at high energy permit a reasonable estimate of that background to be made.

Favorable cases still to be measured are shown in Table 1. The $^{12}C(\pi^-)$ case should be remeasured. The $^{7}Li(\pi^-)$, $^{10}B(\pi^-)$ and $^{13}C(\pi^-)$ cases all have counterparts in π^+, π^o reactions[6]. The lack of agreement with theory is shown in Fig. 5. $^{14}N(\gamma\pi^-)^{14}O$ presents a clean case for study and will be published (DeCarlo, this Symposium). The $^{16}O(\gamma\pi^+)^{16}N$ case should be rejected.

6. Radioactivity in the $\Delta(1232)$ Resonance Region:
Discussion of Results

There is considerable disagreement among the results and much of the older theory is limited to descriptions in terms

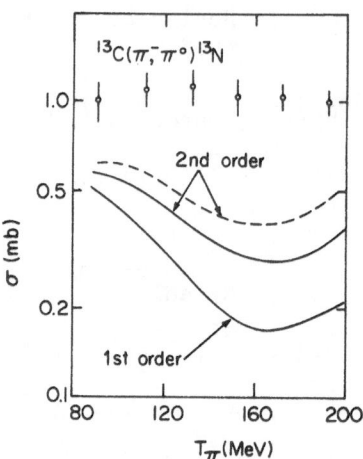

Figure 5. First and second-order DWIA results for the $^{13}C(\pi^+, \pi^\circ)^{13}N$ total cross-section with Pauli correlations only (solid line) and with dynamical intermediate-coupling correlations (dashed line). Data from J. Alster[6].

of surface vs. volume production. Therefore, only the most recent results are presented here, but they are thought to be correct experimentally and they illustrate the points of interest.

The results for $^{12}C(\gamma\pi^-)^{12}N$ are discussed in the Proceedings by Nagl and Überall. There is a factor of two difference in the experimental results found by Bernstein[4] and Epaneshnikov[7]. The $^{12}C(\gamma\pi^-)^{12}N$ and $^{7}Li(\gamma\pi^-)^{7}Be$ cases are now being studied in the resonance region at Bates by Bosted and others.

Recent high energy measurements are those on $^{27}Al(\gamma\pi^+)$ ^{27}Mg, $^{51}V(\gamma\pi^+)^{51}Ti$ and $^{51}V(\gamma\pi^-)^{51}Cr$ by Blomqvist[5]. Six points were taken with about 20% errors from 180 – 400 MeV. The results for ^{27}Al are shown in Fig. 6. The solid curve is the best fit to the data while the lower and upper cross-hatched areas are theoretical calculations with and without the final

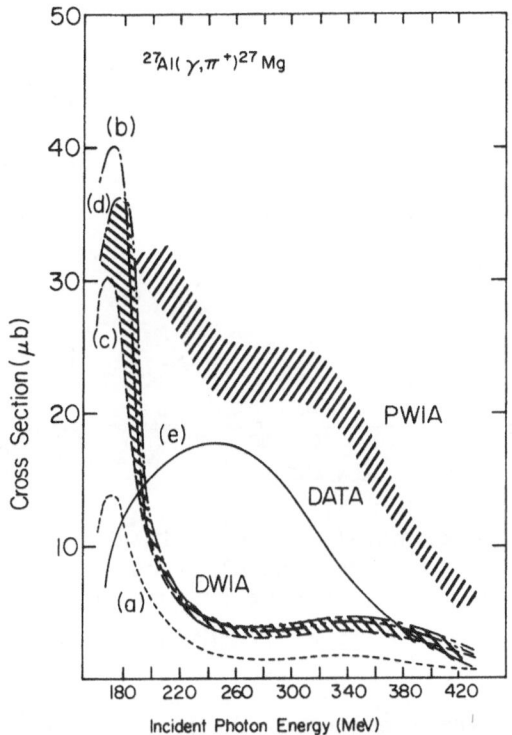

Figure 6. The solid curve is the data for $^{27}Al(\gamma\pi^+)^{27}Mg$, the
lower and upper cross-hatched areas are theory with and without final state interactions[5].

state interaction, labeled DWIA and PWIA. The uncertainty in
in the number of final states to be included in the shell
model calculations which involve transitions from a $d_{5/2}$ pro-
ton hole to the neutron states s 1/2; $|0\boxtimes d_{3/2}>$, $|2\boxtimes d_{3/2}>$,
and some of the $|4\boxtimes d_{3/2}>$. The results are obtained using
the Coulomb plus modified Kisslinger potentials, and the results are similar for the standard Kisslinger and local Laplacian potentials. Similar results are found for $^{51}V(\gamma\pi^+)$
^{51}Ti. Not only the magnitude but the shape of the theoretical calculations is in strong disagreement with experiment.
The blame is placed on the optical potential which gives too
weak an s-wave repulsion and too strong a p-wave absorption,

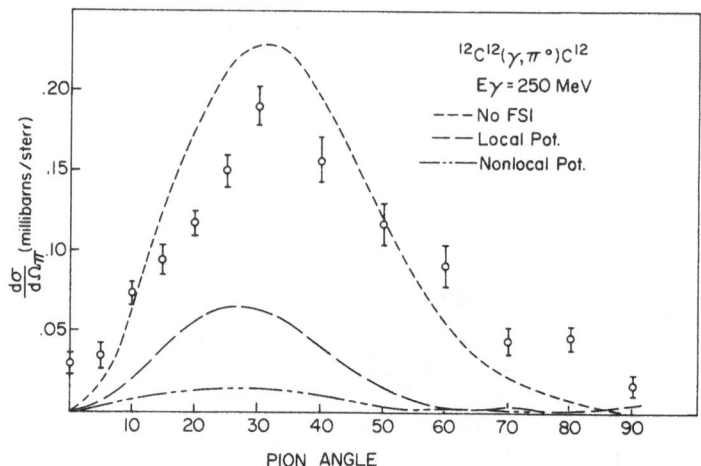

Figure 7. $\gamma\pi^\circ$ differential cross-sections for ^{12}C in the resonance region. The experimental cross-section is larger than that calculated for coherent production using the impulse approximation with final state interactions[8,2].

perhaps from the gradient terms of the potential. It is interesting to compare these results with those for $\gamma\pi^\circ$ production in the resonance region. The experimental data (Fig. 7) by Davidson[8] is compared with the PWIA and DWIA calculations of Saunders[2] for ^{12}C. Similar results were obtained in heavy nuclei. Again, the absorptive part of the potential seems too strong. The same situation may also prevail in the (π^+,π°) reaction as shown in Fig. 5. These results suggest that further study of $\pi^\pm$ and π° photoproduction is needed in the resonance region. There is only one modern radioactivity measurement[9], and this has coarse energy steps, large error bars and considerable uncertainty due to the sum over many states. The $\gamma\pi^\circ$ measurements[8] are at a single energy with poor energy resolution on the π°. The π^+, π° measurements (Alster) may suffer from proton contamination. On the other hand, calculations by DeCarlo reported at this Conference are in agreement with $^{14}N(\gamma\pi^-)^{14}O$ measurements made at Lund (to be published).

7.   Related Reactions

Two reactions related to $\gamma\pi^{\pm}$ production in the resonance region are worth mentioning. These are the $^{12}C(\gamma,n)^{11}C$ and $^{32}S(\gamma,np)^{30}P$ reactions measured by Hylten[10] and Van Hise[11] using the radioactivity of the daughter nucleus. The $^{12}C(\gamma,n)^{11}C$ cross-section (Fig. 8) has a resonance magnitude of a factor of twenty or thirty greater than that for π^{+} production to bound states. The $^{32}S(\gamma,np)$ cross-section has a similar behavior. This may be due to pion absorption with nucleon emission, or it could be π° production to unbound states. Mesonic contributions to photo-nucleon production are discussed by Anderson[12].

The effects of π° production to bound states have been seen by Forkman[13] in measurements through the resonance region of the activity of long-lived isomeric states in ^{89}Y, ^{115}In and ^{197}Au. The maximum cross-section is of order 500 µb, compared to 400 µb typical of the γn and γnp reactions.

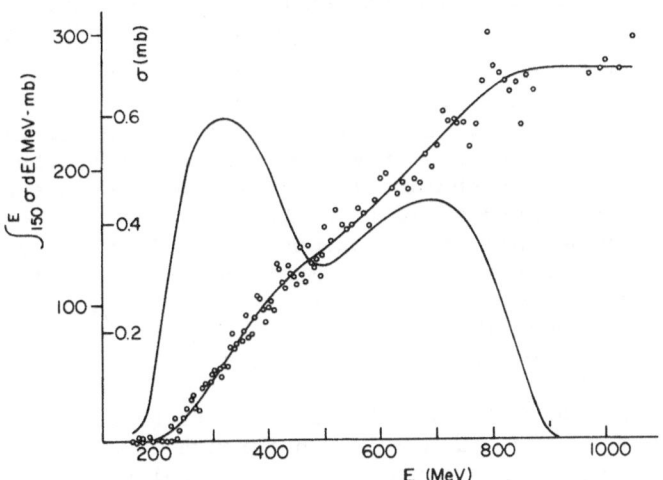

Figure 8. The integrated cross-section and cross-section for $^{12}C(\gamma,n)^{11}C$ as a function of photon energy E.[10]

In view of the interest in the absorptive part of the
nuclear potential and in the π-nucleus interaction, it may
be worthwhile to improve the radioactivity measurements of
this type, since only a few cases have been measured and
those with relatively bad statistical errors.

V. $\pi^+ \to \mu^+ \to e^+$ MEASUREMENTS

1.   Method

The π^+ meson decays in 2.0×10^{-8} sec to a 5 MeV μ^+
which in turn decays to a β^+ with maximum energy 50 MeV in
2.2 μsec. The procedure is to stop the π^+ and the μ^+ in
the target and detect the β^- which escapes the target. The
2.2 μsec lifetime of the μ^+ permits one to count after the
beam pulse which is shortened to 1 or 2 μsec. Pulse rates of
3000 pps and 1000 pps were typical at the Saclay and Bates
linear accelerators where the work was done. The advantages
of measuring a total cross-section by the $\pi^+\mu^+e^+$ method using
the high energy β^+ after the beam pulse thereby eliminating
interferences, the detection system is the same for all cases,
and the cross-section for $p(\gamma\pi^+)n$ is large and well known, so
that the data from each target can be taken relative to that
from CH_2(polyethelene). The problems with the method are

that the Coulomb barrier makes the cross-section low near
threshold, causing difficulty in resolving the contribution
from individual levels, and restricting the cases to light
nuclei, and that pion loss from the targets limits the method
to 20-30 MeV over threshold. The experimental precision is
limited by the different electron absorption in the various
targets, and non-mesic contributions from multiple scattering
events in the detector telescope must be subtracted from the
data. Despite the difficulties and limits of the method,
it has given reproducible results near threshold with an accuracy of a few per cent relative to the proton cross-section.

The experimental situation is summarized in Table 3,
which lists the cases from hydrogen through oxygen. The
results of the measurements are discussed below, but a few
comments should be made about cases for which there is no
data since several of these may be worth doing.

TABLE 3

Possible Cases for $\pi^+ \to \mu^+ \to e^+$ Measurements

Target	% Abund.	GS Spin	Daughter	Energy (MeV)	Daughter Spin	log ft	Comment
^{1}H	99.9	1/2 +	n		1/2 +	3.1	Need thres-data
^{2}H	0.1	1 +	2n				Audit[16] Booth[15]
^{3}He	10^{-4}	1/2 +	^{3}H		1/2 +	3.1	Argan[30]
^{4}He	100	0 +	t,n;d,2n				Phase space; resonances? No data
^{6}Li	7.4	1 +	^{6}He	0	0 +	2.9	Audit[16]
^{7}Li	92.6	3/2 -	2t,n;(^{7}He)	1.8	2 +		Low σ, phase space; resonances? No data.
^{9}Be	100	3/2 -	^{9}Li	0 2.69 4.31 etc.	3/2 - 1/2 -	5.5	Milder[3]
^{10}B	19.6	3 +	^{10}Be	0 3.37 6.0 etc.	0 + 2 +	13.7	No data.
^{12}C	99	0 +	^{12}B	0 0.95 1.70 etc.	1 + 2 +	4.1	Milder[3]
^{13}C	1	1/2 -	^{13}B	0 3.48 etc.	3/2 -	4.0	No data.
^{14}N	99.6	1 +	^{14}C	0 6.09 etc.	0 + 1 -	9.0	Deutsch[28]
^{16}O	99.8	0 +	^{16}N	0-.4 3.36 etc.	2-,0-, 3-,1-	6.7	Milder[3]

2. Cases

^{1}H. There is not data for $p(\gamma \pi^+)n$ at energies below 15 MeV
above threshold. While one may be fairly confident that the
$\bar{\sigma} \cdot \bar{\varepsilon}$ term is well known from higher energy results which give
$\sigma_p = (q/k)201 \pm 7\mu b$ as $q \to 0$, nevertheless it is disturbing
that no measurements exist nearer threshold.

^{4}He, ^{7}Li. Since ^{4}H is unbound, the reaction products are
t + n, d + 2n, etc., resulting in a very small cross-section
near threshold because of phase space considerations. The
cross-section should be a factor of 100 down from the proton
cross-section of few MeV over threshold, although it might be
substantial if resonances exist in the daughter systems. The
cross-sections rise rapidly with energy and should be mea-
surable at 20 MeV above threshold.

The same comments apply to π^+ production from ^{7}Li, except
here the phase space is even smaller with at least three particles in the final state. However, there should be a resonance in the ^{7}He system at 11.07 MeV analogous to the $J = 3/2^-$,
$T = 3/2$ state at 11.25 MeV in ^{7}Li which could produce an
appreciable cross-section.

^{10}B. The $3^+ \to 0^+$ ground state transitions ($T = 0$ to
$T = 1$) is a chance to observe a spin-flip transition which is
highly forbidden for beda decay but may be observable in π^+
production. The first and second excited (2^+) states are well-
separated, giving the possibility of measuring their contributions separately. There is no interference from ^{11}B which has
a threshold about 11 MeV higher than ^{10}B.

^{13}C. The ^{13}C$(\gamma \pi^+)^{13}$B reaction looks very favorable with
a low log ft value, large separation to the first excited
state, and an M1 spin flip transition to $T = 3/2$, $J = 3/2^-$
ground state of ^{13}B from the $T = 1/2$, $J = 1/2^-$ ground state of
^{13}C . A target of separated isotope is required.

3. Results of $\pi^+ \to \mu^+ \to e^+$ Measurements

^{2}H(π^+). Tzara[14] pointed out that the cross-section for ($\gamma\pi^+$)2n is decreased from the p($\gamma\pi^+$)n cross-section near threshold by phase space requirements, but that the cross-section is enhanced by the strong attractive force between the outgoing neutrons in their T = 1 state. The deuteron case provides an ideal testing ground for the reaction mechanism because the wave functions and final state interactions are well known in principle. The cross-section was measured by Booth[15] and found to be in fairly good agreement with theory. More accurate measurements by Audit[15] from 1 - 4 MeV above threshold gave the deuteron cross-section to ± 2.5% relative to the proton. The results are shown in Fig. 9 compared with calculations by Tzara[14], O'Connell[17], and Noble[18] labelled T, O, and N. Audit found that if the scattering length is taken

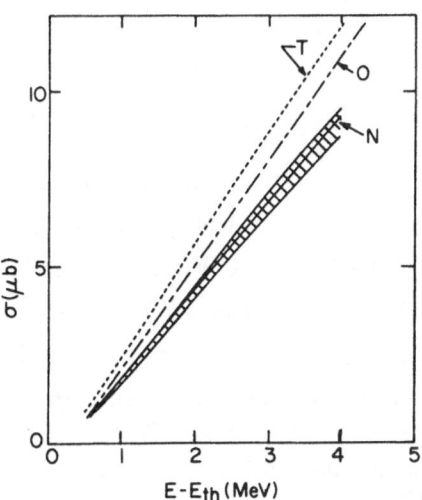

Figure 9. D($\gamma\pi^+$)nn cross-section near threshold[16]. The cross-hatched area is the experimental result with errors. Theory is shown as T (Tzara[14]), O (O'Connell[17])*, and N (Noble[18]).

*See editor's note at the end of this article.

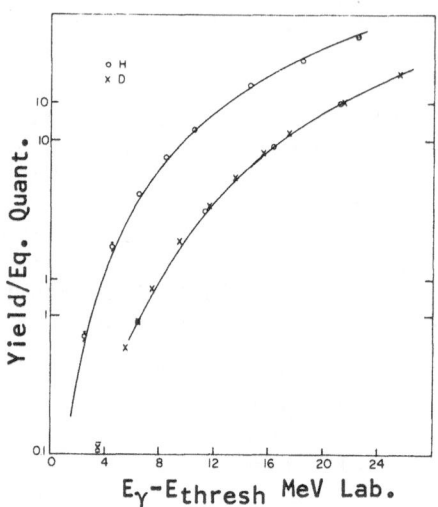

Figure 10. $d(\gamma\pi^+)nn$ yield up to 20 MeV[20]. The solid line through the $p(\gamma\pi^+)n$ data (upper curve) is the theoretical yield from Blomqvist and Laget[9] scaled to fit the data. The solid line near the $d(\gamma\pi^+)nn$ data is the theoretical yield calculated by O'Connell[17] and Laget[21] scaled by the same factor.

as a free parameter in the fit, the result is a_{nn} = -16.3 ± 3.4 compared to the accepted value of a_{nn} = -16.3 ± 0.7. (Baer[19] reported a_{nn} = -18.4 ± 0.4 at Zurich.) Noble's calculation used s- and p-wave pions in the elementary amplitude, the neutrons in both $^3S_0$ and $^1S_0$ states, and all higher partial waves to the n-n subsystem, but did not include the deuteron d state. The energy range was extended to 20 MeV by Booth[20] with results shown in Fig. 10 compared with calculations by O'Connell[17] and Laget[21] using approximations similar to Noble's[18]. There is overall very good agreement between experiment and theory. The experiment cannot resolve the small differences in calculations..

$^3He(\pi^+)$. The reaction $^3He(\gamma\pi^+)^3H$ has been measured from 1 - 5 MeV over threshold with the results shown in Fig. 11,

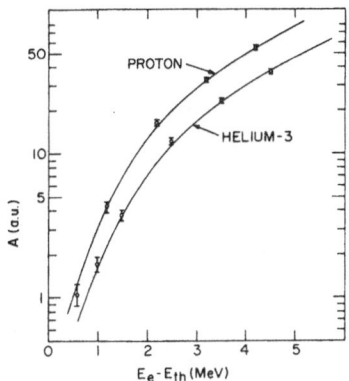

Figure 11. ^{3}He$(\gamma\pi^+)^3$H yield near threshold[30], shown relative to the ^{1}H$(\gamma\pi^+)$n yield. The solid curves are the theoretical yields for reactions.

where the solid line through the ^{3}He data is the theoretical yield relative to the proton, calculated using only the $\bar{\sigma} \cdot \bar{\epsilon}$ term of the Hamiltonian. The measured axial matrix element agrees with a calculation from the Panofsky ratio and disagrees with π^- capture results. This result plus the deuteron result indicates that meson exchange corrections are not needed in $\gamma\pi^+$ reactions, unlike the case for M1 electro-disintigration.

^{6}Li(π^+). The ^{6}Li$(\gamma,\pi^+)^6$He reaction was measured from 1 - 7 MeV over threshold by Audit[22]. The π^+ cross-section to the ground state was measured to within ± 4%. The results are in agreement with calculations by Delorme[23] and Cammarata[24] using a phenomelogical potential obtained from electron scattering[25], using an optical potential obtained from meson capture in light nuclei, and using a gs ^{6}He configuration with a strong L = 1 admixture. The agreement is weakened by neglecting the Fermi motion of the nucleons, by possible many-body effects, and by reliance on a nuclear wave function obtained for the analogue state only. The differences between a strong and weak L = 1 admixture is about 10%, which is not much greater than the experimental and theoretical uncertainties.

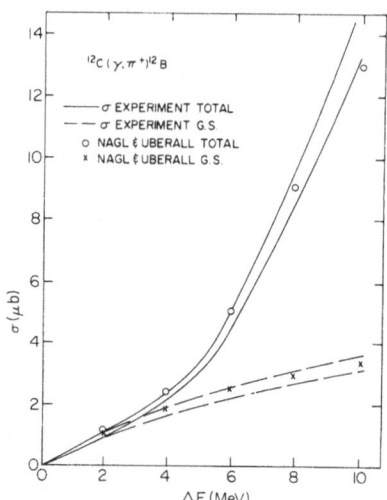

Figure 12. $^{12}C(\gamma\pi^+)^{12}B$ cross-section[3] compared with Helm model calculations[26]. The data lies between the curves. The dashed lines are a fit to the ground state data compared with the calculation (x), while the open circles are compared with the total cross-reaction.

$^{12}C(\pi^+)$. Yields on $^{12}C(\gamma\pi^+)^{12}B$ were taken from 1-20 MeV above threshold with the results relative to the proton shown in Fig. 12. The dashed curves are a best fit to the ground state data, which is given by $\sigma_{gs} = 0.07 \pm .005\ a_p(q/k)S$, while the solid lines encompass the unfolded experimental cross-section. The circles are calculated by Nagl[26] using parameters from electron scattering. Nagl's calculation is about 33% lower than an earlier one by Donnelly[27]. There is some variation in the theoretical cross-section summed over all states depending on the choice of optical model parameters, but the experimental uncertainties and the variations in theoretical cross-section due to experimental errors in the input data preclude a succinct choice of optical models. There are large contributions to the cross-section from transitions to the 1^+ ground state, to the 2^- giant quadropole resonance (4.37 MeV) and to the El spin-flip resonance strength spread through 8 - 10 MeV. Similar

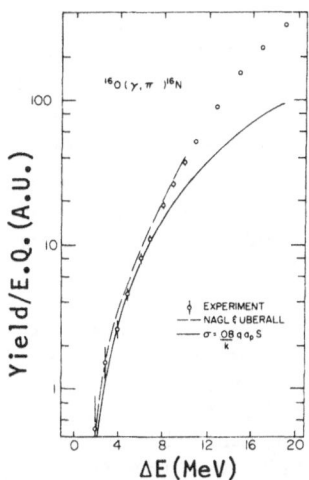

Figure 13. $^{16}O(\gamma\pi^+)^{16}N$ Yield[3]. The dashed curve is a Helm model calculation[26] for all states while the solid curve is a fit to the low energy points.

strength to these states is shown in the π^- capture data reported by Alder et al. at this Symposium.

$^{16}O(\pi^+)$. The $^{16}O(\gamma\pi^+)^{16}N$ reaction has been carried out by Milder[3] at energies from 2 - 16 MeV over threshold, with the yield shown in Fig. 13. The four low-lying states of ^{16}N at 0 - 0.4 MeV cannot be resolved, but should be separable as a group from the next level at 3.36 MeV. The data is compared with a synthesized yield using Helm model cross-sections calculated by Nagl[26] with parameters obtained from electron scattering. The data near threshold suffered from the need to make a large (30%) correction for water contamination in the BeO target, causing the substantial error bars shown in the figure. The ground state quartet is fitted by a cross-section $\sigma = .08 \pm .005$ a_p (q/k)S. The yield curve clearly shows the presence of the El spin-flip giant dipole states located at 7 - 8 MeV.

$^{14}N(\pi^+)$, $^9Be(\pi^+)$. Measurements on $^{14}N(\gamma\pi^+)^{14}C$ and $^9Be(\gamma\pi^+)^9Li$ have been reported briefly by Deutsch[28] and

Milder[3]. The large log ft value for β decay in $^{14}N \rightarrow {}^{14}C$ is consistent with the fact that the π^+ cross-section to the ground state is a factor of 100 lower than that to the 7.01 MeV and 8.32 MeV levels, whose analogues are strongly excited by M1 electron scattering and π^- capture (Alder et al., this Symposium). The $^9Be \rightarrow {}^9Li$ ground state transition does not have the usual shape for the pure G - T transition, and has a small cross-section consistent with a small $\bar{\sigma} \cdot \bar{\epsilon}$ contribution.

In summary, the results of π^+ photoproduction measurements using the $\pi^+ \rightarrow \mu^+ \rightarrow e^+$ method have sufficient experimental accuracy, of order $\pm$ 5%, to make a reasonable test of the combined effects of the photoproduction mechanism, the final state interactions, and the nuclear structure aspects for energies up to 20 MeV over the threshold in light nuclei. The work on ^2H and ^3He gives confidence in the production mechanism, while that on ^6Li and ^{12}C gives some confidence in the treatment of the nuclear structure information and of the final state interactions. It is true that the π^+ wave function is less sensitive than the π^- to conditions inside the nucleus[1] due to the Coulomb repulsion, so these results do not necessarily mean that the theory would have done equally well in calculating π^- cross-sections. In fact, the π^+ total cross-section measurements have produced only a limited amount of new nuclear structure information and, as Table 3 shows, there are few cases left to explore. The real question is this: can one proceed to use differential photo-pion cross-section measurements as a nuclear structure probe? At 10 - 20 MeV pion energy, the Coulomb barrier is no longer a problem to π^+ production and a pion spectrometer can determine the energy of the excited states to within a few hundered KeV, so the experimental situation is much better. The total cross-section π^+ measurements reasonably may be said to have confirmed our models for the production mechanism, which should work equally well in the π^- case. The situation is less clear for the final state interaction, particularly since the pion mean free path is increasing at higher energies, the optical potential appropriate for π^- capture is less useful as the pion energy increases, and the usual pion optical potentials may give the wrong results in the $\Delta(1232)$ resonance region. This suggests that work will be necessary with pion spectrometers on targets with well known nuclear structure in order to determine the appropriate parameters for the pion-nucleus optical potential. Once those parameters are determined, pion spectroscopy should be useful as a tool for nuclear structure studies, especially in identifying states with large spin flip components.

REFERENCES

1. J. Koch and T. Donnelly, Nucl. Phys. B64, 478 (1973).
    J. Koch, Proc. Int. Conf. on Meson Nuclear Physics, Pittsburgh (1976).
    J. Koch, private communication.
2. L. Saunders, Nucl. Phys. B7, 293 (1968) and Thesis M.I.T. (1967).
3. F. Milder et al., Bull. Am. Phys. Soc. 23, 611 (1978).
4. A. Bernstein, et al., Phys. Rev. Lett. 37, 819 (1976).
5. I. Blomqvist, et al., Phys. Rev. C15, 988 (1977)a.
    I. Blomqvist, et al. Lund; LUNFD6/(NFFR-3016) (1977)a.
6. J. Warzawski, et al., Proc. of VII Int. Conf. on High Energy Physics and Nuclear Structures, Zurich (1977).
7. V. Epaneshnikov, Sov. Journ. Nucl. Phys. 19, 242 (1974).
8. G. Davidson, Thesis, M.I.T. (1959).
9. I. Blomqvist and J. Laget, Nucl. Phys. A280, 405 (1977).
10. G. Hylten, Nucl. Phys. A158, 225 (1970).
11. T. Van Hise, R. Meyer, and J. Hummel, Phys. Rev. 139B, 554 (1965).
12. G. Anderson, B. Forkman and B. Friberg, Nucl. Phys. A171, 551 (1971).
13. B. Forkman, et al., Nucl. Phys. A211, 310 (1973).
14. C. Tzara, Nucl. Phys. A256, 381 (1975).
15. E. Booth, et al., Phys. Lett. 66B, 236 (1977).
16. G. Audit, et al., Phys. Rev. C16, 1517 (1977)a.
17. J. O'Connell, private communication.
18. J. Noble, Phys. Lett. 67B, 39 (1977).
19. H. Baer, Int. Conf. on High Energy and Nuclear Structure, Zurich 1977.
20. E. Booth, et al., Bull. Am. Phys. Soc. 23, 601 (1978).
21. J. Laget, private communication.
22. G. Audit, et al., Phys. Rev. C15, 1415 (1977)b.
23. J. Delorme and A. Figureaux, private communication (1977).
24. J.B. Cammarata and T.W. Donnelly, Nucl. Phys. A267, 365 (1976).
25. J. Bergstrom, L. Auer, and R. Hicks, Nucl. Phys. A251, 401 (1975).

26. A. Nagl, private communication.
27. T.W. Donnelly, private communications.
28. J. Deutsch et al., Proc. V. Int. Conf. on High Energy
     Physics and Nuclear Structure, Santa Fe (1975).
29. J. Vergados, Phys. Rev. C12, 1278 (1975).
30. P. Argan, et al., Rap. Int. Saclay DPh-N/HE/78/05 (1978).
31. I. Hughes and P. March, Proc. Phys. Soc. 72, 895 (1958).
32. P. March and T. Walker, Proc. Phys. Soc. 77, 293 (1961).
33. P. Dyal and J. Hummel, Phys. Rev. 127, 2217 (1967).
34. A. Masaike, J. Phys. Soc. Japan 19, 427 (1964).
35. A. Yavin and G. dePasquali, Phys. Rev. B2, 431 (1963).
36. R. Meyer, W. Walters and J. Hummel, Phys. Rev. 138B,
     1421 (1965).
37. W. Walters and J. Hummel, Phys. Rev. 143, 833 (1966).
38. I. Blomqvist, G. Nydahl and B. Forkman, Nucl. Phys.
     A162, 193 (1971).
39. H. Medicus et al., Bull. Am. Phys. Soc. 22, 546 (1977).
40. V. Noga, et al., Yad Fiz (U.S.S.R.) 14, 904 (1973).
41. K. Min, et al., Phys. Rev. C14, 807 (1976).

*Editor's Note: A more recent calculation by E.T. Dressler,
W.M. MacDonald and J.S. O'Connell, to be published, gives
much better agreement with the data than the curve by the
same authors in Figure 9.

THEORY OF PION PHOTOPRODUCTION IN ^{12}C AND ^{16}O

A. Nagl* and H. Überall*[†]

Department of Physics, Catholic University of

America, Washington, D.C.

In a previous communication,[1] we presented a calculation of the total cross section for π^{-} photoproduction in ^{12}C leading to the ground state of ^{12}N. This theory was based on the elementary interaction Hamiltonian of Berends et al.,[2] with the nuclear transition density being described by a Helm model fitted to inelastic electron scattering data. The distortion of the pion wave function was obtained from an optical potential as available at that time. The results were compared with experimental data near threshold by Bernstein et al.,[3] and at higher energies by Epaneshnikov et al.,[4] and were found too low compared to the former, and too high compared to the latter. Here we present improved calculations which lead to considerably better agreement with both the threshold and the high-energy data.

Our results are close to those obtained by two other recent calculations[5,6] in the threshold region, and are presented by curves no. 1 and 2 of Fig. 1, where the data of Ref. 4 are shown as points and of Ref. 3 by solid curves I, II representing the data uncertainties. The improvement in the theoretical curves consists mainly in the use of an optical potential with parameters determined by a recent phase shift analysis of π N scattering data,[7] with isovector terms included, and the Lorenz-Lorentz effect[8] taken into account (parameter $\xi = 1$).

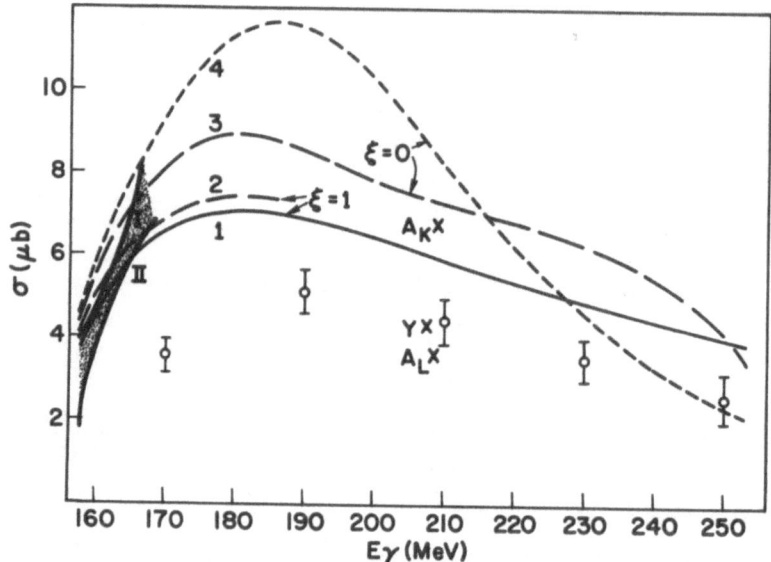

Figure 1.   Total cross section of $^{12}C (\gamma,\pi^-)^{12}$ N.   Data from
Ref. 3 (shaded area), Ref. 4 (points).

The optical potential used is for $^{\pm}$ mesons:

$$V_{opt} = - (2\pi/E_\pi)\{(1 + \frac{E_\pi}{m}) [b_0\rho(r) \pm b_1(\rho_n-\rho_p)] + \frac{E_\pi}{2m} \nabla^2 c'(r)$$

$$+ B(1 + \frac{E_\pi}{m})\rho^2(r) - (1 + \frac{E_\pi}{m})^{-1} \vec{\nabla}\cdot \frac{c'(r)}{1 + \frac{4\pi}{3} \xi c'(r)} \vec{\nabla}$$

$$- c(1 + \frac{E_\pi}{m})^{-1} \vec{\nabla}\cdot\rho^2(r)\vec{\nabla} \}$$

where

$$c'(r) = c_0\rho(r) \pm c_1 [\rho_n(r) - \rho_p(r)],$$

$\rho(r)$ being the sum of the proton (ρ_p) and neutron (ρ_n) densities.   Here E_π = total pion energy, m = nucleon mass.   The
complex values of b_0, b_1, c_0 and c_1 were obtained from meas-

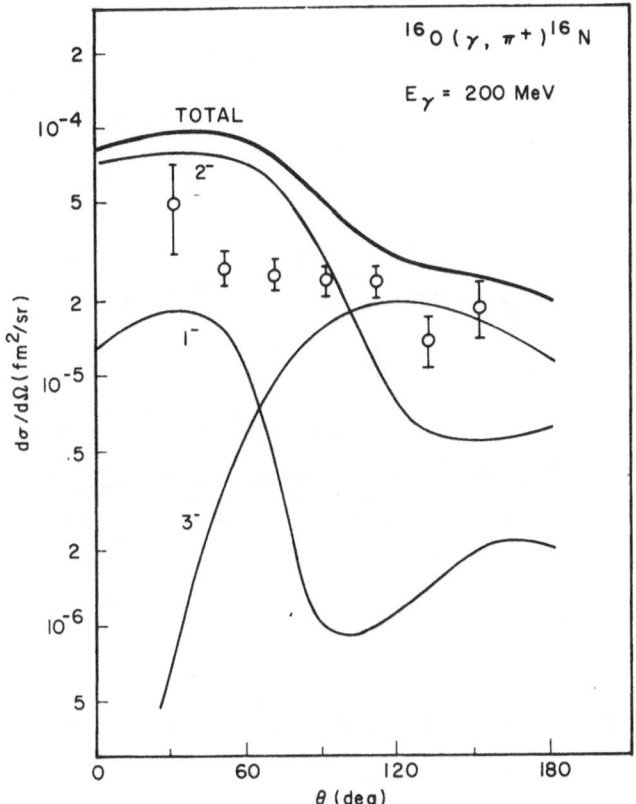

Figure 2.   Differential Cross Section of ^{16}O$(\gamma,\pi^{+})^{16}$N.
Data from Ref. 10.

ured π-N scattering, except that ReB$_o$ = -0.003 m$_\pi^{-1}$ was from
pionic-atom data, as were ImB = 0.042 m$_\pi^{-4}$, ImC = 0.076 m$_\pi^{-6}$
while ReB, ReC are either (1) taken equal and opposite[9] to
ImB, ImC, or (2) zero. Curves 1 and 2 correspond to these
two choices.

As an indication for the sensitivity to the optical potential, we also show curve no. 4 which corresponds to curve
2 with ξ = 0, and no. 3 with potential parameters fitted
empirically to π^- nucleus scattering. Also shown are individual theoretical points A_k, A_L using two versions of an optical

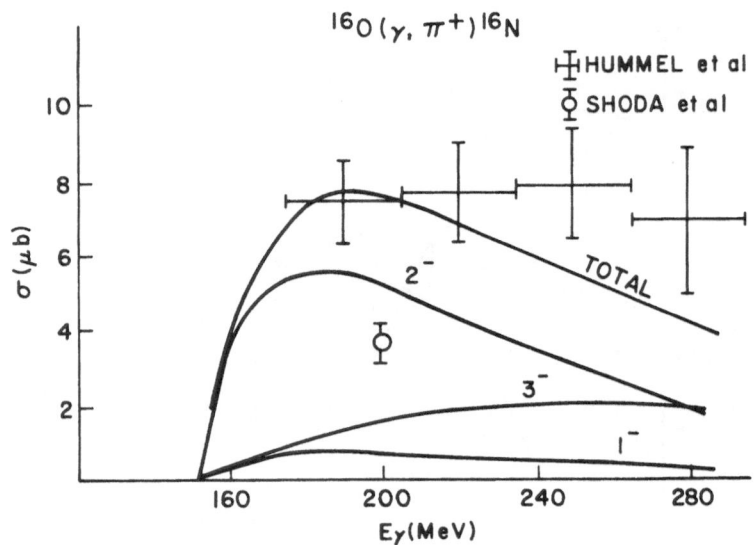

Figure 3.   Total Cross Section of $^{16}O(\gamma,\pi^+)^{16}N$.   Data
from Ref. 11.

potential by Amann et al., and Y using another by Yoo and
Sternheim.

We have also calculated π^+ photoproduction in ^{16}O leading
to the ground (2^-) and low-lying $(3^-,1^-)$ excited states in
^{16}N using the same optical potential (with corresponding
nucleon densities).   The differential cross section data in
Fig. 2(a) are those of Sung, Shoda et al.[10], and the total
cross section data in Fig. 2(b) those of Meyer et al.[11]
The point at 200 MeV in Fig. 2(b) was obtained by integrating
the data of Fig. 2(a).   These calculations may be compared to
our previous results obtained with plane-wave pions.[12]

REFERENCES

* Supported by the National Science Foundation.
† Also at the Naval Research Laboratory, Washington, D.C. 20375

1. A. Nagl and H. Überall, Phys. Letters <u>63B</u>, 291 (1976).
2. F.A. Berends, A. Donnachie and D.L. Weaver, Nucl. Phys. <u>B4</u>, 1 (1967).
3. A.M. Bernstein et al., Phys. Rev. Letters <u>37</u>, 819 (1976).
4. V.D. Epaneshnikov et al., Sov. J. Nucl. Phys. <u>19</u>, 242 (1974).
5. G.N. Epstein et al., Univ. of Pittsburgh preprint (1977).
6. W.C. Haxton, Los Alamos report LA-UR-77-2898 (1977).
7. G. Rowe, M. Salomon and R. Landau, Proc. 7th Int. Conf. High Energy Phys. Nucl. Struct., Zurich, Sept. 1977.
8. M. Ericson and T.E.O. Ericson, Ann. Phys. (NY) 36, 327 (1966).
9. J. Hüfner, Phys. Rep. <u>21C</u>, 1 (1975).
10. B.N. Sung, K. Shoda et al., Proc. Int. Conf. on Nucl. Struct., Tokyo, Sept. 1977.
11. R.A. Meyer, W.B. Walters, and J.B. Hummel, Phys. Rev. <u>138</u>, B1421 (1965).
12. A. Nagl, F. Cannata and H. Überall, Acta Physica Austriaca <u>48</u>, 267 (1978).

NEGATIVE PION PHOTOPRODUCTION FROM ^{28}Si

K. Srinivasa Rao and S. Susila

MATSCIENCE, The Institute of Mathematical
Sciences
Madras-600020, India

We present here a preliminary report on an impulse
approximation calculation of ^{28}Si$(\gamma,\pi^-)^{28}$P, from the region
just above the photopion threshold to the first pion-nucle-
on resonance region. The theoretical results obtained in
the volume and surface production models are compared with
the experimental data of Epaneshnikov et al.[1].

We assume the ground state of ^{28}Si to be spherical.
There exists seven low-lying particle stable states[2] of ^{28}P
below 1.6 MeV, which would have contributed to the production
process, since the activation method has been employed[1] to
measure the cross sections for ^{28}Si$(\gamma,\pi^-)^{28}$P. In the absence of experimental spin-parity assignments to these states[2],
except for the ground state of ^{28}P, which is a 3^+ state, we
assume that these are the isobaric analogues of the lowest
T = 1 states of ^{28}Si. Following the study of Donnelly and
Walker[3], these seven states of ^{28}P are assigned to the spin-
parities = 3^+, 2^+, 3^+, 2^+, 4^+, 5^- and 4^-. When the residual
interaction is absent, the ground state of ^{28}P and the first
2^+ state may be assigned to the $(2\,s_{1/2})(1\,d_{5/2}^{-1})$ particle-hole
configuration, while the second 3^+, the second 2^+ and the

first 4^+ state may be assigned to the $(1d_{3/2})(1d_{5/2}^{-1})$ configuration. The two odd-parity states of ^{28}P $(5^-, 4^-)$ may be
assigned to the $(1f_{7/2})(1d_{5/2}^{-1})$ configuration. We refer to
this scheme as the Independent Particle Model (IPM), for the
low-lying $T = 1$ states of ^{28}P.

Donnelly and Walker[3] performed a particle-hole configuration mixing calculation for ^{28}Si, in the Tamm-Dancoff Approximation, and we use these wave functions in our study
and denote this description for the states of ^{28}P by DW.

As in the case of our earlier work[4], the photoproduction
of π^- from $^{28}Si(0^+, g.s.)$ leading to the seven bound states of
^{28}P, is studied by means of a nuclear transition amplitude obtained in the impulse approximation, using the C.G.L.N. amplitudes[5] for single nucleon pion photoproduction.

For the radial wave functions, we take the harmonic
oscillator forms with the oscillator strength parameter:
$b = 2.038$ fm, which is in agreement with high energy elastic
electron scattering data[6]. The cross sections obtained, denoted by IPM(V) and DW(V) in Fig. 1, as a function of incident photon energy, are very large when compared to the experimental data, which is consistent with our theoretical
studies on charged pion photoproduction from complex nuclei.

We have found the phenomenological surface production
model to be successful in accounting for the experimental
data in our earlier studies[1,7]. Freed et al.[8] have shown that
this model simulates the effect of the Final State Interaction
(FSI) of the outgoing pion with the residual nucleus, provided the optical potential contains only an s-wave part.
Since the use of the complete optical potential, having both
s- and p-wave parts, suppresses the cross sections drastically[8,9] at the first pion-nucleon resonance region to values
for below the experimental data, the phenomenological surface
production model alone is used here to simulate the effect of
FSI.

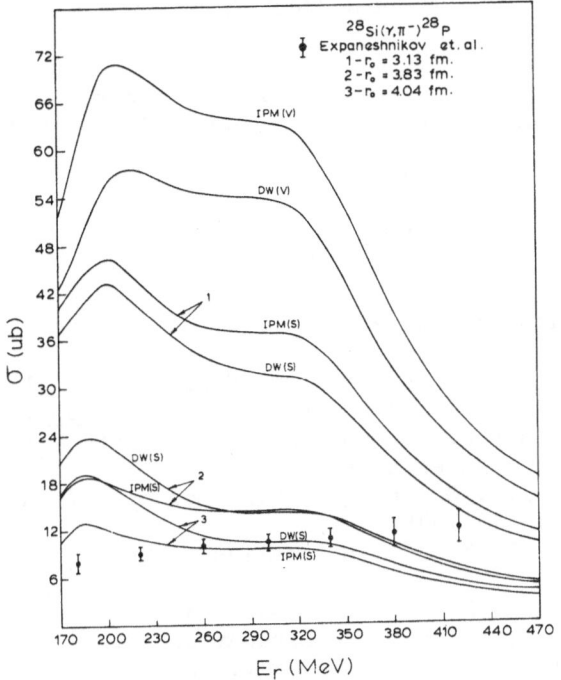

FIGURE 1

   In Fig. 1, the curves denoted by IPM(S) and DW(S) have
been obtained using the surface production model for three
different values of the cut-off parameter: r_o = 3.13 fm.,
3.83 fm and 4.04 fm. The value of 3.13 fm. corresponds to
the r.m.s. radius[6] of ^{28}Si, while the value of 4.04 fm. corresponds to the charge radius of ^{28}Si. Unlike in the case of
our earlier studies[4,7], a value larger than the r.m.s. radius
of the nucleus is required in this case, for the cut-off parameter, to account for the experimental data. An analysis
of our results suggests that the use of an energy dependent
cut-off parameter - with a larger value in the threshold
region and a smaller value in the resonance region - will
provide a better fit to the experimental data for ^{28}Si(γ,π^-)
^{28}P, than an energy independent cut-off parameter.

ACKNOWLEDGEMENT

The authors thank Professor Alladi Ramakrishnan for encouragement and one of us (S.S.) acknowledges with thanks the award of a U.G.C. Teacher Fellowship.

REFERENCES

1. V.D. Epanishnikov et al., Sov. J. Nucl. Phys. $\underline{23}$ (1976) 129.
2. P.M. Endt and C. Van der Leun, Nucl. Phys. $\underline{A214}$ (1973) 1.
3. T.W. Donnelly and G.E. Walker, Ann. Phys. $\underline{60}$ (1970) 209.
4. K. Srinivasa Rao, J. Tamil Nadu Acad. Sci. Vol. 1 (1978) 127, and references therein.
5. G.F. Chew, M.L. Goldberger, F.E. Low and Y. Nambu, Phys. Rev. 106 (1957) 1345.
6. G.G. Li, I. Sick and M.R. Yearian, Phys. Letts. $\underline{37B}$ (1971) 282.
7. V. Devanathan, K. Srinivasa Rao and R. Sridhar, Phys. Letts. $\underline{25B}$ (1967) 456, V. Devanathan, G.N.S. Prasad and K. Srinivasa Rao, Phys. Rev. $\underline{C8}$ (1973) 188.
8. N. Freed and P. Ostrander, Phys. Rev. $\underline{C11}$ (1975) 805, Phys. Letts. $\underline{61B}$ (1976) 449.
9. R. Handel and W. Weisse, Lett. al. Nuo. Cim $\underline{14}$ (1975) 592.

PHOTOPRODUCTION OF π^- FROM ^{14}N

V. DeCarlo, N. Freed, W. Rhodes

Department of Physics

The Pennsylvania State University

University Park, Pennsylvania 16802

We have carried out calculations on the reaction ^{14}N(γ,π^-)^{14}O$_{g.s.}$ and compared our results to those of activation measurements recently completed at Lund[1]. Since the nuclear wave functions are well known, this reaction should provide a stringent test of the dependence of the cross section on the various optical potentials used to describe the π-nucleus final state interactions.

Calculations were carried out in the DWIA using three sets of nuclear wave functions: Cohen-Kurath[2], Baer, et al[3], and pure j-j coupling. The Cohen-Kurath wave functions are consistent with a wide range of electromagnetic properties in the p-shell, including the anomalously slow ^{14}C$\rightarrow$ ^{14}N Gamow-Teller beta decay. The Baer wave functions give good results for radiative pion capture and electron scattering in the p-shell. Finally, for comparison purposes, we used j-j coupled wave functions which give poor fits to almost all properties in the upper p-shell. Harmonic oscillator wave functions with an oscillator length b = 1.70 fm were used throughout.

Final state interactions (fsi) were simulated by a variety of π-nucleus optical potentials plus a Coulomb potential. The specific forms used were the standard Kisslinger[4], local Laplacian[5], and modified Kisslinger[6]

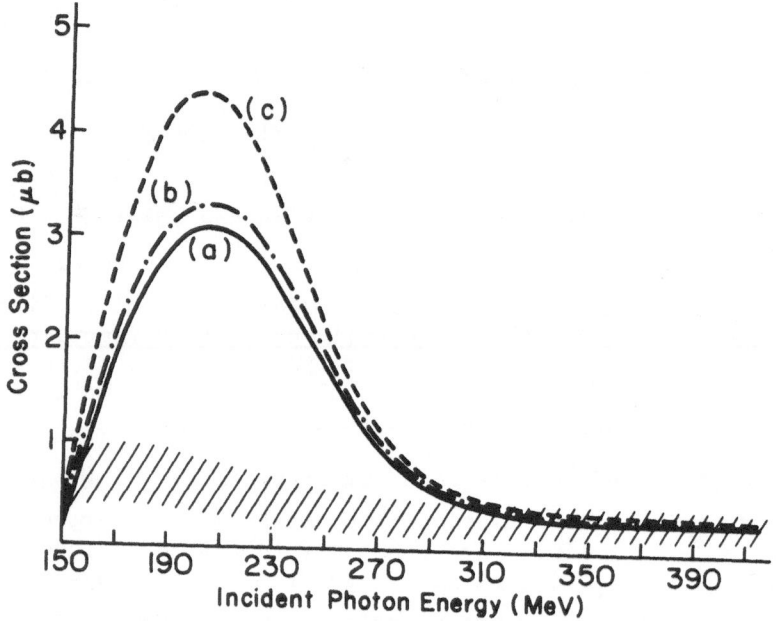

Figure 1. Comparison of theory and experiment for the local Laplacian potential. Curves (a), (b), and (c) are the theoretical curves for Cohen-Kurath, Baer, and j-j coupled wave functions, respectively. The hatched area represents the results of experiment.

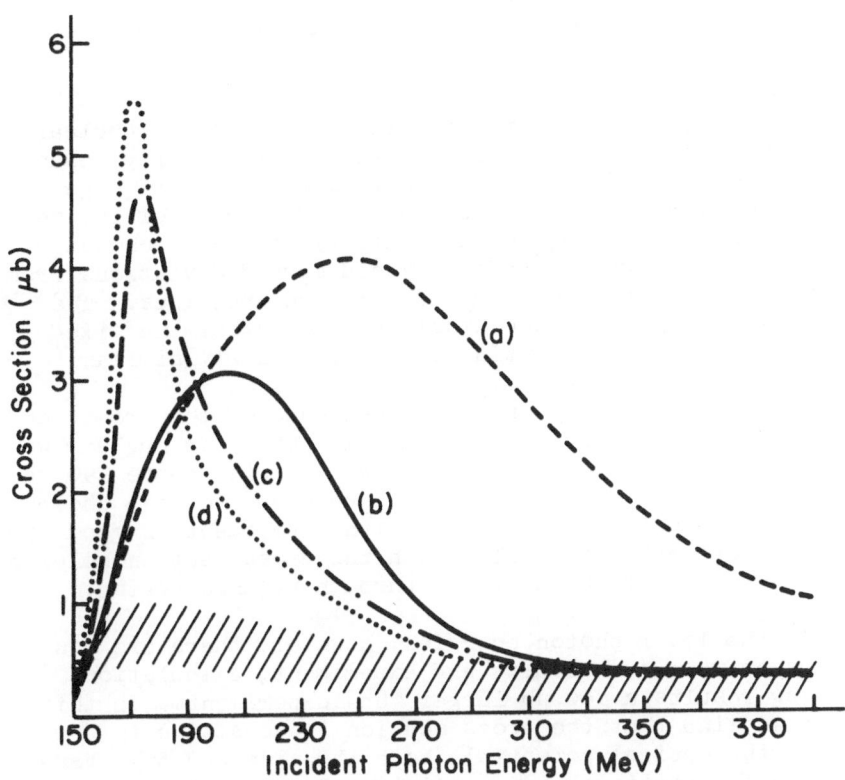

Figure 2:   Comparison of theory and experiment for Cohen-
Kurath wave functions. Curve (a) is the result of no fsi
calculations. Curves (b), (c), and (d) are the theoretical
curves for the local Laplacian, modified Kisslinger, and
Kisslinger potentials, respectively. The hatched area
represents the results of experiment.

potentials. The optical potential strength coefficients were derived from a parametrization by Salomon[7], who fit an analytic function to recent πN phase shift data[8]; the resulting scattering amplitudes are smooth functions of energy in the range 0-250 MeV. The outgoing distorted pion wave functions used in the evaluation of the radial integrals were generated by the program PIRK[9].

In Figure 1 we show the result of varying the nuclear wave functions with the optical potential (local Laplacian, in this case) fixed. Our calculations show that at high photon energies the cross section is independent of the set of wave functions used; in the vicinity of the Δ resonance, however, the j-j wave functions yield marked deviations from the results of the other two sets of wave functions. The curves obtained from fixed Kisslinger potential and fixed modified Kisslinger potential are qualitatively similar.

In Figure 2 we examine the dependence of the cross section on the various optical potentials for fixed nuclear wave functions (Cohen-Kurath). For photon energies above 290 MeV we find that the cross section is independent of the choice of optical potential. Agreement with experiment is good for all three optical potentials, with the cross section holding fairly constant at about 0.3 μb throughout this region.

At the lower photon energies, however, there are discrepancies between our results and experiment, calculations yielding results a few μb larger than experiment. In this region we find that the cross section is sensitive to the form of the optical potential; both the size and the shape of the cross sections vary significantly as we pass from one optical potential to another. This disagreement with experiment could arise from the failure of these optical potentials to reproduce π-nucleus elastic scattering data for $T_{\pi-} < 50$ MeV.

REFERENCES

1. V. DeCarlo, N. Freed, W. Rhodes, B. Bulow, G. G. Jonsson, K. Lindgren, and R. Pettersson, to be published.
2. S. Cohen and D. Kurath, Nucl. Phys. <u>73</u>, 1 (1965); H. J. Rose, O. Hausser, and E. K. Warburton, Rev. Mod. Phys. <u>40</u>, 591 (1968).
3. Helmut W. Baer, et al., Phys. Rev. <u>C12</u>, 921 (1975).
4. L. S. Kisslinger, Phys. Rev. <u>98</u>, 761 (1955).
5. H. K. Lee and H. McManus, Nucl. Phys. <u>A167</u>, 257 (1971).
6. G. A. Miller, Phys. Rev. <u>C10</u>, 1242 (1974).
7. M. Salomon, TRIUMF Rpt. TR1-74-2, Low Energy (0-250 MeV) Pion-Nucleon Phase Shift Fits (1974).
8. S. Almehed and C. Lovelace, CERN preprint TH 1408 (1971); P. W. Coulter, Phys. Rev. Lett. <u>29</u>, 450 (1972); J. R. Carter, D. V. Bugg, and A. A. Carter, Nucl. Phys. <u>B58</u>, 378 (1973).
9. R. A. Eisenstein and G. A. Miller, Comp. Phys. Comm. <u>8</u>, 130 (1974).

A METHOD FOR MEASUREMENTS OF THE TOTAL ^{2}H(γ,π^-)PP

CROSS SECTIONS AT THRESHOLD

B. Schoch, F. Klein and G. Lührs

Institut fur Kernphysik

Universität Mainz

In recent years measurements of the total cross section for the reactions ^{2}H$(\gamma,\pi^+$nn$)$ and ^{2}H$(\gamma,\pi^0)^2$H close to the production threshold were performed[1]. For the ^{2}H(γ,π^-)pp reaction data of similar good quality doesn't exist. In this paper a method is prescribed and the feasibility for the reaction ^{2}H(γ,π^-)pp is demonstrated. Figure 1 shows the principle: The photon beam hits a liquid or high pressured gas target of ^{2}H. π^- are created on the deuteron and stopped in the same target cell. The major part of the π^- are captured by deuterons via the reaction ^{2}H$(\pi^-,$n$)$n. The energy is shared equally by the two neutrons (E_n = 68.2 MeV). One of these monochromatic neutrons is detected by a neutron detector using the time of flight method.

The actual set up is shown in Figure 2. A short electron beam burst ($\simeq$ 2 M sec) with a rep. rate of 300 cps is deflected by a bending magnet and hits a radiator. The bremsstrahlung beam is collimated on the target. A cleaing magnet sweepts out the electron beam.

If the endpoint energy of the photon spectrum exceeds the pion production threshold, neutrons are produced by the described mechanism as well as from the photo disintegration of the deuteron. By choosing suitable kinematics even 9 MeV above pion threshold the most energetic neutrons originating from the photodisintegration are separated. A time of flight spectrum measured with a bremsstrahlung spectrum with an

171

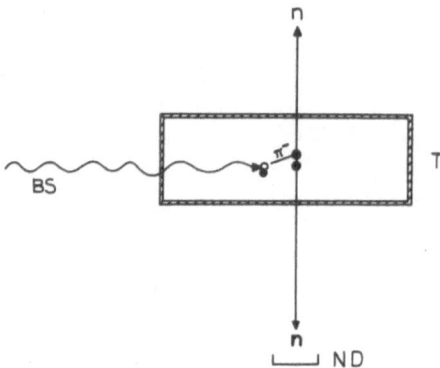

Figure 1:   Principle of the method used - BS:Bremsstrahlung
beam, T:Target filled with liquid or high pressured H,
ND:Neutron detector

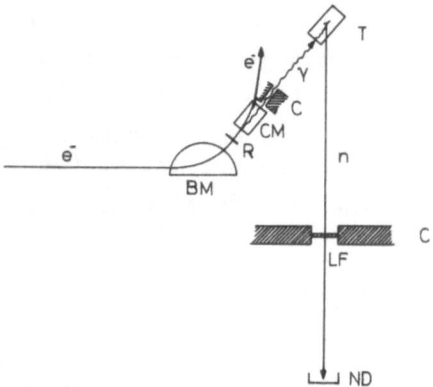

Figure 2:   Setup used in the measurement - BM:Bending magnet,
R:Radiator, CM:Cleaning magnet, T:Target, N:Neutron detector,
C:Collimator, Lf:Lead filter

endpoint energy $E\gamma^{o}$ = 9 MeV above threshold is shown in
Figure 3.

     The monochromatic peak is clearly seen separated from
the continuous neutron spectrum coming from the photodisintegration. In this experiment the total cross section for

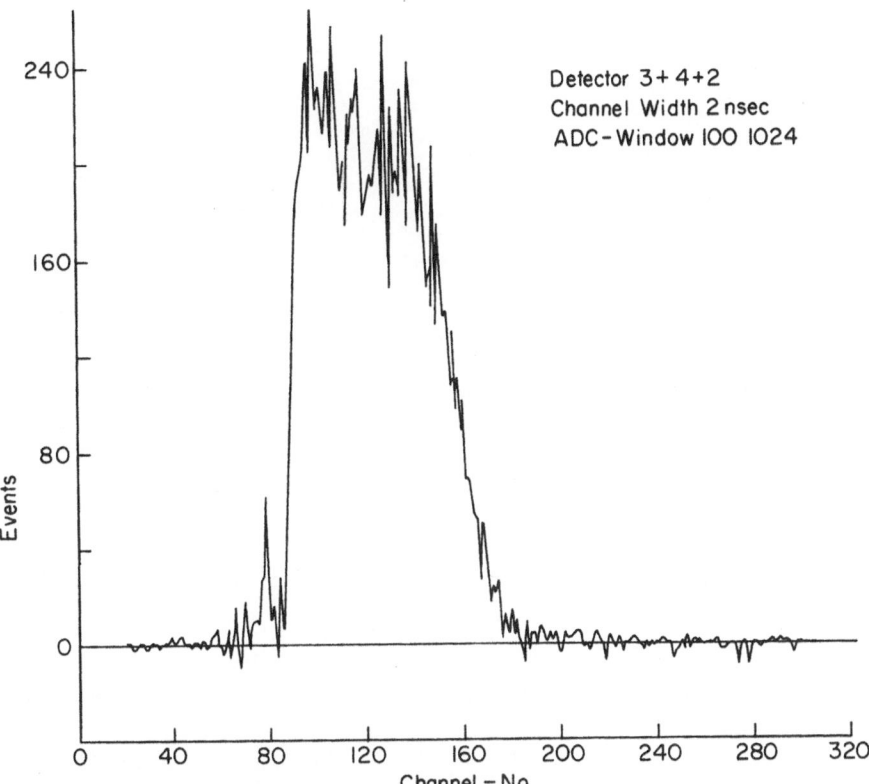

Figure 3: Neutron time of flight spectrum. The endpoint energy of the bremsstrahlung spectrum is 9 MeV above pion production threshold.

^{2}H(γ,π^-)pp is determined relative to the differential cross section of the photodisintegration of the deuteron, which can be done with rather good accuracy.

REFERENCES

1. C. Tzara, Meson-Nuclear Physics 1976 (Carnegie-Mellon Conference) p. 566.

CHARGED PION SPECTROSCOPY

K. Shoda

Laboratory of Nuclear Science, Tohoku University

Tomizawa, Sendai 982, Japan

1. INTRODUCTION

Photopion production from nuclei is usually described in the impulse approximation by adding elementary photopion production over all constituents in nuclei. The elementary photopion amplitude on nucleon is given by[1]

$$= i\vec{\sigma}\cdot\vec{\epsilon}\ F_1 + \vec{\sigma}\cdot\vec{p}\vec{\sigma}\cdot(\vec{k}\times\vec{\sigma})\ (F_2/pk)$$

$$+ i\vec{\sigma}\cdot\vec{k}\vec{p}\cdot\vec{\epsilon}(F_3/pk) + i\vec{\sigma}\cdot\vec{p}\vec{p}\cdot\vec{\epsilon}(F_4/p^2). \qquad (1)$$

In case of low momentum photopion production, Kroll-Ruderman term $\vec{\epsilon}\cdot\vec{\sigma}$ is known to be dominant[2] and the cross section of photopion production from nuclei is expressed by

$$\left(\frac{d\sigma}{d\Omega}\right)_{(\gamma,\pi)} = \frac{p}{km_\pi^2}\ e^2 f^2\ \frac{1}{2J_i+1} \sum_{M_i M_f S_\gamma} |\vec{\epsilon}\cdot\vec{m}|^2 \qquad (2)$$

$$\vec{m} \approx \langle J_f M_f | \phi_\pi^* \sum_{n=1}^{A} \vec{\sigma}_n \tau_n^\pm \exp(i\vec{k}\cdot\vec{r}_n) | J_i M_i \rangle. \qquad (3)$$

This approximation is suitable for threshold photopion production and radiative pion capture but it seems necessary to include the other terms of Eq. 1 for high momentum photopion production.

175

Since the photopion cross section depends on the nuclear transition matrix element and the pion wave in nuclei as shown in Eqs. 2 and 3, the study of this cross section can be a useful tool in nuclear spectroscopy to learn about the spin-flip type interaction as in the case of radiative pion capture.

In the radiative capture experiment the direction of mommentum transfer is averaged, while in the threshold photopion experiment using the residual activity the detected decay electron loses the memory of the original pion direction. Therefore angular dependence of the reaction cannot be studied in these experiments.

Initial information in nuclear physics is generally obtained from angular distributions as well as the energy distributions. Therefore the photopion angular distribution provides important information in addition to the energy distribution results.

Since the pion counting rate is not large in most experiments, the reduction of background is an important problem. The magnetic spectrometer will be suitable for the charged photopion experiment with low background and good resolution. Pion loss due to its short life is related to the path length of the spectrometer and the pion energy. This limits the photopion experiment to somewhat large pion-energy and rather high excitation above the threshold.

The momentum transfer involved in various types of experiments is shown in Fig. 1 using ^{12}C target as an example, where comparison is made with electron scattering and π, μ capture. Angular distribution experiments will yield information about q-dependence in the reaction. In such a study, information on the single state and also on the giant resonance state could be obtained.

The experiments of the charged photopion spectroscopy started only quite recently. I would like to report on the present situation of this new field of study.

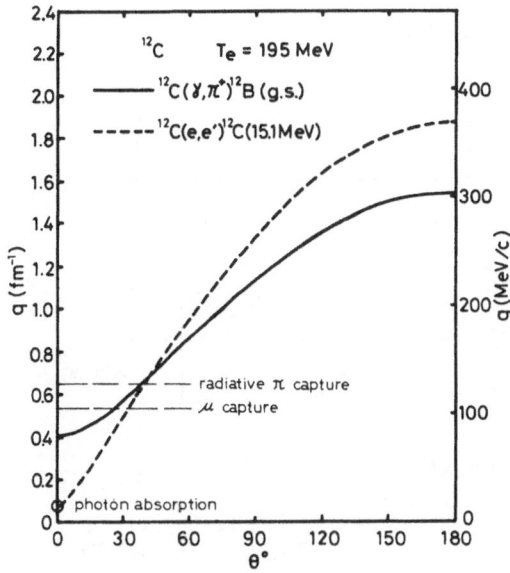

Figure 1.   Diagram of momentum transfer in various reactions on ^{12}C.

2.   EXPERIMENTAL METHOD AND ARRANGEMENT

Three experimental methods are applicable for photopion spectroscopy using bremsstrahlung (the end point energy E_e), or an electron beam (kinetic energy E_e).   These are to measure the pion yield curves, the isochromats and the pion spectra at various emission angles as described in Fig. 2.   The photon spectrum in the reaction is also shown in the figure.

In the case of photopion yield curve experiments, the total yield of produced pion is counted at successive incident energies.   A break in the yield curve shows a sharp resonance in the (γ, π) cross section or a particular strong channel feeding to a residual state in the reaction.   Since no fine structures are expected in the (γ, π) cross section, the breaks indicate strong channels feeding to particular residual states.   Experiments of this type have been made at Saclay and Bates[3].

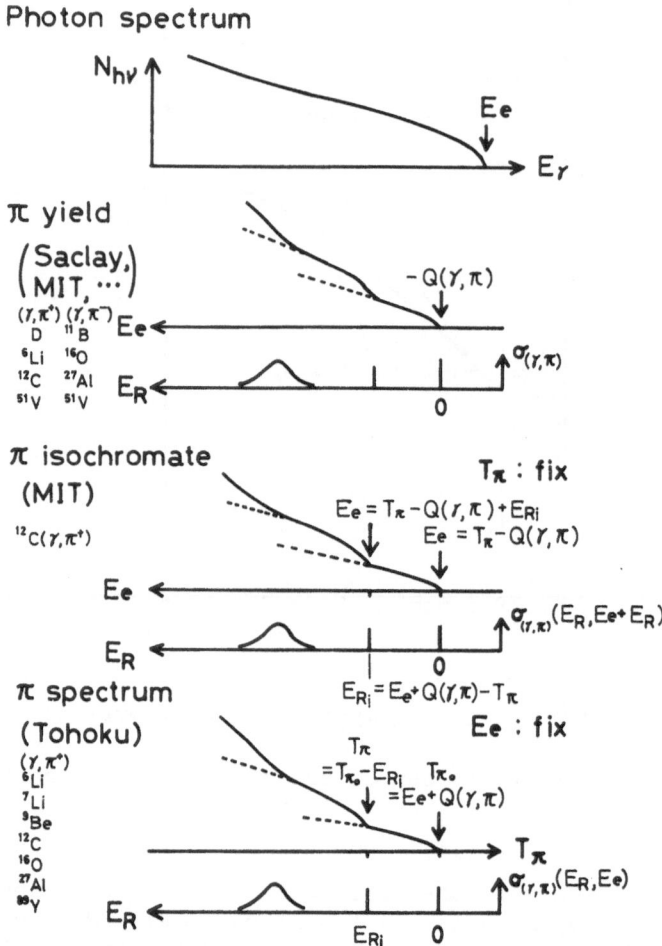

Figure 2. Description of experimental methods for pion spectroscopy with bremsstrahlung or virtual photon in (e,e'π)
reactions.

        Isochromats are given by plotting the counts of photo-
pion of fixed energy T_π. When a strong break is found in
the isochromat, energy of the residual state relating to
the break is given by

$$E_R = E_e + Q - T \tag{4}$$

Similar studies are also possible for (e,e') reactions.

An example of the isochromat results obtained by the Rensselaer-MIT group[4] is shown in Fig. 1 of the following contribution by Paras et al. Breaks are found in the curve. Corresponding residual energies obtained by Eq. 4 and the transition strength to each residual state can be obtained by analysis of the curve.

Photopion energy distributions from bremsstrahlung or (e,e'π) reactions should show breaks corresponding to the opening of a pion decay channel leaving the residual state given by Eq. 4 as shown in Fig. 2. Analysis can be made similarly to the case of isochromats.

Pion energy distributions from (e,e'π$^+$) reaction have been measured with the Tohoku Linac using a 169.8° deflecting double focussing magnetic spectrometer ($\sim$ 5m path length). Detectors are 10 mm x 2 mm, 1 mm thick Si(Li) solid state detectors mounted on 1 mm thick ceramic bases and compose a 33 array of 3-fold coincidence system (energy resolution is (0.05%).

In this system for positive photopion experiments, charged nuclear particles are stopped in the first detector and eliminated by the coincidence requirement. A single pulse height distribution from the second detector is shown in Fig. 3a. A strong coincident background is expected from the positrons. The pulse height of the positron is estimated equal to that of the electron as shown in Fig. 3b. Background can be mostly discriminated by suitable bias setting. However, the energy loss pulse height distribution shows a Landau shape which extends above the discrimination bias. Therefore the positron background still exists after three-fold coincidence and is given by

$$N_{BG} = N_e + \eta^3, \tag{5}$$

where η is a ratio of single background positron counts above the bias setting to the total. At normal bias setting in the present experiment, η is about 0.1 and it leads to a very small background after taking coincidence.

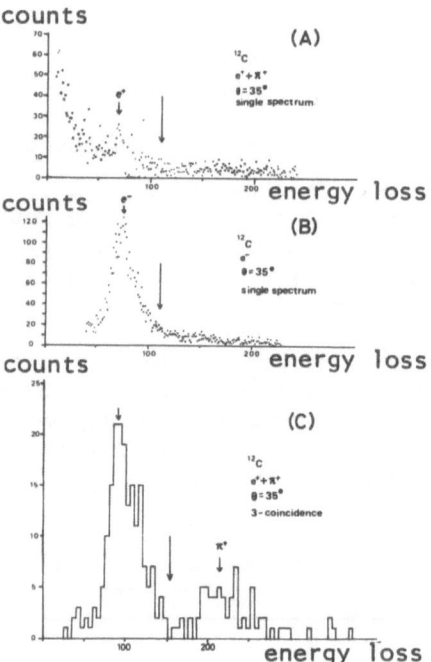

Figure 3. Pulse height distribution from the second detector
for $(e,e'\pi^+)$ experiment. Arrows indicate the discriminator
bias height. (a) Single pulse height distribution. (b) Single pulse height distribution of e . Pulse height distribution for positive charged particles gated with coincidence
pulses among the three detectors.

An example of a pulse height distribution of the output
from the second detector gated by the threefold coincidence
output is shown in Fig. 3c. Separation of $^+$ is good and positron background is small enough except at forward angles.

In order to improve the resolution of the result the
target orientation is chosen so that the total sum energy loss
of the electrons and the pion in the target is equal for any
pion production point as shown in Fig. 42. In this case, the
energy spread is due only to the straggling of the loss energy
in the target. The spread is about 300 keV for a thick carbon
target (180 mg/cm^2) for detection angles smaller than 150°.
This method is not applicable for large angle experiments
($\theta \geqslant 150°$) and photon bombarding experiments, where the ener-

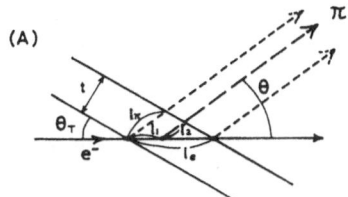

(A)

$$\Delta E = \left(\frac{dE}{dx}\right)_e l_1 + \left(\frac{dE}{dx}\right)_\pi l_2 \approx \text{constant}$$

solution: $\left(\frac{dE}{dx}\right)_e / \left(\frac{dE}{dx}\right)_\pi = \sin(\theta_T + \theta)/\sin\theta_T$

energy spread: ε = straggling of ΔE

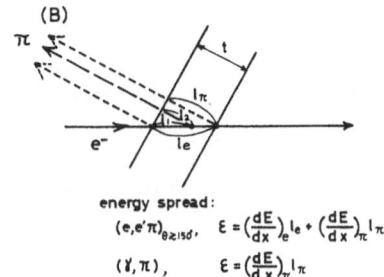

(B)

energy spread:

$(e, e'\pi)_{\theta \gtrsim 15\delta}, \quad \varepsilon = \left(\frac{dE}{dx}\right)_e l_e + \left(\frac{dE}{dx}\right)_\pi l_\pi$

$(\gamma, \pi), \quad \varepsilon = \left(\frac{dE}{dx}\right)_\pi l_\pi$

Figure 4. Description of target setting. (a) Optimum resolution experiment. (b) Other condition experiments.

gy spread is equal to the full energy loss in the target (see
Fig. 4b).

A similar detector system has been used at the University
of Saskatchewan[5]. They used four telescopes of three (2-ΔE,
1-E, Si surface barrier detectors each set at the focal plane.
Figure 5 shows their result for ^{12}C at $\theta = 90°$. The figure
also shows Tohoky data at $\theta = 50°$, $90°$. Both data are quite
reasonable.

In order to check the absolute value, electroproduction
of π^+ from H has been measured using LiH target at Tohoku.
The result of energy distribution of π^+ is shown in Fig. 6a.
This result was analyzed with the virtual photon spectra calculated by Dalitz and Yennie[6]. The resulting (γ, π^+) cross
section is plotted in Fig. 6b in comparison with a previous
H(γ, π^+) result[7]. Good agreement is found between them showing

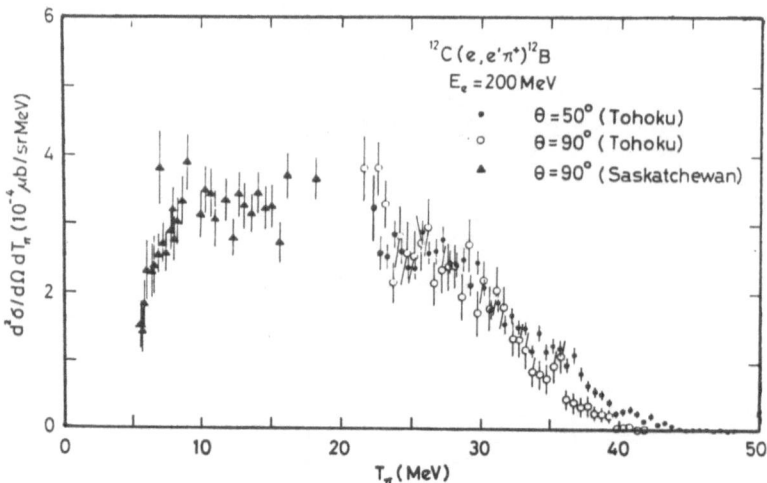

Figure 5.   Photopion energy distributions measured at Saskatchewan and Tohoku.

that the present analysis is reasonable for hydrogen.

3.   EXPERIMENTAL RESULTS AND DISCUSSIONS

3.1   Low Lying Residual States

Energy distributions of π^+ have been measured in the end point region at various angles on ^{6}Li, ^{9}Be, ^{12}C, and ^{16}O. Experiments have also been tried for ^{10}B, ^{11}B, ^{14}N, ^{27}Al and ^{31}P but the yields were not enough for separation from the background.

Examples of energy distributions are shown for ^{12}C in Fig. 7 together with theoretical estimates. Solid curves in the figure are obtained by folding the theoretical cross section with a virtual photon spectrum[6]. The (γ, π^+) cross section was derived from analogous electron scattering data with the Helm model and with the pion plane wave impulse approximation (PWIA)[8]. The dashed lines are also the theoretical results calculated using a nuclear shell model and pion wave derived with an optical potential consisting of Coulomb and strong-interaction as given by the modified Kisslinger type[9], that is pion distorted wave impulse approximation (DWIA)[9].

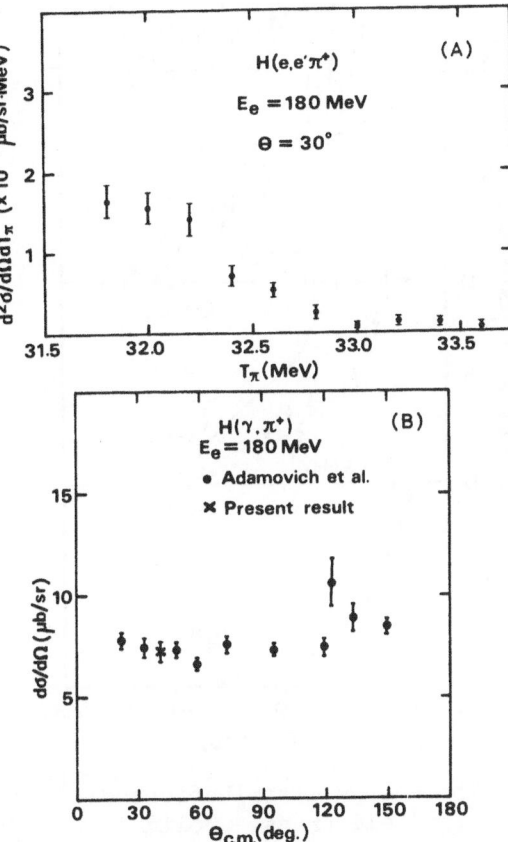

Figure 6. (a) Pion energy distribution of H(e,e'π^+). (b) ($\gamma \pi^+$) cross section deduced from the energy distribution using a virtual photon spectrum calculated from Dalitz and Yennie's formula[6]. Previous result[7] is compared.

The residual states corresponding to the breaks in the energy distributions for ^{12}C are the ground (1^+), 0.9 MeV (2^+) and 4.4 MeV states in ^{12}B. These agree with the isochromat results in Fig. 3 and also with the result of radiative π^- capture.

The photopion energy distribution of (e,e'π^+) reactions and (γ,π^+) cross section can be expressed as

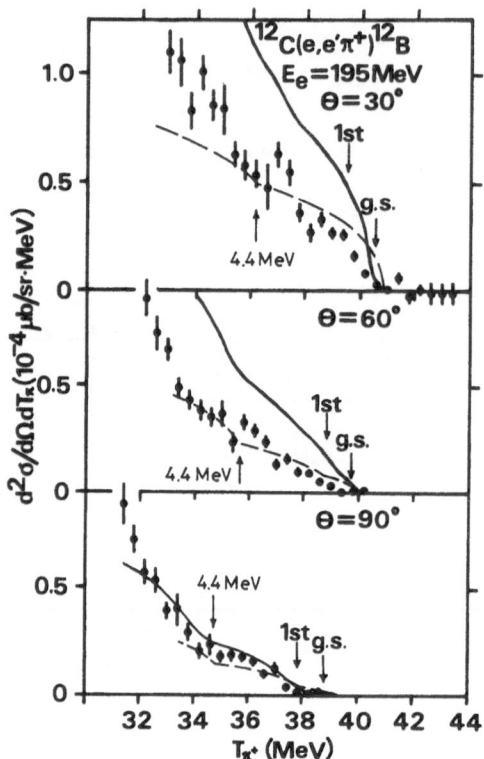

Figure 7. Photopion energy distributions of $^{12}C(e,e'\pi^+)$ with 195 MeV (total energy) electrons. Solid curve: theoretical estimate by generalized Helm model fitted to electron scattering data (PWIA)[8]. The folding is made with virtual photon spectra[6]. Dashed curve: shell model calculation result (DWIA)[9]

$$\frac{d^2\sigma_{(e,e'\pi^+)}(E_{Ri})}{d\Omega dT} = \sum_i \frac{d\sigma_{(\gamma,\pi^+)}(E_{Ri})}{d\Omega} N_{h\nu}(E_e, E_\gamma), \quad (6)$$

where E_{Ri} is the excitation energy of the residual states and $N_{h\nu}(E_e, E_\gamma)$ is the virtual photon spectrum. By fitting this equation to the $(e,e'\pi^+)$ data, the (γ,π^+) cross section can be deduced. The cross sections so obtained should not be directly compared with the radiative pion capture and isochromat results because the momentum transfer is not equal in these reactions.

Angular distribution of (γ, π^+) are given for these cross sections at various angles. The results are obtained for ^{6}Li, ^{9}Be, ^{12}C and ^{16}O. Examples are shown for ^{12}C and ^{16}O in Figure 4 and 5 in the following contribution of Ohashi et al. For this analysis the virtual photon spectra derived by Dalitz and Yennie's formula[6] were used.

There are two methods for theoretical estimation of photopion cross sections by Eqs. 2 and 3, that is phenomenological estimation of the nuclear matrix element and calculation by nuclear models.

When the pion wave function is approximated by a plane wave, $\vec{m}$ defined by Eq. 3 can be expressed as

$$\vec{m} \approx \langle J_f M_f | \sum_{n=1} \vec{\sigma}_n \tau_n^\pm \exp(i\vec{q}\cdot\vec{r}_n) | J_i M_i \rangle, \qquad (7)$$

$$= \mp \sqrt{2(T+1)} \int d^3 r \, e^{i\vec{q}\cdot\vec{r}} \rho(\vec{r}), \qquad (8)$$

where $\vec{q}$ is the momentum transfer and $\rho(\vec{r})$ is radial spin flip density. Überall's group has estimated $\rho(\vec{r})$ from electron scattering data using the generalized Helm Model[10] and calculated the cross sections by this phenomenological calculation[8]. Their results[8,11] are shown in Figs. 4 and 5 of the contribution of Ohashi et al. In these calculations, the momentum dependent terms in the interaction of Eq. 1 and the final state interactions are not included. Another result on ^{16}O which has an improved treatment for final state interactions[12] is also compared. The shell model calculation by the same group[13] is shown in Fig. 4 of Ohashi et al. The patterns seem in qualitative agreement with the experimental results, but not always complete.

Precise theoretical calculations have been made using

the shell model in various degrees of approximations. For
example Furui calculated the energy and angular distibutions
of (e,e'π^+) reactions in connection with the Tohoku experiments[9,14]. He also estimated the virtual photon spectrum and
concluded that it is 2 ∿ 3 times as small as that from Dalitz
and Yennie's formula[9,14]. The main origin of this difference
is the fact that the assumption of $<J^2> \propto k^2$ in Dalitz and
Yennie's formula is not valid and the applicability of the
formula to the region of the final momentum $p'_e \simeq 0$ is not
evident.

The (e,e'π^+) experimental results are recalculated with
his virtual photon and compared with theoretical results in
Figs. 9 and 10. The Furui's results in the figures were calculated by DWIA and shell model with suitable normalization.
Agreement seems much improved.

Recently various approximations have been applied in
theoretical calculations. A conclusion to be drawn from the
theoretical studies is the importance of final state interaction between the pion and nucleus and also correct virtual
photon estimation for the (e,e'π) experiment. A general theoretical study of photopion reactions has been given by Epstein
et al.[15]. After these problems are settled, the photopion
study will give a good method to study the isovector spin-flip
contribution in nuclear transition for individual states.

3.2 Wide Range Photopion Energy Distribution and Giant Resonances

Photopion energy distributions in an energy range from
about 25 MeV up to 50 MeV have been measured with the (e,e'π^+)
reaction on ^{6}Li, ^{7}Li, ^{9}Be, ^{12}C, ^{27}Al and ^{89}Y using 180 MeV or
200 MeV (total energy) electrons from the Tohoku Linac. The
results are shown in Fig. 1 of the following contribution by
Shoda et al., where arrows indicate the pion maximum energy
leaving the residual ground state.

Dotted bars in the figure indicate the position corresponding to the neutron separation energy of the residual nuclei. As shown in the figure, the photopion yield strongly

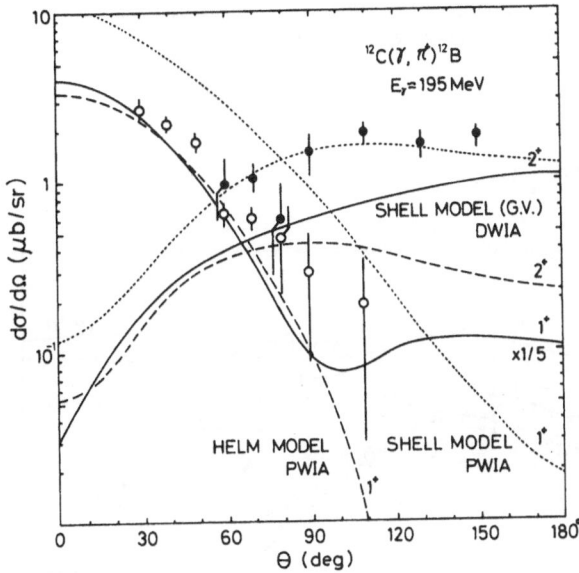

Figure 8. The photopion angular distribution deduced from $^{12}C(e,e'\pi^+)^{12}B$ using a virtual photon spectrum derived by Furui[9,14]. Solid curve: theoretical result by shell model calculation (DWIA)[14]. Dashed curve: theoretical result by generalized Helm model (PWIA)[8]. Dotted curve: shell model calculation (PWIA)[13]. (Editor's note: the experimental points have been increased by a factor of 2.5 compared to those shown in Fig. 4 of the contribution by Ohashi et al.)

increased in the region that the neutron is unbound in the residual nucleus. This suggests that photopion production leads to neutron emission based on the elementary process $\gamma + p \rightarrow n + \pi^+$ in the reaction.

On the basis of virtual photon theory the (e,e'π) spectra can be expressed in terms of the (γ,π) cross section feeding a residual state, and the virtual photon spectra as

$$I(T_\pi) = \int_0^{E_e+Q-T} \sigma(E_\gamma, E_R) N_{h\nu}(E_e, E_\gamma = -Q + T_\pi + E_R) dE_R. \tag{9}$$

As suggested by theory, the momentum and energy dependence of
the cross section is expressed by

$$\sigma(E_\gamma, E_R) = \frac{p_\pi}{E_\gamma} M(E_R, q).$$

(10

Using Eq. 10 to Eq. 9,

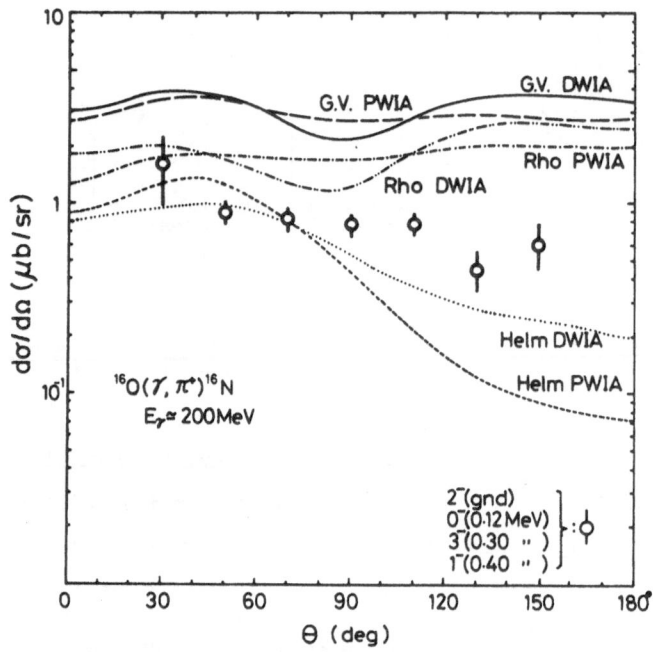

Figure 9. The photopion angular distribution deduced from
$^{16}O(e,e'\pi^+)^{16}N$ (sum of the lowest 4 levels) using a virtual
photon spectrum derived by Furui[9,14]. Solid curve: theoretical result by shell model (Gillet-VinhMau wave function[21]
(DWIA)[14]. Broken curve: same as above but PWIA[14]. 2-dot-dash
curve: result by shell model (Migdal model wave function of
Rho[22] (DWIA)[14]. Dot-dash curve: same as above but PWIA[14].
Dashed curve: result by generalized Helm model but improved
for final state interaction[12].

$$\frac{I(T_\pi)}{p_\pi} = \int_0^{E_e+Q-T_\pi} M(E_R,q)\frac{1}{E_\gamma} N_{h\nu}(E_e,-Q+T_\pi+E_R)dE_R. \tag{11}$$

This relation is analogous to the bremsstrahlung yield curve shown as

$$Y(E_e) = \int_0^{E_e} \sigma(E_\gamma)N_{h\nu}(E_e,E_\gamma)dE_\gamma. \tag{12}$$

When the correspondence is taken between $I(T_\pi)/p_\pi$, T_π, E_R, $\frac{1}{E_\gamma} N_{h\nu}(E_e, -Q+T_\pi+E_R)$ in Eq. 11 and $Y(E_e)$, E_e, E_γ, $N_{h\nu}(E_e,E_\gamma)$ in Eq. 12, respectively, $M(q,E_R)$ can be analyzed from an energy distribution by Eq. 11 using a photon difference method similar to that for Eq. 12. The cross sections $d^2\sigma/d\Omega \, dE_R(E_R,E_\gamma \approx E_e)$ are calculated from the spectra in Fig. 13 using a variable bin Penfold-Leiss method[16] and the result is shown in Fig. 2 of the contribution of Shoda et al.

The results are compared with the radiative π^- capture result[17] which is analogous to the present data except that the relevent momentum transfer is smaller. Though the resolution is not good in the present result because of poor statistics in the energy distribution data, some prominent strengths appear in both results suggesting spin-flip type giant resonances. It is hoped that future experiments with improved statistics and angular distribution studies will provide more complete information.

Double-differential cross sections for the electroproduction of pions have been measured by the Mainz group[18]. Their result is for a fixed energy pion and theoretical comparison was made with a nuclear sum rule description or integration over a range of virtual photon and residual state energies in the shell model.

Nuclear structure and photopion production mechanisms in highly excited states of nuclei have also been studied by the photon difference method and tagging method over the resonance region by the INS (Tokyo) group[19,20]. These experiments are natural extensions to higher excitation of pion spectroscopy which was discussed in the present report focussing mainly on low lying states.

4. CONCLUSIONS

The present situation of photopion spectroscopy was discussed here for π^+ photoproduction from nuclei. Similar studies will be possible for π^- photoproduction when some caution is made for discrimination of π^- from the strong e^- background. Cerenkov back up counters will be useful in this case.

The main conclusions in π^+ photopion spectroscopy are as follows.

    i.   Photopion transition strengths can be obtained using conventional magnetic spectrometers but more effective spectrometers are desired.

   ii.   From the theoretical point of view, final state interactions will be most important in order to separate nuclear information correctly. Momentum dependent terms of the interaction will become important for precise study. Virtual photon spectra are also important for the (e,e'π) analysis. These problems should be solved with a good interplay between theoretical and experimental study on the nuclei for which nuclear property is well known.

  iii.   Spin-flip type giant resonances will be studied using pion spectroscopy. High efficiency, variable angle and broad range spectrometer will be effective.

The virtual photon problem is being tackled at Bates and Tohoku. For the purpose of more effective study, improved spectrometers are under construction at MIT, Tohoku and other institutes. Many new developments in charged photopion spectroscopy are expected in the very near future.

REFERENCES

1. F.A. Berends, A. Donnachie and D.L. Weaver, Nucl. Phys. B4 (1967) 1, 54.
2. M. Ericson and M. Rho, Physics Reports 5 (2) (1972) 57.
3. G. Audit et al., Phys. Revi 15C (1977) 1415; K. Min et al., Phys. Revi 14C (1976) 807.

4. N. Paras, A.M. Bernstein, K.I. Blomqvist, G. Franklin,
    M. Pauli, B. Schoch, J. LeRose, K. Min, D. Rowley, P.
    Stoler, E.J. Winhold, and P.F. Yergin, contribution to
    this symposium 1978.

5. R.M. Sealock, H.S. Caplan, M.K. Leung, G.J. Lolos and
    S. Hontzeas, Preprint 1978.

6. R.H. Dalitz and D.R. Yennie, Phys. Rev. 105 (1957) 1598.

7. M.I. Acamovich, V.G. Larionova, A.V. Lebedev, S.P.
    Kharlamov and F.R. Yagudina, Photomesic and Photonuclear
    Processes, Proceeding (Truday) of the P.N. Lebedev
    Physics Institute Vol. 34, ed. Academian D.V. Skobel'tsyn
    (Translated from Russian, Consultants Bureau, N.Y., 1967)
    p. 49.

8. F. Cannata, B.A. Lamers, C.W. Lucas, A. Nagl, H. Überall,
    C. Werntz and F.J. Kelly, Can. J. Phys. 52 (1974) 1405.

9. S. Furui, Nucl. Phys. A300 (1978) 385.

10. M. Rosen, R. Raphael and H. Überall, Phys. Rev. 163
    (1967) 927; H. Überall and B.A. Lamers, J.B. Langowrthy
    and F.J. Kelly, Phys. Rev. 6C (1972) 1911.

11. A. Nagl, F. Cannata and H. Überall, Acta Physica Austri-
    aca 48 (1978) 267.

12. A. Nagl and H. Überall, contribution to this symposium
    1978.

13. J.S. Seaborn, V. Devanathan and H. Überall, Nucl. Phys.
    A219 (1974) 461.

14. S. Furui, private communication; contribution to this
    symposium 1978.

15. G.N. Epstein, M.K. Singham and F. Tabakin, Phys. Rev.
    17C (1978) 702.

16. E. Bramanis, T.K. Deague, R.S. Hicks, R.J. Hughes, E.G.
    Muirhead, R.H. Sambell and R.J.J. Stewart, Nucl. Instr.
    Meth. 100 (1972) 59.

17. J.C. Alder et al., Meson-Nuclear Physics (Carnegie-
    Melone Conference 1976) p. 624, p. 626; P. Truöl, ibid.
    p. 581.

18. F. Borkowski, C. Schmitt, G.G. Simon, V. Walther, D.
    Drechsel, W. Haxton, R. Rosenfelder, Phys. Rev. Lett.
    38 (1977) 742.

19. K. Baba et al., preprint 1978; I. Endo, Phys. Lett.
    47B (1973) 469.

20. S. Homma, contribution to this symposium 1978; and private
    communication.

21. V. Gillet and N. VinhMau, Nucl. Phys. 54 (1964) 321.

22. M. Rho, Phys. Rev. 161 (1967) 955.

STUDY ON THE LOW-LYING STATES VIA THE PHOTOPION PRODUCTION

ON SEVERAL NUCLEI

H. Ohashi, K. Nakahara, M. Yamazaki and K. Shoda

Laboratory of Nuclear Science, Tohoku University

Tomizawa, Sendai 982, Japan

B. N. Sung

Department of Physics, College of Natural Science

Seoul National University, Seoul, Korea

Photopion production from complex nuclei will reveal it-self as a new probe which sheds light on nuclear structure. Similar results will be obtainable by the inverse reaction, i.e., radiative capture reaction, yet photoproduction is superior because of a possibility of changing the momentum transferred to the nucleus by changing the detection angle of photo-produced pions, whereas in the case of radiative pion capture, the momentum transfer is fixed.

The photopion production amplitude from a nucleon near the threshold is dominated by a spin-flip type of transition $\vec{\sigma} \cdot \vec{\epsilon}$, where $\vec{\sigma}$ is the nucleon spin and $\vec{\epsilon}$ is the photo polarization vector. The photopion cross section on complex nuclei, under the PWIA, is given by[1]:

$$\left(\frac{d\sigma}{d\Omega} \right)(\gamma,\pi) = \frac{p}{km_\pi^2} e^2 f^2 \frac{1}{\hat{J}_i^2} \cdot \sum_{M_i} \sum_{M_f} \sum_{S_\gamma} |\vec{\epsilon} \cdot \vec{\mu}|^2 \qquad (1)$$

$$\vec{\mu} = \langle J_f M_f | \sum_{n=1}^{A} \vec{\sigma}_n \tau^+ -\exp(i\vec{q} \cdot \vec{r}_n) | J_i M_i \rangle, \qquad (2)$$

where m_π is the pion mass, $e^2 = 1/137$, $f^2 = 0.08$, $\tau^\pm$ is the isospin raising and lowering operator and $\vec{q}$ is the momentum transferred to the nucleus.

Energy and angular distributions of electro-produced pions were measured for ^{6}Li at the incident total energy of 180 MeV and for ^{9}Be at 200 MeV, in addition to the previous data on ^{12}C (195 MeV) and ^{16}O (200 MeV)[2]. An electron beam from Tohoku University linear accelerator was used and emitted pions were momentum analyzed with a double-focusing electromagnetic spectrometer, and were detected with a 33-array of triple coincidence system of Si(Li) solid state detectors set on the focal plane. All charged nuclear particles are stopped in the first detectors of the coincidence system. Pions were discriminated from background positrons with the discriminator bias setting as shown in Fig. 1.

After correcting for the decay of pions during their passage through the apparatus, photoproduction cross sections leading to the states of the residual nucleus were deduced from the electroproduction cross section using the relation,

$$\frac{d^2\sigma_{(e,e'\pi^+)}(T_\pi, E_{Ri})}{d\Omega dT_\pi} = \sum_i \frac{d\sigma_{(\gamma,\pi^+)}(E_\gamma, E_{Ri})}{d\Omega} N_{h\nu}(E_e, E_\gamma) \quad (3)$$

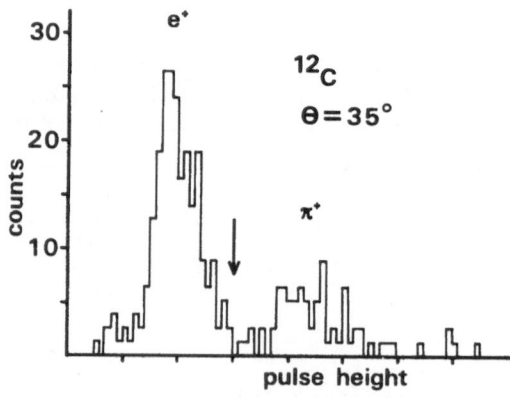

Figure 1. An example of the pulse height distribution from a single Si(Li) solid state detector in the triple coincidence system.

where $(d\sigma(E_\gamma, E_{Ri})/d\Omega)_{(\gamma,\pi^+)}$ is the (γ,π^+) differential cross section leading to the residual energy E_{Ri}, and $N_{h\nu}(E_e, E_\gamma)$ is the virtual photon spectrum associated with an electron of energy E_e. Differential cross sections analyzed by Eq. 3 with the virtual photon spectrum derived by Dalitz and Yennie[9] are shown in Figs. 2 - 5 for ^{6}Li, ^{9}Be, ^{12}C and ^{16}O, respectively. Theoretical estimates using the intermediate-coupling shell model of Barker[3] for ^{6}Li[4] and of Boyarkina[5] for ^{9}Be, generalized Helm model for ^{12}C[1] and ^{16}O[6] are also shown in these figures.

In order to check the absolute value, the cross sections of π^+ electro-production from hydrogen were measured at E_e = 180 MeV by using a LiH target. In Fig. 6, the cross section of $H(\gamma,\pi^+)$ deduced from Eq. 1 is shown by a cross and compared with the previous results obtained by real photons at E_γ = 180 MeV[7] which are shown by circles. The agreement between both results are satisfactory.

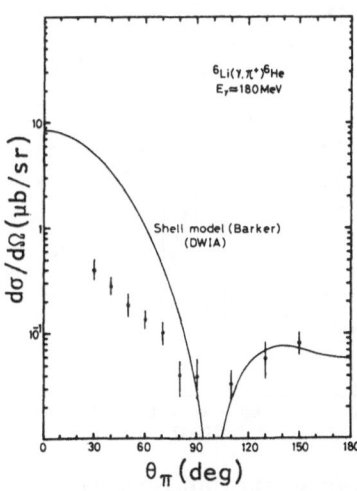

Figure 2. Pion angular distributions in ^{6}Li(γ,π^+) ^{6}He at $E_\gamma \simeq$ 179 MeV. Solid curves are the theoretical estimates of the DWIA results[4] by using Barker's wave function[3].

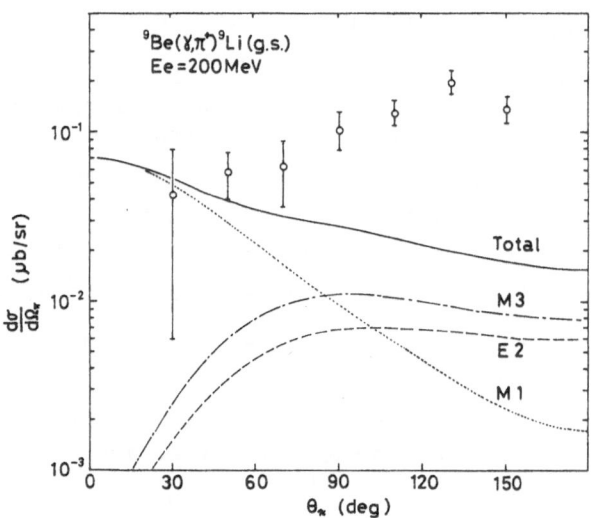

Figure 3. Pion angular distributions in ^{9}Be$(\gamma,\pi^+)^9$Li at $E_\gamma \simeq$
199 MeV. Solid curves are the theoretical estimates of
the PWIA result by using Boyarkina's wave function[5].

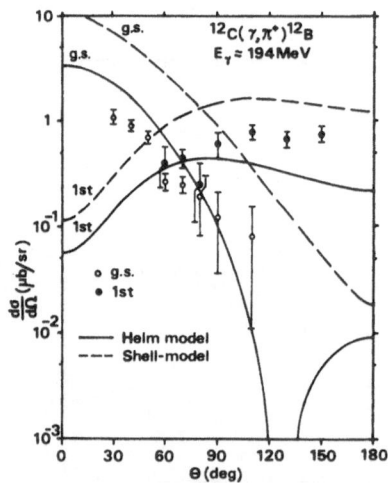

Figure 4. Pion angular distributions in ^{12}C$(\gamma,\pi^+)^{12}$B at $E_\gamma \simeq$
194 MeV. Curves are the theoretical estimates of the
PWIA result by using the generalized Helm model[1] and shell
model.[11] Open circles and closed circles show cross sections
leading to the ground and first excited state, respectively.

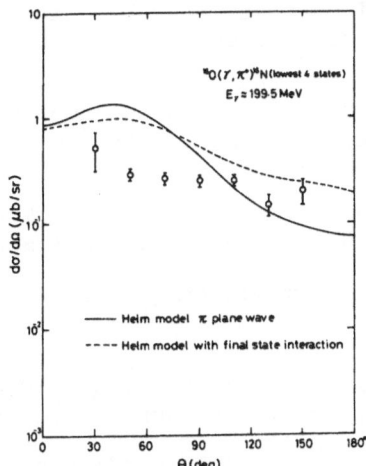

Figure 5.   Pion angular distributions in $^{16}O(\gamma,\pi^+)^{16}N$ at $E_\gamma \simeq$ 199 MeV. Solid curves are the theoretical estimates of the PWIA result by using the generalized Helm model[6]. Dashed curves are theoretical results taking into account final state interactions[10].

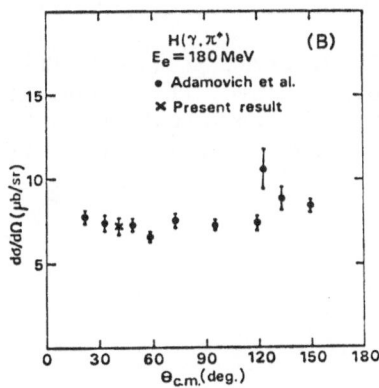

Figure 6.   The cross section of $H(\gamma,\pi^+)$ at $E_\gamma \simeq$ 179 MeV.   The present result deduced by using the virtual photon theory shown by a cross is compared with the previous result shown by circles[7].

With DWIA, Furui[4] has shown much better agreement between the experiments and calculations for the nuclei mentioned above. He also suggests[8] that the number of the virtual photons in the $(e,e'\pi^+)$ reaction of about two or three times as small as that obtained from the Dalitz-Yennie formula for the light nuclei. In order to solve this problem, bremsstrahlung experiments on ^{12}C and ^{51}V have been performed at this laboratory, and are in the process of being analyzed.

REFERENCES

1. F. Cannata et al., Can. J. Phys. 52 (1974) 1405.
2. K. Shoda et al., Proc. Int. Conf. Nucl. Str. (Tokyo, 1977) 482.
3. F.C. Barker, Nucl. Phys. 83 (1966) 418.
4. S. Furui, Res. Rept., Lab. of Nuclear Sci., Tohoku Univ., Sendai, Japan 11 (1978) 20.
5. A.N. Boyarkina, Izv. Akad. Nauk, SSSR, Ser Fiz. 28 (1964) 337.
6. A. Nagl et al., Acta Physica Austriaca 48 (1978) 267.
7. M.I. Adamovich et al., Photomesic and Photonuclear Processes, Proceedings (Trudy) of the P.N. Lebedev Physics Institute, Vol. 34, ed. Academician D. V. Skobel'tsyn (translated from Russian, Consultants Bureau N.Y., 1967), 49.
8. S. Furui, Prog. Teor. Phys. 58 (1977) 864.
9. R.H. Dalitz and D.R. Yennie, Phys. Rev. 105 (1957) 1598.
10. A. Nagl and H. Uberall, contribution to these Proceedings.
11. J.S. Seaborn, V. Devanathan and H. Uberall, Nucl. Phys. A219 (1974) 461.

CHARGED PION ELECTROPRODUCTION ON LIGHT NUCLEI NEAR THE THRESHOLD

S. Furui

National Laboratory for High Energy Physics

Oho-machi, Tsukuba-gun, Ibaraki, 300-32, Japan

In order to study pion fields in nuclei, it is necessary to find probes sensitive to the detailed behavior of real and virtual pions inside nucleus. Pion photoproduction or electroproduction above threshold can be useful since the angular distribution of pions allows us to obtain the momentum transfer dependence of the form factor. Shoda et al. have measured the kinetic energy spectra of π^+ produced from ^{12}C, ^{16}O, ^{9}Be, ^{7}Li, ^{6}Li, ^{27}Al and ^{89}Yb by electrons of laboratory kinetic energy about 200 MeV[1,2]. Near the tip of the spectrum, nuclear excitations are restricted to the low lying states and we can obtain contribution of each level if the level spacings are relatively large. They have analyzed their data in the plane wave impulse approximation (PWIA) assuming the Helm model for the nucleus.

We have analyzed their data in the distorted wave impulse approximation (DWIA) using the shell model wave functions which reproduce electron scattering, muon capture and/ or β-decay[3,4]. We have adopted the "peaking approximation", in shich we have assumed the contribution of virtual photons emitted in the direction of the incident electrons is dominant.

In the PWIA we have obtained the double differential cross section for pions of laboratory kinetic energy T_π and the production angle θ as

$$\frac{d^2\sigma}{2\pi d(\cos\theta)dT_\pi} = \frac{\alpha}{\pi}\,\frac{M^2}{8\pi^2}\,\sum_f - \frac{1}{b^2}\,\ln\left|\frac{b-c}{b+c}\right|$$

$$\times\left[\frac{|\vec{p}|^2+|\vec{p}'|^2}{|\vec{Q}|^2}\,\left(\frac{|\vec{p}'|}{E_{\vec{p}'}}\right)^3\,\frac{1}{|\vec{Q}|}\,\frac{1}{(2J_i+1)}\left(\frac{\sqrt{2g}}{2M(1+m_\pi/M_f)}\right.\right.$$

$$\left.\left.<J_f||\Sigma_1'(\vec{q})\sqrt{\tfrac{1}{2}}\,\tau_{-1}||J_i>[\phi_\pi])^2\right]_{\text{peak}}$$

where we took the Kroll-Ruderman term only. The momentum transfer is denoted by $\vec{q}-\vec{Q}-\vec{k}$ where $\vec{Q}$ and $\vec{k}$ are the momenta of the photon and the pion respectively and $\Sigma_1'(\vec{q})$ is the Fourier transform of the operator

$$\Sigma_1'(\vec{x}) = -i\,\frac{1}{|\vec{q}|}\,[\vec{\nabla}\times\vec{M}_{11}(\vec{x})]\cdot\vec{\sigma}\;.$$

We denote the masses of nucleon, pion and final nucleus by M, m_π and M_f respectively, the momenta of incident electron and final electron by $\vec{p}$ and $\vec{p}'$ respectively and the final electron energy by E'. J_i and J_f are the spins of the initial and the final nucleus, τ_{-1} is the spherical component of the ispspin operator and $[\phi_\pi]$ is the distortion factor of the pion. We parameterize the final electron momentum scattered at an angle θ as

$$|\vec{p}'(\theta)| = A/(B+C\cos\theta)$$

and define

$$b = 2m_e^2\,Q_0^2 BC/|\vec{p}|,\quad c = m_e^2 Q_0^2 B^2/(|\vec{p}|A^2) + 2|\vec{p}|$$

where m_e is the mass of the electron and Q_0 is the photon energy. The suffix "peak" signifies that the peaking approximation is adopted.

The model can be extended to DWIA and the photoproduction vertex can be replaced by that of Berends et al.[5] We consider the terms proportional to $\tau_{-1}\sigma_\pm$ and the terms proportional to

$\tau_{-1}\sigma_3$. The latter is dominated by the pion pole contribution and smaller than the former. By analogy with the result of the soft pion limit[6], where the photoproduction vertices reduce to the Gamow-Teller term and the induced pseudoscalar term, we expect the coherent virtual pion exchange between nucleons in the nucleus to modify the vertices appreciably.

Using the equivalent photon approximation, we can obtain the photoproduction cross section from the electroproduction cross section. We have estimated the number of equivalent photons by performing the integration over the phase space of the final electrons with the assumption of the dominance of the Kroll-Ruderman term. The number depends on the momentum dependence of the photoproduction vertices and the distortion of the wave function of the final electrons. There is a factor of about 2.5 difference between our estimate and that of Shoda et al.

Our DWIA result for π^+ photoproduction from ^{12}C is shown as the solid curve in Fig. 9 in the preceding review by K. Shoda. We have adopted Kurath's intermediate coupling shell model wave function* with the oscillator length parameter b=1.64 fm. Experimental points are taken from Ref. 1, but multiplied with the factor 2.5**due to the difference of the number of equivalent photons. The term proportional to $\tau_{-1}\sigma_3$ gives a small contribution for this nucleus.

In Fig. 1 we show the DWIA result for ^9Be. We have adopted Barker's intermediate coupling shell model wave function which reproduces the inelastic electron scattering to the isobaric analog state well. In the case of M1 and M3 transitions, we observe a slight enhancement around 90° due to the term proportional to $\tau_{-1}\sigma_3$. We compare our DWIA result and PWIA result to the experimental data in Fig. 2. The data of Ref. 2 is multiplied by the factor of 2.5.** The PWIA calculated with Boyarkina's intermediate coupling shell model wave function gives results similar to that of Barker's wave function. Enhancement of the longitudinal spin and isospin mode due to coherent virtual pion exchange between nucleons in the nucleus is the most natural possibility to explain the broad peak around 120° shown in the data.

We show our PWIA results and DWIA results for ^{16}O cal-

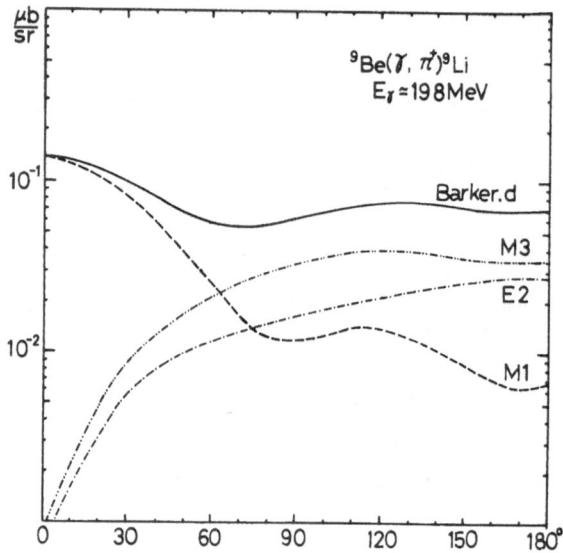

Figure 1. The angular distribution of π^+ photoproduction from ^{9}Be calculted in DWIA with Barker's wave function. Experimental points from Ref. 2 are multiplied by factor 2.5**.

culated with Gillet-VinhMau's Tamm-Dancoff wave functions and Rho's Migdal model wave functions in Fig. 10 of the preceding review by Shoda. Since the low lying 2^-, 0^-, 3^- and 1^- states of ^{16}N are close together, the sum of the four contributions is shown. The experimental data of Ref. 2 is multiplied by the factor of 3.0**. Our results are larger than the data especially at large angles. It may be due to the deficiency of the wave function of the 3^- state.

In the case of ^{6}Li, the experimental data at 30° is about a factor of 2 smaller than the DWIA result calculated with the Saskachewan-A wave function.

In the case of ^{7}Li, the Gamow-Teller matrix element is small and the transition is dominated by the M3 component. Our DWIA result is about a factor of 2.5 larger than the PWIA result at large angles. We expect measurements of the angular distribution would clarify renormalization effects on the vertices.

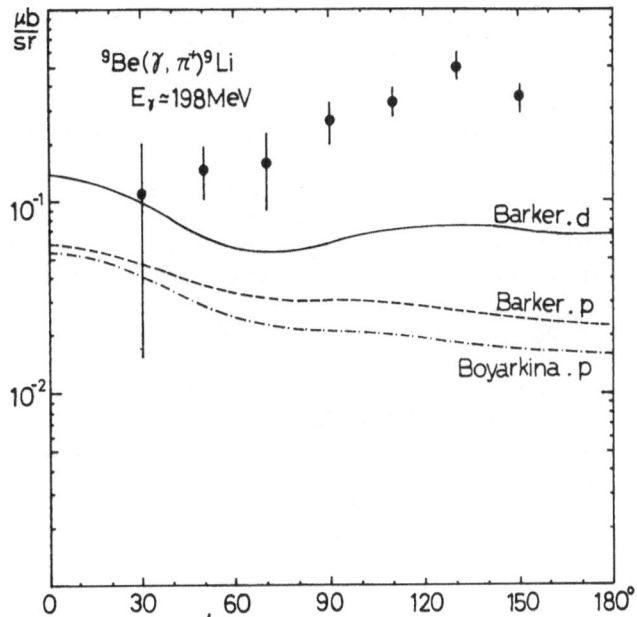

Figure 2. The angular distribution of π^+ photoproduction from ^{9}Be calculated in DWIA (d) and PWIA (p).

    In summary, charged pion electroproduction near the threshold is a useful tool for studying nuclear structure and it compliments studies of electron scattering, muon capture, β-decay and radiative pion capture. Although the accuracy of the experimental data is as yet insufficient for detailed studies, we expect a substantial improvement in the very near future. If comparison of π^- photoproduction and π^+ photoproduction near the threshold is performed, useful information on the isospin structure of mirror nuclei would be obtained.

NOTES AND REFERENCES

*Thanks are due to Prof. D. Kurath for kindly supplying us with tables of his wave function.

**Editor's note: According to the author, the number of effective virtual photons used in Ref. 1 is about a factor of 2.5 greater than the author's estimate. The correction

made to the data does not appear to take into account the real
photon contribution, which presumably is not in question.
Thus, the correction factor of 2.5 is too large.

1. K. Shoda, H. Ohashi and K. Nakahara, Phys. Rev. Lett.
    39 (1977) 1131.
    H. Ohashi, K. Shoda and K. Nanahara, these Proceedings.
2. K. Shoda et al., Research Report of Laboratory of Nuclear
    Science, Tohoku University, 10 (1977) 188 (unpublished);
    Proc. of Int. Conf. Nucl. Structure (Tokyo, 1977) 482.
    H. Ohashi et al., these Proceedings.
3. S. Furui, Prog. Theor. Phys. 58 (1977) 864; Nucl. Phys.
    A300 (1978) 385.
4. S. Furui, to be published in Nucl. Phys. A.
5. I. Blomqvist et al., Phys. Rev. C15(1977) 988.
6. M. Rho, Proc. of Int. School of Phys. "Ettore Majorana",
    1976, Erice, Italy (to be published).

PHOTOPION ENERGY DISTRIBUTIONS AND SPIN-ISOSPIN GIANT RESONANCES

K. Shoda, M. Yamazaki, K. Nakahara and H. Ohashi

Laboratory of Nuclear Science, Tohoku University

Tomizawa, Sendai 982, Japan

Positive charged photopion production from nuclei includes a dominant spin-isospin type interaction, $\vec{\epsilon} \cdot \vec{\sigma} \, \tau$, where ϵ is the photon polarization vector. The (γ, π^+) cross section is approximately given by this interaction as

$$\left(\frac{d\sigma}{d\Omega} \right)_{(\gamma, \pi^+)} = \frac{P_\pi}{km_\pi^2} e^2 f^2 \frac{1}{2J_i + 1} \sum_{M_i M_f} \sum_{S_\gamma} |\vec{\epsilon} \cdot \vec{m}|^2 \tag{1}$$

$$\vec{m} \approx \langle J_f M_f | \phi_\pi^* \sum_{n=1}^{A} \vec{\sigma}_n \, \tau_n^- \exp(i\vec{k} \cdot \vec{r}_n) | J_i M_i \rangle \tag{2}$$

When the pion wave function is approximated by a plane wave, the matrix element is reduced to

$$\vec{m} \approx \langle J_f M_f | \sum_{n=1}^{A} \vec{\sigma}_n \, \tau_n^- \exp(i\vec{q} \cdot \vec{r}_n) | J_i M_i \rangle, \tag{3}$$

where $\vec{q}$ is the momentum transfer.

This matrix element is quite similar to that of the spin-flip isovector electromagnetic type transition. Measurement of this type transition strength has been difficult in the usual - and -decay experiments, because a very strong El type transition background makes the study impossible. In the photopion production, the spin-flip type transition is the leading process; therefore no other type of stronger background exists.

Distributions of spin-flip type transition strengths in residual nuclei can be studied by the photopion reaction as well as radiative π^- capture. This will give one of the best methods to study spin-flip giant resonances.

The photopion energy distribution in $(e,e'\pi)$ reactions can be expressed as

$$I(T_\pi) = \int_0^{E_e+Q-T} \sigma(E_\gamma,E_R)N_{h\nu}(E_e,E_\gamma=-Q+T_\pi+E_R)dE_R \qquad (4)$$

As suggested by Eq. 1, the energy dependence of the cross section is given by

$$\sigma(E_\gamma,E_R) = \frac{P_\pi}{E_\gamma} M(E_R,q). \qquad (5)$$

When this expression is used in Eq. 4, we obtain

$$\frac{I(T_\pi)}{P_\pi} = \int_0^{E_e+Q-T} M(E_R,q) \frac{1}{E_\gamma} N_{h\nu}(E_e,-Q+T_\pi+E_R)dE_R. \qquad (6)$$

Using this relation, $M(E_R,q)$ can be analyzed from the energy distribution $I(T_\pi)/P_\pi$ by an unfolding method using virtual photon spectra.

In order to study the giant resonance, the photopion energy distirbutions have been measured with the $(e,e\pi^+)$ reaction. About 200 mg/cm^2 thick targets of ^{6}Li, ^{7}Li, ^{9}Be, ^{12}C, ^{27}Al and ^{89}Y have been bombarded by an electron beam of 180 MeV or 200 MeV (total energy) using the Tohoku linear accelerator. The produced pion energy distributions have been measured with a double focussing magnetic spectrometer at $\theta=50°$ or $60°$ with dynamic range $\Delta p/p = 3.3\%$ and an array of 33 triple coincidence systems of Si(Li) solid state detectors. Several field settings of the spectrometer have been used in order to measure the pion in an energy range of about 25 MeV up to 50 MeV. The data were corrected for the detector efficiency, pion

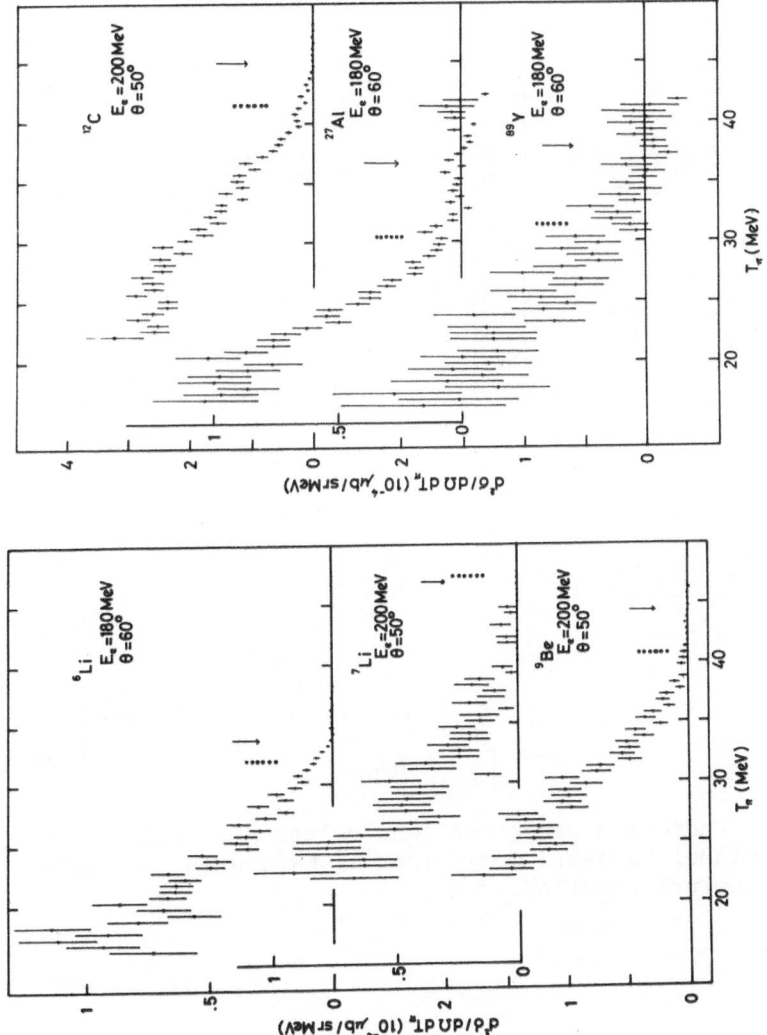

Figure 1. Photopion energy distribution from the (e,e'π⁺) reaction. Solid arrows: calculated maximum pion energy. Dotted bars: energy corresponding to the neutron separation of the residual nuclei.

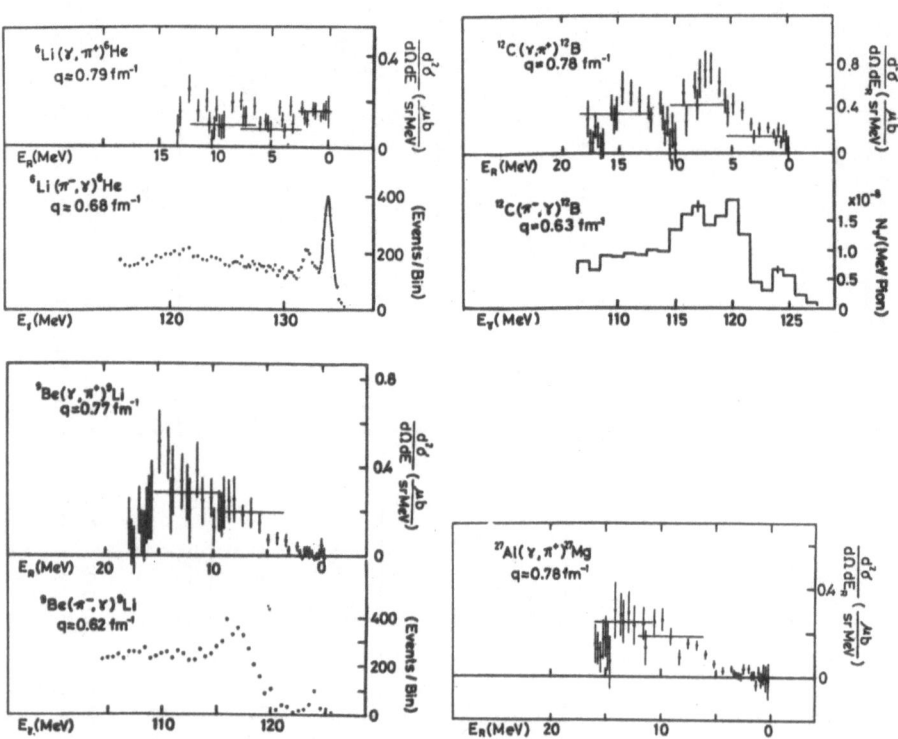

Figure 2. Photopion cross sections relating to the energy of the residual nuclei. Comparison is made with the radiative π⁻ capture results[2].

decay during passage through the spectrometer and the contribution of bremsstrahlung produced by the beam passing through
the target.

The results are shown in Fig. 1 where the arrows indicate the calculated pion maximum energy leaving the residual
nucleus in the ground state. The apparent end point energy of
the experimental result is often much smaller than the calculated maximum energy. The dotted bars in the figure indicate
the neutron separation energy of the residual nucleus. Agreement can be seen between this energy and the apparent end
point energy in the spectra. This fact suggests a strong contribution of the elementary process

$$p + \gamma \to n + \pi^{+},$$

where n is unbound in the nuclear photopion production.

The energy distributions are unfolded with Eq. 6 by a
variable bin Penfold-Leiss method[1]. The results are shown
in Fig. 2. These results show rather broad resonances (the
resonance widths may reflect the resolution in the present
analysis) and indicate an existence of isovector spin-flip
giant resonances.

A comparison with the results of radiative π^{-} capture[2]
is also made in the same figure. Though the momentum transfer
is different, both results show a good correspondence with
each other. More precise experiments including angular distributions are being prepared with a broad-range magnetic
spectrometer now under construction at this laboratory.

REFERENCES

1. E. Bramanis, T.K. Deague, R.S. Hicks, R.J. Hughes, E.G.
   Muirhead, R.H. Sambell and R.J.J. Stewart, Nucl. Instr.
   Meth. 100 (1972) 59.
2. J.C. Alder et al., Meson-Nuclear Physics (Carnegie-Mellon
   Conference 1976) p. 624, p. 626; P. Truöl ibid p. 581.

PION SPECTRA TO INDIVIDUAL FINAL STATES

IN THE REACTIONS $^{12}C(\gamma,\pi^{\pm})^{12}B,^{12}N$

N. Paras, A. M. Bernstein, K. I. Blomqvist,

G. Franklin, J. LeRose, K. Min, M. Pauli,

D. Rowley, B. Schoch, P. Stoler, E. J. Winhold

and P. F. Yergin - MIT-RPI Collaboration

The cross sections for the reactions $^{12}C(\gamma,\pi^{+})^{12}B$ and $^{12}C(\gamma,\pi^{-})^{12}N$ to various states of the final nucleus have been measured at a pion energy around 29 MeV for electron energies from 187 to 193 MeV, and at a pion energy around 17 MeV with electron energies from 174 to 181 MeV, at a pion angle 90° with respect to the incident photon beam. The observed states in ^{12}B and ^{12}N are the analogues of T = 1 states in ^{12}C which are excited by inelastic electron scattering[1,2]. Furthermore, the (γ,π) one-body operator tends to strongly excite spin-flip components of the transitions to these states.

EXPERIMENTAL PROCEDURE

The experimental arrangement is briefly discussed and shown in Figure 9 of the review by Winhold in these Proceedings. Electrons from the Bates linear accelerator irradiated a 0.2 gm/cm^2 graphite target oriented at an appropriate angle to the electron beam as discussed in the review of K. Shoda. The pions were produced primarily via electro production with some contribution from the real bremsstrahlung produced in the target itself.

The pions were detected and momentum-analyzed at an angle of 90° with respect to the incident electron beam using a magnetic spectrometer with a focal plane multiwire proportional counter. The pions were separated from electrons and other background by requiring coincidence with a backup array consisting of three scintillation counters and a Cerenkov counter for electron suppression. Further reductions in the background were accomplished by putting appropriate software cuts on the scintillator pulse height spectra, as well as on the event time between scintillator signals. The wire chamber spectrum for 29 MeV positive pions which covers the pion energy from 25 to 32 MeV is shown in Figure 1. This spectrum is constructed by taking runs at electron beam energies from 185 to 191 MeV in 1 MeV intervals, while keeping the pion energy interval fixed, and then combining the data from all the runs.

In order to extract cross sections from such spectra it is necessary to determine the detector's efficiency times solid angle. This is obtained by normalization to hydrogen, using a cryogenic pressurized gas target at 230 MeV electron energy. In the data analysis the effective virtual photon spectrum for pion photoproduction used was that of Dalitz and Yennie[3], but increased by a factor of 1.27 to reflect the real-to-virtual pion production measurements which we have made for several light nuclei. For the real photons, the

Bethe-Heitler bremsstrahlung spectrum was used. The normalization was checked by direct calculation using the known characteristics of the spectrometer, and the agreement was within 10%, which lends a degree of confidence to the virtual photon spectrum[*].

In Figure 1, the solid curves represent a least square fit to the data using the photon spectrum as discussed above.

RESULTS

The results of the experiments are shown in Figures 2-4. In Figure 2 and 3 the experimental cross sections for transitions to the various states in ^{12}B for positive pion energies around 29 and 17 MeV respectively are shown with error

---

[*]See the contribution by Furuii for another discussion of this point.

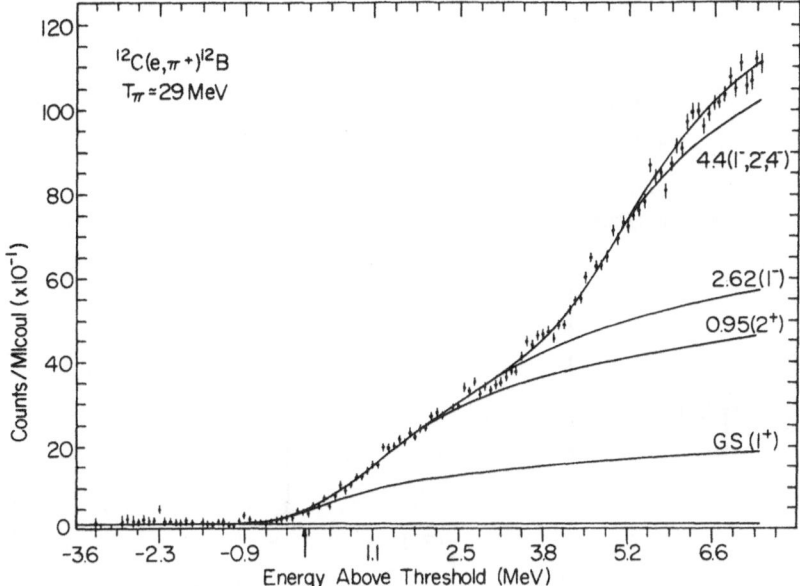

Figure 1:   Spectrum of positive pions of energy around 29
MeV, constructed as discussed by E. J. Winhold in these
Proceedings.   The solid curves are the results of a least
square fit.   The zero of energy is the limit of pions with
ideal zero overall resolution.

bars which represent a folding together of purely statistical
errors as well as those due to the uncertainty in pion endpoint channel.   In addition, systematic errors are estimated
to be about 10%.   The main sources of systematic errors are
detector solid angle and efficiency, and the ratio of virtual
to real photons.   Muon contamination was estimated by a Monte
Carlo procedure to be less than a few percent.   The energy
scale also indicates the energies of the ^{12}C analogue states.

Transitions to the ground state and 0.95 MeV levels in
^{12}B, which are analogues to the 15.11 and 16.1 MeV spin flip
$(1P^{3/2})^{-1}(1P^{1/2})$ states in ^{12}C, have previously been seen in
$(e,e'\ \pi^+)$ reactions at somewhat higher electron energy by the
Tohoku group[4].   A striking feature of the spectrum is a

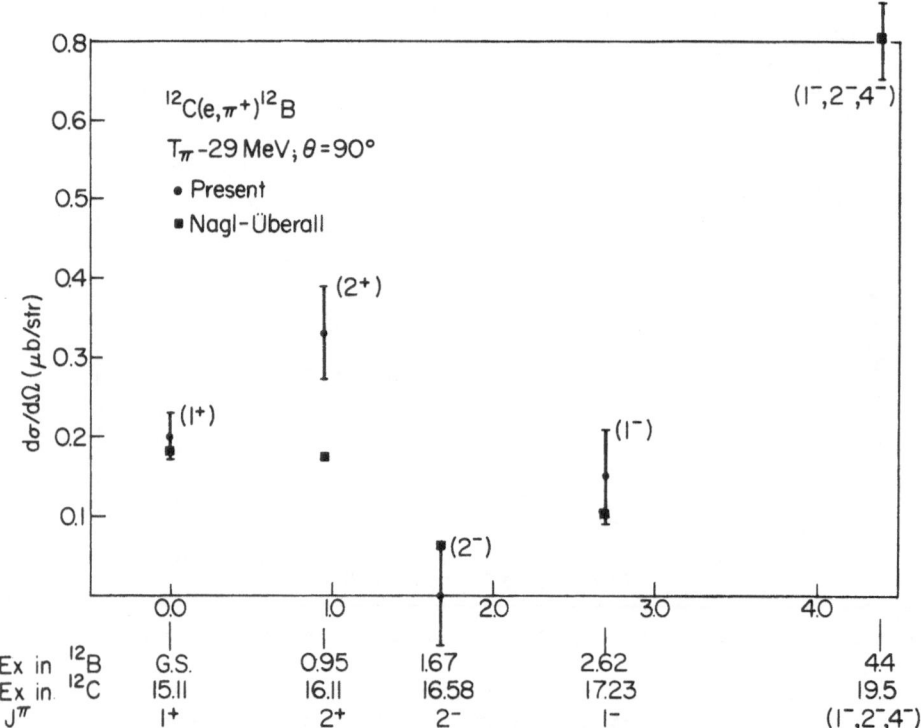

Figure 2. Extracted π^+ differential cross sections at 90°
to states in ^{12}B for pion kinetic energy of 29 MeV. The
black squares are calculations by Nagl and Uberall[5].

complex of strongly excited states beginning at about 4.5
MeV, which correspond to the negative parity complex at about
19.5 MeV in ^{12}C with dominant structure $(1P^{3/2})^{-1}(1d^{5/2})$.
This shows that these transitions have very strong spin-flip
components. Also shown in Figures 2-3 is the result of a
Helm model calculation by Nagl and Überall[5].

    In Figure 4 is shown the π^- spectrum to ^{12}N at a pion
energy around 17 MeV. The shape is almost identical with the
π^+ spectrum to the analogue states in ^{12}B as shown in Figure 5
In particular, we point out the strong spin flip-transition

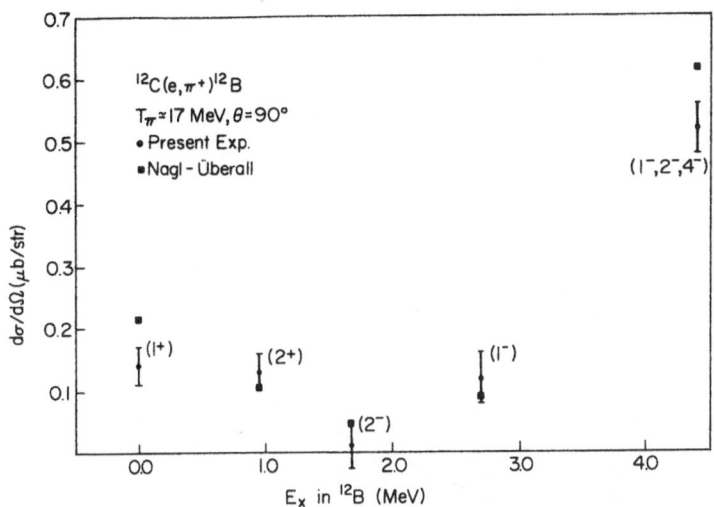

Figure 3: Same as Figure 2, except for pion energy of 17 MeV.

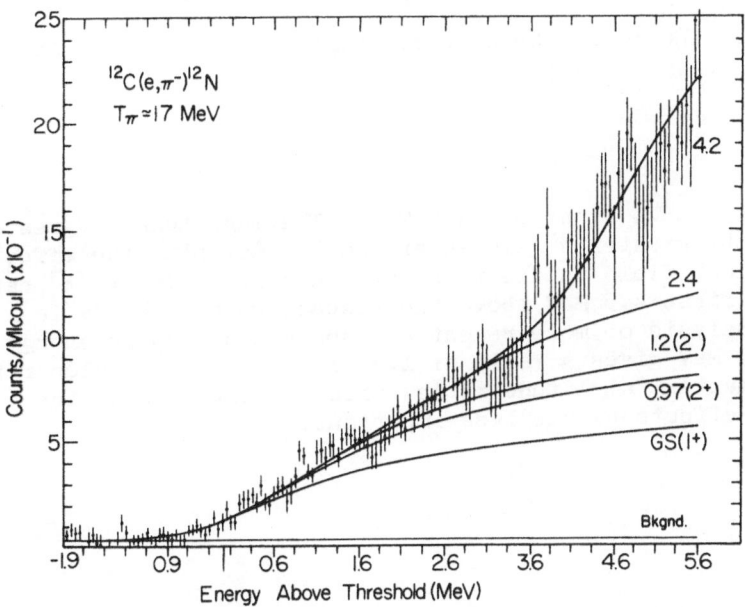

Figure 4: Spectrum of negative pions with energy of 17 MeV.

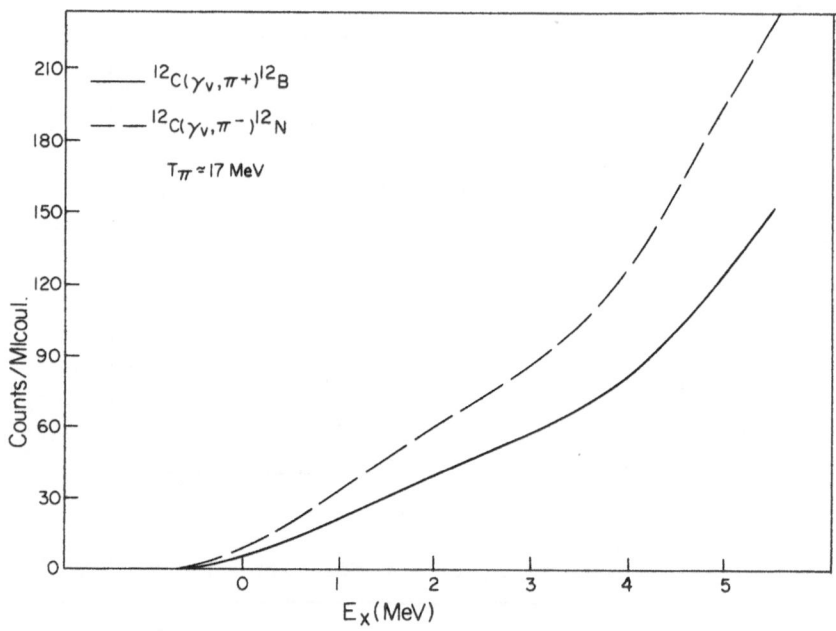

Figure 5:   Ratio of the best fit curves for π^- and π^+ spectra.

strength in ^{12}N at around 4.5 MeV. The magnitude is greater
due to the greater π^- elementary amplitude, plus the effect of
the Coulomb field on the pion wave function. The π^-/π^+ ratio
for the first 4.5 MeV above the respective thresholds is 1.55 ±
.08. Analysis of more recent spectra taken at pion energy a-
round 29 MeV gives a ratio of 1.37 ± .07. This is closer to
the elementary amplitude ratio, which is expected as the
Coulomb effects become less important.

REFERENCES

1. H. Überall, B. A. Lamars, C. W. Lucas and A. Nagl. Phys. Lett. 44B, 4 (1973) 324.
2. Yamaguchi, T. Terasawa, K. Nakahara and Y. Torizuka, Phys. Rev. C 3, 5 (1971) 1750.
3. R. H. Dalitz and D. R. Yennie, Phys. Rev. 105 (1957) 1598.
4. K. Shoda, H. Ohashi, and K. Nakahara, Phys. Rev. Letts., 39, 1131 (1977). Also, K. Shoda, these Proceedings.
5. A. Nagl and H. Überall, private communication.

NEUTRAL PION PHOTOPRODUCTION ON NUCLEI NEAR THRESHOLD

N. de Botton

Département de Physique Nucléaire CEN Saclay

B.P. 2, 91190 Gif-sur-Yvette, France

1. INTRODUCTION

Historically, the first π^0 photoproduction near threshold measurements on nuclei were performed twenty-five years ago. With almost the same experimental procedure we are using presently, data on nuclei ranging from deuterium to lead were collected until 1965. For the last ten years there has been no significant development in the field. The renewal of interest in the subject, attested by the present session of our meeting, comes primarily because of the advent of a new generation of linear electron accelerators providing significantly higher photon fluxes with a larger duty-cycle. This improved capability allows both better accuracy in the measurements and the exploration of the very near-threshold region which could not be reached by our predecessors because of the nearly vanishing cross sections.

In discussing the frame of interpretation of the experiments, I shall try to explain the essential motivations of this investigation and show that one can learn about basic pion photoproduction on nucleons, photoproduction mechanism in nuclei and nuclear matter distribution. The main difficulty will obviously be to disentangle the contribution of these different factors by studying a set of judiciously chosen nuclei. Next, I shall sketch the experimental procedure utilized and discuss the difficult problem of the absolute calibration of the measurements. I shall conclude by showing the data on deuterium taken at the Saclay linac which will illustrate these points. Preliminary data will be presented by my

colleagues of Bates on complex nuclei, and by Eric Vincent of
Saclay on ^{3}He and ^{4}He. This will give you an overall view of
the most recent developments in the field and an insight of
the possibilities and limitations of these experiments.

2. DISCUSSION OF THE INTERPRETATION FRAME

The theoretical framework used to study pion photoproduction is the distorted wave impulse approximation. The
effects of the various nucleons in the nucleus are assumed to
be additive and an effective elementary amplitude of the
process on the nucleon is used. The pion wave is distorted
by its interaction with the nucleus; this distortion is usually moderate because of the weakness of the pion nucleus interaction at low energy. To the extent that the pion momentum is small, the pion wave function will be constant inside
the nucleus and we will probe the transition form factor of
the nuclear initial and final states at momentum transfer m_π.

2.1 Neutral Pion Production on the Nucleons

The photoproduction amplitude on a free nucleon is calculated using the operator:

$$O_\pi = \vec{K} \cdot \vec{\sigma} + L$$

In the vicinity of threshold the quantities $\vec{K}$ and L can be
accounted for using the two multipolar amplitudes E_{o+} and
M_{1+} which are both real in this energy region. The dipole E_{o+}
is almost constant with energy whereas M_{1+} which dominates the
$\Delta(1236)$ resonance region varies like qk, the product of the
pion momentum by the photon momentum in the center of mass
system; O_π can be expressed in terms of E_{o+} and M_{1+}[1].

$$O_\pi = [E_{o+} \, \hat{\epsilon} + M_{1+} \, \hat{\epsilon}(\hat{q}\cdot\hat{k}) - M_{1+} \, \hat{k}(\hat{\epsilon}\cdot\hat{q})]\cdot\vec{\sigma} + 2M_{1+} \, \hat{q}\cdot(\hat{k}\times\hat{\epsilon})$$

$\hat{\epsilon}$ is the photon polarization.

The differential cross section on the nucleon reads:

$$\frac{d\sigma}{d\Omega} = \frac{q}{k} \{E_{o+}^2 + 2E_{o+} M_{1+} \cos\theta + M_{1+}^2 (\cos^2\theta + \frac{5}{2}\sin^2\theta)\},$$

and the integrated cross section

$$\sigma = 4\pi \frac{q}{k} [E_{o+}^2 + 2M_{1+}^2]$$

From the experimental values of the multipoles E_{o+} of Table 1 one can see that the proton and neutron photoproduction cross sections at threshold are comparable and two orders of magnitude smaller than the charged pion photoproduction ones. In the energy region we investigate, the nucleonic π^o cross sections are typically a few tenths of a microbarn.

Threshold photoproduction of π^o is a topic interesting on its own right. $E_{o+}(p\pi^o)$ and $E_{o+}(n\pi^o)$ are important parameters of s-wave pion physics; however, they are experimentally known with a very poor accuracy. Values of neutral pion channels E_{o+} can discriminate between theoretical models which agree on charged pion channels. The experimental value of $E_{o+}(\pi^o)$ for instance rules out the pseudo scalar Born approximation results (see Table 1).

The determination of the threshold value

$$E_{o+}(p\pi^o) = (-0.0018 \pm 0.0006)m_\pi^{-1}$$

is coming from the analysis of two measurements by Govorkov et al.[2] who investigated the energy dependence of the 90° center of mass cross section in the range 160-200 MeV using a conventional experimental set-up. To illustrate the poor accuracy of these measurements let us notice that this value proceeds from the combination of the two following parameter determinations extracted from the experiment: 0.06 ± 0.06 and 0.09 ± 0.08.

The Govorkov value agrees with the analysis by Mullensiefen of the Hitzeroth[3] experiment performed with nuclear emulsion to detect the recoil protons inside a hydrogen gas

TABLE 1

The experimental and theoretical values of the dipole pion
photoproduction amplitudes E_{o+} in units $10^{-3} m_\pi^{-1}$

Channel	Experiment[a]	Born PV[b]	Born PS[c]	Current[d] Algebra
$\gamma p \rightarrow n\pi^+$	28.6 ± 0.14	28.	28.1	29.0
$\gamma n \rightarrow p\pi^-$	-31.5 ± 1.0	-32.2	-31.7	-33.2
$\gamma p \rightarrow p\pi^o$	-1.8 ± 0.6	-2.5	-8.1	-2.4
$\gamma n \rightarrow n\pi^o$	0.3 ± 0.9[e]	0.4	5.6	0.3

a) from Ref. 7.
b) from Ref. 13
c) from Ref. 23
d) from Ref. 24
e) deduced from the three other channels values using relation (4).

target. It is also in good agreement with the multipole analysis in the $\Delta(1236)$ region by Noelle[4] which yields $E_{o+}(p\pi^o) =$
-0.0019.

Depending on the type of analysis the M_{1+} multipole
varies from $M_{1+}(\pi^o p) = 0.008 \ qk/m_\pi^2$ to $M_{1+}(\pi^o p) = 0.010 \ qk/m_\pi^2$
(see Ref. 2).

We notice that in contrast with charged pion production
where E_{o+} dominates even at 200 MeV photon energy, in the case
of neutral pion production the M_{1+} contribution to the total
cross section is equal to the E_{o+} one, 2 MeV above threshold.
From this constation, it follows that non spin-flip terms are
important very close to the threshold.

At this point, one must observe that no absolute measure-

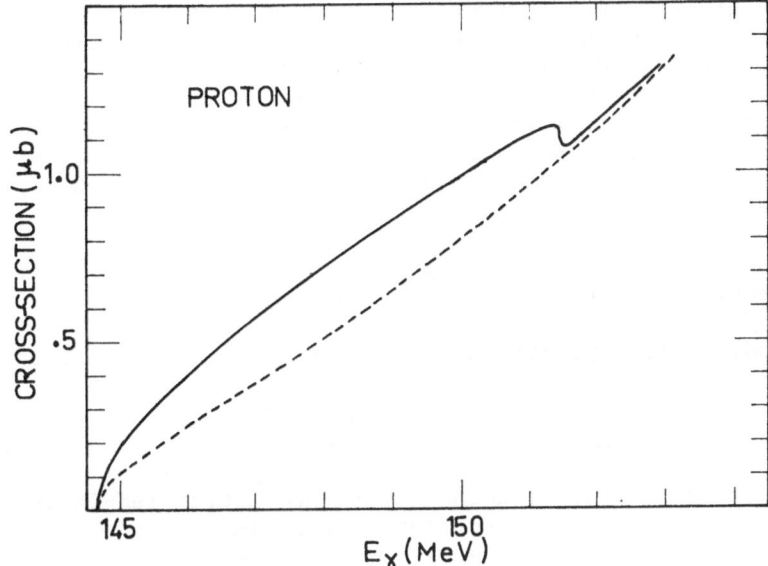

Figure 1. The proton π^o photoproduction cross section. The dotted line corresponds to the multipole amplitudes E_{o+} = 0.0025 and M_{1+} = 0.008 qk/m_π^2. The solid line includes the cusp effect produced by the coupling with the $\pi^+ n$ channel as predicted by J.M. Laget[6].

ment closer than 13 MeV from threshold has been performed so far on the reaction:

$$\gamma + p \rightarrow \pi^o + p; \quad E_{\gamma \text{ threshold}} = 144.67 \text{ MeV}. \tag{1}$$

This remark has an importance in view of the singularity which is expected in the cross section of Eq. 1 at the inception of the competing reaction

$$\gamma + p \rightarrow \pi^+ + n; \quad E_{\gamma \text{ threshold}} = 151.44 \text{ MeV}. \tag{2}$$

This effect has been theoretically investigated by Baldin et al.[5] who conclude to a maximum 20% addition to the multipole $E_{o+}(p\pi^o)$ value. A more recent calculation by Laget[6] confirms this estimation (see Fig. 1).

The effect would enhance the cross section close to threshold by as much as 50% as compared to the prediction based on the quoted values of the multipoles. Let us stress that the mentioned calculations based on the unitarity relations for the S matrix and a q dependence at threshold of the reaction (Eq. 2), do not deal with the dynamical properties of the pion nucleon system in an energy region where the isotopic multiplet mass differences are important.

The implications of this peculiar circumstance for our measurements will be discussed later.

Needless to say, direct experimental observation of reaction:

$$\gamma + n \rightarrow \pi^\circ + n \tag{3}$$

is impossible. However, using invariance properties in isospin space of strong interactions - isoscalar, and electromagnetic interactions - isoscaler + third component isovector - , multipolar amplitudes of the four photoproduction channels can be expressed in terms of three independent amplitudes. The knowledge of the three other channel multipoles allows determination of the reaction (Eq. 3) multipoles by the relation

$$\mu(n\pi^\circ) = \mu(p\pi^\circ) - [\mu(n\pi^+) + \mu(p\pi^-)]/\sqrt{2} \tag{4}$$

Unfortunately, the value[7]

$$E_{o+}(n\pi^\circ) = (0.0003 \pm 0.0009)m_\pi^{-1}$$

deduced using (Eq. 4) is affected by a large uncertainty which stems equally from the poorly known $E_{o+}(p\pi^\circ)$ and from the absolute uncertainty of $E_{o+}(p\pi^-)$ which is seven times larger than that of $E_o(n\pi^+)$ As for $M_{1+}(n\pi^\circ)$, all estimations agree and yield approximately $M_{1+}(n\pi^\circ) \simeq 0.9\ M_{1+}(p\pi^\circ)$.

It appeared to the Saclay group[8] that an improvement of our knowledge of the nucleonic amplitudes was possible by studying π° photoproduction very near threshold on the lightest nuclei: d and ^{3}He.

Indeed, at threshold, in the plane wave impulse approximation and assuming frozen nucleons, elastic π° photoproduction cross sections on these two nuclei are such that:

$$\sigma_d \sim |E_{o+}(p\pi^\circ) + E_{o+}(n\pi^\circ)|^2$$

$$\sigma_{3_{He}} \sim |E_{o+}(n\pi^\circ)|^2$$

Comparison with hydrogen would at least put some constraints on $E_{o+}(n\pi^\circ)$.

2.2   The Nuclear Process

Using the nucleonic amplitudes and assuming a plane wave pion the nuclear matrix element for a transition between nuclear states ψ_i and ψ_f will be

$$< \psi_f \ | \sum_{i=1}^{A} (\vec{K}\cdot\vec{\sigma}_i + L) \ e^{i(\vec{k}-\vec{q})\cdot\vec{r}_i} |\psi_i>$$

The spin independent term contributions of all nucleons will add coherently, whereas only a few nucleons can contribute in the spin dependent production. Except for light non zero spin nuclei, even in the vicinity of threshold, the coherent photoproduction will dominate. For instance, the elastic coherent π° photoproduction cross section will read:

$$\frac{d\sigma}{d\Omega} = \frac{q}{k} \frac{(1+m_\pi/M)^2}{(1+m_\pi/M_A)^2} \ 2|Z \ M_{1+}(\pi^\circ p)+N \ M_{1+}(\pi^\circ n)|^2 \ F^2(t) \ \sin^2\theta$$

$F(t)$ is the matter body form factor (form factor of the nucleons positions) at transfer $t = (k-q)^2$, since neutrons as well as protons contribute almost equally to the process. Neutral pion photoproduction on complex nuclei can then be considered as a tool for investigating the matter density of the nucleus and hopefully the neutron density by using the electron scattering charge density information.

So far we have neglected the pion nucleus interaction; we shall now try to understand how it will affect the results.

2.3   The Pion Nucleus Interaction

At low energy the pion nucleus interaction is very weak, especially for neutral pions, as shown by the scattering lengths $a(\pi,\pi)$ displayed in Table II: they are of the order of 0.1 fm: two orders of magnitude smaller than the nucleon nucleon ones.

The emitted π° can be described by a distorted wave calculation using an optical potential in a Schrodinger equation. However, since we have been using effective nucleonic amplitudes, the correct procedure consists in calculating the pion multiple scattering series, because the interaction of the pion with the nucleon on which it was produced is already contained in the photoproduction amplitude. This remark is particularly important in the case of very light nuclei (as witnessed by π^{+} photoproduction results on d and ^{3}He). Anyway, distortion of the π° wave inside the nucleus is expected to be moderate in the threshold region.

TABLE II

The pion nucleon scattering lengths in units 10^{-3} m_{π}^{-1}.

a) From Ref. 25.

Channel	Scattering Length
$\pi^{\circ} n \to \pi^{\circ} n$	$-$ 4. $\pm$ 3.
$\pi^{-} p \to \pi^{-} p$	83. $\pm$ 3.
$\pi^{-} n \to \pi^{-} n$	$-$ 92. $\pm$ 2.
$\pi^{-} p \to \pi^{\circ} n$	$-$124. $\pm$ 3.

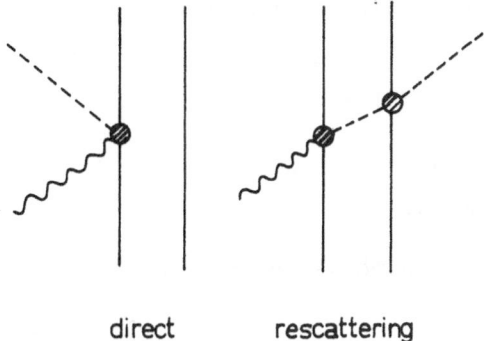

direct   rescattering

Figure 2. The direct and the rescattering processes in π°
photoproduction on deuterium.

         However, the coupling of the various nucleon π photopro-
duction and π scattering channels will induce a very special
feature, first pointed out by Koch and Woloshyn[9], for the E_{o+}
non coherent part of the amplitude. Because $E_{o+}(\pi^{\circ})$ are
lower than $E_{o+}(\pi^{\pm})$ by more than one order of magnitude, the
two step process charged pion photoproduction and virtual
charge exchange will compete with the direct process (see
Fig. 2). Since non coherent production can principally be
observed in very light nuclei, we will be sensitive to the
process, in the nuclei from which we were expecting to learn
about $E_{o+}(n\pi^{\circ})$.

         To evaluate, in first approximation, this effect at
threshold, we will use a very simple model[10] which gives
reasonable predictions of the $\pi^{-}d$ and $\pi^{-}{}^{3}$He scattering lengths
deduced from the experimental energy shifts of pionic d and
^{3}He. In this model, the nucleons are taken as fixed scatter-
ers, and the pion in the intermediate state considered as
free; the nuclear wave functions are pure S states. The rescattering amplitude

$$A_r = \langle\psi_f| \sum_{ij} a(\pi^{\mp},\pi^{\circ})(j) \frac{1}{|\vec{r}_i - \vec{r}_j|} e^{i\vec{k}\vec{r}_i} E_{o+}^{\pi^{\mp}}(i)|\psi_i\rangle$$

which describes s wave photoproduction on nucleon i and

charge exchange scattering on nucleon j, simplifies to

$$A_r = <\phi_f| \frac{1}{|\vec{r}_i - \vec{r}_j|} e^{i\vec{k}\cdot\vec{r}_i}|\phi_i><S_f\ T_f| \sum_{ij} a(j)E_{o+}(i)|S_i\ T_i>$$

where $\phi_{i,f}$ are radial wave functions, S and T spin and isospin of the nuclear states.

The first term in the A_r expression will be noted $<\frac{1}{r}>_{m_\pi}$.

For elastic π°, photoproduction on deuterium, using a Reid soft core deuteron s wave function we find

$$<\frac{1}{r}>_{m_\pi} = 0.52\ fm^{-1}\ (0.74\ m_\pi).$$

At this point we should note that the quantity $<\frac{1}{r}>_{m_\pi}$ is very dependent of the central density of the nucleus; the use of a non-realistic wave function is dangerous because it enhances the rescattering effect. For instance $<\frac{1}{r}>_{m_\pi}$ (Hulthen) 0.65 fm^{-1}. The total amplitude becomes

$$A = [E_{o+}(p\pi^\circ) + E_{o+}(n\pi^\circ)]$$

$$-(1+m_\pi/M) <\frac{1}{r}>_{m_\pi}\ a(\pi^-,\pi^\circ)[E_{o+}(p\pi^-)-E_{o+}(n\pi^+)]$$

$$= (-0.0016 \pm 0.0014) - (0.0063)$$

The rescattering amplitude is at least two times larger than the direct amplitude.

For ^{3}He elastic π° photoproduction, the rescattering term happens to have exactly the same expression as for deuterium. The quantity $<\frac{1}{r}>_{m_\pi}$ has been deduced starting from the value of the ^{3}He Coulomb energy for the S.S.C. Lavergne and Gignoux[11] wave function.

$$< \frac{1}{r} >_{m_\pi} = 0.44 \text{ fm}^{-1}.$$

The direct amplitude is $E_{o+}(n\pi^\circ) = 0.0003 \pm 0.0009$ and the rescattering one $A_r = -0.0054$.

Second order scattering can be neglected because it introduces an additional factor

$$a(\pi,\pi) \text{ x} < \frac{1}{r} >_{m_\pi} \text{ which is typically 1/10.}$$

More refined calculations on deuterium[12,18] and ^{3}He[14] taking into account the complete photoproduction elementary nucleonic amplitudes and the Fermi motion of the nucleons have been developed. However, these estimations still suffer from ambiguities and uncertainties related to the off shell behavior of the elementary amplitudes. Anyhow, the importance of the rescattering term makes of these reactions a very sensitive testing ground for the photoproduction mechanism description; this aspect is important for instance in view of the precise calculation required for the much smaller rescattering effects met in π^+ photoproduction[15]. In Figs. 3 and 4 the various predicted cross sections for d and ^{3}He are displayed.

Pion rescattering in light complex nuclear targets has been investigated by J. Vergados and R.M. Woloshyn[16] and found to dominate the coherent production very near threshold for non-zero spin targets; in the case of ^{4}He they find ~ 20% increase in the cross section at all energies from threshold up to 8 MeV above threshold.

3.   THE EXPERIMENTAL PROCEDURE

Experimental procedure has barely changed since the first measurements performed in the early fifties. The photon source is bremsstrahlung. The photon flux is monitored by a quantameter. Near threshold, the knowledge of the absolute energy scale is crucial; it can be established through the study of the π^+ photoproduction on the proton which yields the threshold

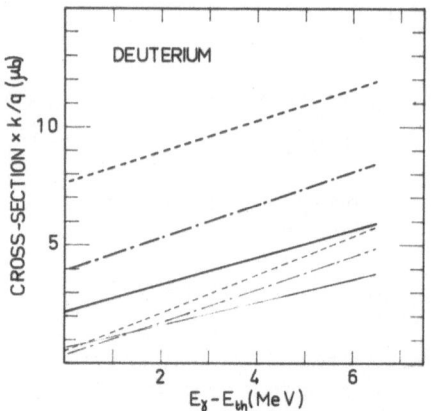

Figure 3. The π° photoproduction reduced cross section on deuterium as a function of the excess energy above threshold. The light lines represent the direct term contribution only; the heavy lines include the rescattering term. Solid lines correspond to Ref. 13, dashed-dotted lines to Ref. 12, and the dotted lines to the simple impulse approximation with frozen nucleons model discussed in the text. All estimates use approximately the same proton cross section (E_{0+} = -0.0025 M_{1+} = 0.008 qk/m_{π}^{2}).

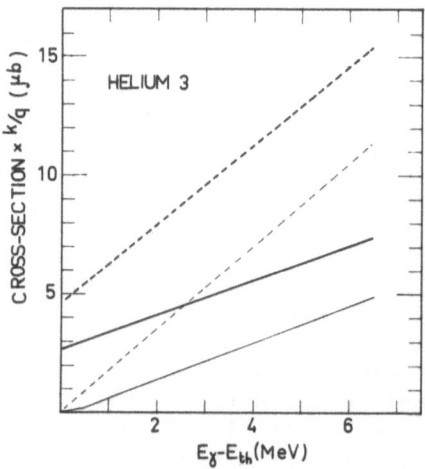

Figure 4. The π° photoproduction reduced cross section on ^{3}He labelled as Fig. 3. Solid lines correspond to Ref. 14.

energy of the process (151.44 MeV) with a $\pm$ 30 keV accuracy[18].

   The two gammas from the pion decay are converted and subsequently detected in two counter telescopes placed symmetri-
cally about the photon beam direction. The angle between the
two telescopes is varied with the π° energy to account for the
decay kinematics. The rotation of the plane of the two detectors, with respect to an axis perpendicular to the photon
beam allows the investigation of the π° angular distribution.
The angular resolution obtained in this way improves with increasing π° energy. The π° detection efficiency is usually
estimated by a Monte Carlo simulation taking into account
exact kinematics, detector geometry, π° decay gammas conver-
sion and propagation in the telescopes.

   Below-threshold measurements give the background level
related to electromagnetic reactions (large angle pairs,...).
Even for high Z nuclei, there is no sizable contribution to
the π° yield in the threshold region from the Primakoff effect[17] (production of a π° through the interaction of the photon with the Coulomb field of the nucleus).

   The π° yield

$$Y(E_e) = \int_{E_{th.}}^{E_e} \frac{d\sigma}{d\Omega} \, \epsilon(E_{\pi^\circ}, \theta) \, \frac{dN}{dk} \, (E, E_e) \, dE$$

is measured as a function fo the end point energy E_e of the
bremsstrahlung spectrum, and compared to "theoretical" yields
obtained by folding theoretical cross sections with the bremsstrahlung spectrum dN/dk and the Monte Carlo calculated efficiency ϵ.

   Absolute normalization of the yield is a major problem;
it requires the calibration of the π° detector and the knowledge of the normalized photon spectrum per unit quantameter.
Hydrogen, whose cross section is inaccurately known cannot play
the calibration role it had in π^+ threshold photoproduction
studies[18]

   There are some plans in Saclay in order to improve the
present situation.

We contemplate calibrating our detection system with the 2.9 MeV π° of stopped π^- charge exchange capture in hydrogen, using the Saclay Linac pion facility. The pion charge exchange branching ratio in hydrogen is deduced from the Panofsky ratio[19] to be 0.61 ± 0.01; the main uncertainty in the efficiency measurement will stem from the determination of the number of stopped pions in the target which can be estimated with an accuracy better than 10%.

Taking advantage of our monochromatic annihilation photon beam, we can consider even with the current operating conditions (50 nA average e^+ current corresponding to ~ 10^7 photons per sec. at 150 MeV), the measurement of the π° photoproduction on the proton, 3 MeV above threshold (expected cross section 0.5 μb) with an overall accuracy of 10%. A larger positron intensity would be desirable to bring into light, through the measurement of the forward differential cross section ($d\sigma/d\Omega$ (0°) ~ $|M_{1+} + E_{o+}|^2$), the predicted cusp behavior induced by the π^+ channel.

4.   EXPERIMENTAL DATA

4.1   Photoproduction of π° on Very Light Nuclei (d, ^{3}He, ^{4}He)

In 1953 in an early attempt to measure the ratio of the cross sections from deuterium and hydrogen at threshold, C. André[20] concluded that for ~ 10 MeV π°, σ_d/σ_p = 4.2 ± 1.3. There were no other data in the threshold region before the Saclay experiment[21]. In the latter experiment we studied the energy range up to 8 MeV above threshold. The π° detection system (two telescopes at 90° from the beam direction) was devised in order to be sensitive to the total cross section; the Monte Carlo calculated efficiency vs. π° energy is displayed in Fig. 5.

The measured yields for deuterium and hydrogen are shown in Fig. 6. They are compared to two theoretical estimations by J. Koch and R. Woloshyn[12] (KW) and by P. Bosted and J.M. Laget[13] (BL). Complete photoproduction operators and re-

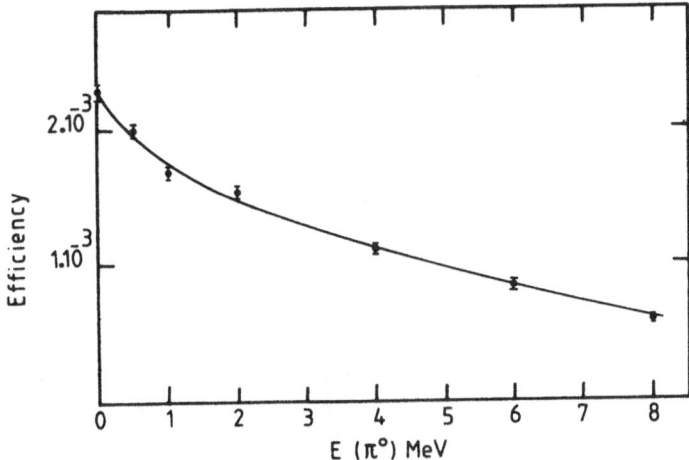

Figure 5. Efficiency of the detection telescope, used in the deuterium Saclay experiment[21], as a function of the π° energy.

scattering amplitudes are used in both papers which differ only by the choice of the deuterium wave function: BL use a parameterization of Reid soft core, whereas KW utilize a Hulthen wave function.

In order to produce "theoretical" yield curves, we have folded the hydrogen predicted cross section with the Monte Carlo calculated efficiency and the bremsstrahlung shape for the different values of the photon spectrum end point energy selected in the experiment. By fitting these values to the experimental hydrogen yields we deduced a normalization factor for the π° detection efficiency which was then used to construct the deuterium theoretical yield curve.

As explained in the discussion on rescattering, the agreement of the experimental data with KW theory is probably fortuitous since the Hulthen wave function enhances artificially the rescattering amplitude.

No definite conclusions other than the importance of the rescattering process can be drawn from the inspection of the only deuterium data. Preliminary results of recent experiments on ^{3}He and ^{4}He are presented by E. Vincent[26], to-

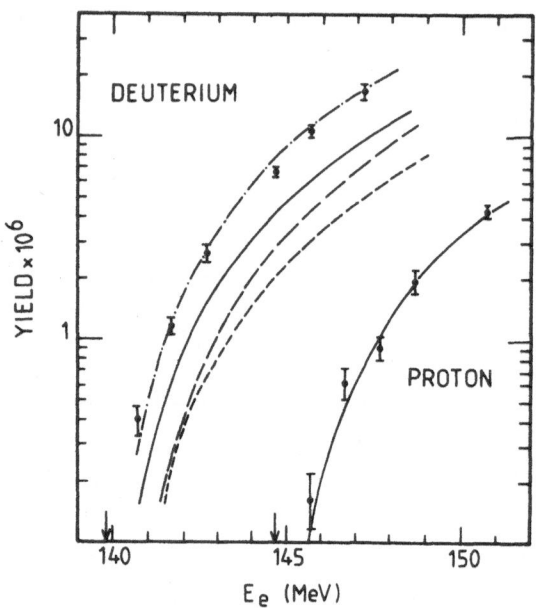

Figure 6.   The π° photoproduction on deuterium from Ref. 21.
The measured photoproduction yields as a function of the
bremsstrahlung end point energy E_e are compared to theoretical estimates for deuterium without rescattering (short dashes
for Ref. 12 and long dashes for Ref. 13) and include rescattering (dash-dots for Ref. 12, and solid line for Ref. 13).
These two theoretical estimates use the same proton cross
section; the corresponding yield has been adjusted to the
hydrogen data.   Arrows indicate the threshold energies.   The
yields are given in microbarn for equivalent quantum.

gether with a detailed yield curve on hydrogen.   The complete
set of data on the four nuclei has not yet been analyzed as
a whole.   A first attempt is to try to understand the four
nuclei in the frame of an impulse approximation approach including the very simple model of rescattering developed in
2.3, and leaving the nucleon multipoles as free parameters to
be adjusted to the data.

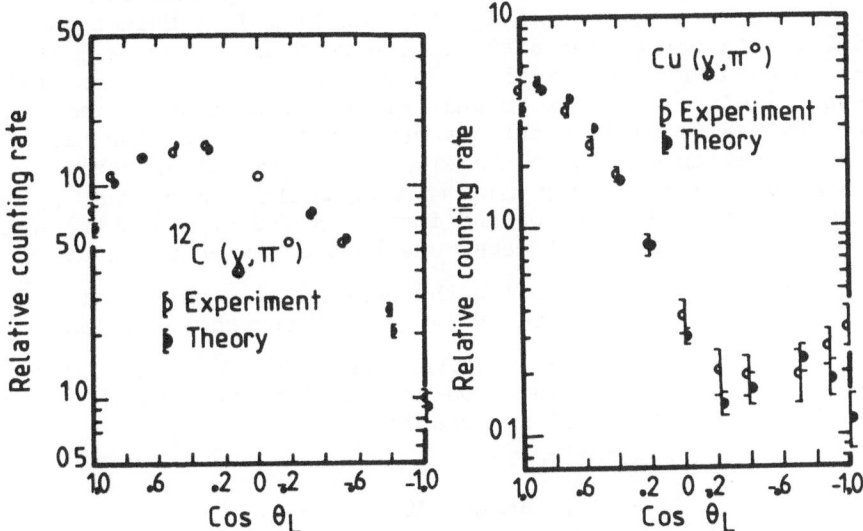

Figure 7. The π° photoproduction on carbon and copper from Ref. 22. Experimental data and best shape fit of the data by a Monte Carlo synthesis. The ordinate is the relative yield and the abscissa the cosine of the colatitude angle of the counter system with respect to the incident photon beam. The end point energy of the bremsstrahlung spectrum was 166 MeV.

4.2   Coherent π° Photoproduction

Extensive measurements of π° photoproduction differential cross section on various complex nuclei targets were performed by Schrack et al.[22] at 166 MeV and by Govorkov et al.[2] at 180 MeV bremsstrahlung end-point energy. A generally good agreement with coherent elastic photoproduction theory is observed for nuclei A < 40 (see Fig. 7). Mean-square radii of nuclear matter distribution were extracted from the data in good agreement with those obtained from other experiments. Inelastic processes and failure of Born approximation were invoked to explain the discrepancies observed for backward production on high A nuclei. More detailed investigation of the nucleonic density distribution would require taking into account pion distortion and improving the experimental accuracy. In their recent measurements undertaken at Bates, J.L. Milder et al.[27] minimize the non-elastic contributions by choosing to investigate the very near-threshold region.

When confronted with the Vergados and Woloshyn[16] calculations which treat completely the pion-nucleus interaction, their results can bring useful information.

REFERENCES

1. G.F. Chew, M.L. Goldberger, F.E. Low and Y. Nambu, Phys. Rev. 106, 1345 (1957).
2. B.B. Govorkov, S.P. Denisov and E.V. Minarik, Proc. (Trudy) of the P.N. Lebedev, Physics Institute 54, 1 (1974).
3. W. Hitzeroth, Nuovo Cimento LXA, 467 (1969).
4. P. Noelle, W. Pfeil and D. Schwela, Nucl. Phys. B26, 461 (1971) and B31, 1 (1971).
5. A.M. Baldin, B.B. Govorkov, S.P. Denisov and A.I. Lebedev, Soviet Journal of Nucl. Phys. 1, 62 (1965).
6. J-M. Laget, Private Communication.
7. M.I. Adamovich, Proceedings (Trudy) of the P.N. Lebedev, Physics Institute 71, 119 (1976).
8. Saclay group: P. Argan, G. Audit, A. Bloch, N. de Botton, J-C. Faure, C. Schuhl, G. Tamas, C. Tzara, E. Vincent (CEN Saclay) and J. Deutsch, D. Favart, R. Prieels and B. Van Oystaeyen (Louvain-la-Neuve).

9.   J.H. Koch and R.M. Woloshyn, Phys. Lett. 60B, 221 (1976).
10. N. de Botton and C. Tzara, Rapport interne DPh-N/HE/ 78/06.
11. A. Laverne and C. Gignoux, Nucl. Phys. A203, 597 (1973).
12. J.H. Koch and R.M. Woloshyn, Phys. Rev. C16, 1968 (1977).
13. P. Bosted and J-M. Laget, Nucl. Phys. A296, 413 (1978).
14. P. Bosted and J-M. Laget, Nucl. Phys. (to be published).
15. P. Argan et al., Rapport interne DPh-N/HE/78/05 and to be published.
16. J.D. Vergados and R.M. Woloshyn, Phys. Rev. C16, 292 (1977).
17. H. Primakoff, Phys. Rev. 81, 899 (1951).
18. G. Audit et al., Phys. Rev. C15, 1415 (1977).
19. V.T. Cocconi et al., Nuovo Cimento 22, 494 (1961).
20. C.G. Andre, UCRL-2425 (1953).
21. P. Argan et al. (to be published).
22. R.A. Schrack, J.E. Leiss and S. Penner, Phys. Rev. 127, 1772 (1962) and Phys. Rev. 140B, 897 (1965).
23. P. de Baenst, Nucl. Phys. B24, (1970) 633.
24. R.D. Peccei, Phys. Rev. 181, 1902 (1969).
25. D.V. Bugg, A.A. Carter and J.R. Carter, Phys. Lett. 44B, 278 (1973).
26. P. Argan et al. (contributed paper to this meeting).
27. J.L. Milder et al. (contributed paper to this meeting).

A MEASUREMENT OF π° PHOTOPRODUCTION ON ^{3}He AND ^{4}He NEAR THRESHOLD

P. Argan, G. Audit, A. Bloch, N. de Botton,

J-L. Faure, C. Schuhl, G. Tamas, C. Tzara and

E. Vincent

DPh-N/HE, CEN Saclay, BP 2, 91190 Gif-sur-Yvette,

France

The study of π° photoproduction from light nuclei is a way to obtain information about the $\gamma p \to p\pi^\circ$ and $\gamma n \to n\pi^\circ$ elementary amplitudes, although near threshold the production of a virtual $\pi^\pm$ and its scattering with charge exchange on another nucleon is expected to be of the same order of magnitude as the direct process.

We have measured the π° photoproduction yields on deuterium, helium-3 and helium-4 in comparison with hydrogen, at energies ranging from around 1 to 8 MeV above threshold. The photon source was a Bremsstrahlung beam derived from the electron beam of the Saclay Linear Accelerator (ALS). The photon intensity was measured by a Wilson type quantameter. We used cryogenic targets filled with liquid gas, and an empty target for control measurements; the walls contribution appeared to be negligible. The photons from π° decay emitted almost back to back were converted in two 0.6 cm thick lead foils placed at 90° on each side of the target. Each electron-positron pair was detected in an array of two lucite detectors (4 cm thick) sandwiching one plastic scintillator (0.5 cm thick). An event was defined by a large coincidence of the six counters; the timing and the pulse height of each individual signal being recorded, we varied off-line the coincidence width to determine the contribution of the accidentals in the time window retained for the good events. In the worst case, the true to accidental ratio was 4. We also made cuts in the amplitudes spectra of the four Cerenkov counters, which give a rough estimate of the energy of the detected particles. The background measured below threshold was assumed constant in the small energy range explored and substracted from the yields above threshold.

The experimental yields, normalized to one target nucleus and to the photon intensity, are listed in Table 1 and presented in Figs. 1a - 1b in the case of H, ^{3}He, ^{4}He (the deuterium results[1] are discussed in another contributed paper). They are related to the π^0 photoproduction differential cross-section $d\sigma/d\Omega$ by:

$$A(E_e) = \int_{E_o}^{E_e} \int_{\Omega_d} B(E,E_e) \, \varepsilon(E,\Omega) \, \frac{d\sigma}{d\Omega} \, d\Omega \, dE$$

$B(E,E_e)$ is the photon spectrum for an incident electron energy E_e, $\varepsilon(E,\Omega)$ is the detection efficiency, E_o is the threshold energy and Ω_d is the solid angle viewed by the detection system.

In order to compare these results with theoretical cross-sections, we have computed the efficiency $\varepsilon(E,\Omega)$ by a Monte Carlo method, taking in account the exact geometry and a simple model for the pair production and its propagation in the counters. So $\varepsilon(E,\Omega)$ is known up to a constant factor only depending on the experimental set-up, which has been slightly modified between our D and ^{3}He-^{4}He measurements. We found $\varepsilon(E,\Omega)$ ranging from around 2×10^{-3} to 1×10^{-3} for 1 to 8 MeV above threshold; it is very little dependent on the π^0 emission angle, so that the detection system is essentially sensitive to the total cross-section only.

The ^{3}He results cannot be reproduced by an impulse approximation calculation; they show the importance at threshold of virtual charged π production and charge exchange rescattering. The ^{4}He results are in good agreement with impulse approximation; in this case, there is no spin-flip production and the rescattering is small. Comparing the ^{4}He data with H should

Table 1. Experimental yields $A(E_e) = \int_{E_o}^{E_e} \int_{\Omega_d} B(E,E_e)\,\varepsilon(E,\Omega)\,\dfrac{d\sigma}{d\Omega}\,d\Omega\,dE$ in microbarns, normalized to one target nucleus and one equivalent quantum, for different values of the Bremsstrahlung end-point energy E_e above threshold E_o.

HYDROGEN		HELIUM-3		HELIUM-4	
$E_e - E_o$ (MeV)	$10^6 \cdot A_H$	$E_e - E_o$ (MeV)	$10^6 \cdot A_{^3He}$	$E_e - E_o$ (MeV)	$10^6 \cdot A_{^4He}$
1.41	0.14 ± 0.05	1.87	0.61 ± 0.19	1.72	0.51 ± 0.20
1.91	0.14 ± 0.06	2.87	1.88 ± 0.24	2.72	1.59 ± 0.24
2.41	0.39 ± 0.05	3.87	2.94 ± 0.28	3.72	3.25 ± 0.42
2.91	0.40 ± 0.06	4.87	6.08 ± 0.47	5.22	8.73 ± 0.68
3.41	0.65 ± 0.07	5.87	9.86 ± 0.83	6.22	12.4 ± 1.0
3.91	0.85 ± 0.10	9.37	23.7 ± 1.2	8.72	29.3 ± 1.8
4.41	1.39 ± 0.17	9.87	24.4 ± 1.8		
4.91	1.60 ± 0.15				
5.41	2.00 ± 0.16				
6.41	3.15 ± 0.25				
7.91	4.40 ± 0.28				

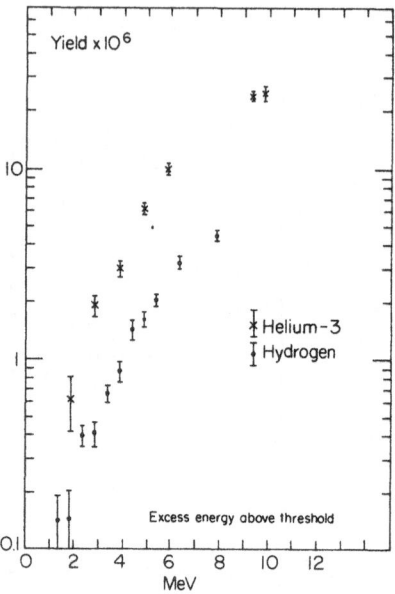

Figure 1(a).   Hydrogen and Helium-3 experimental yields.

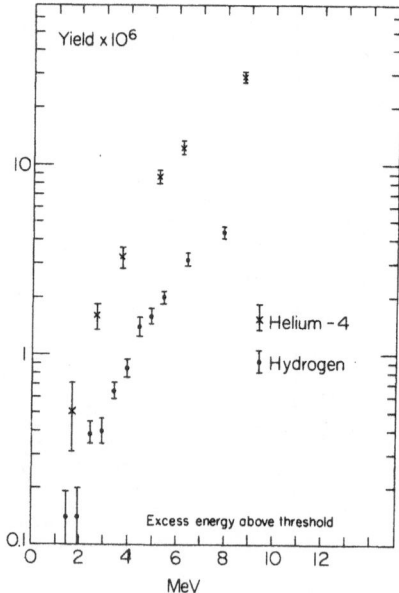

Figure 1(b).   Hydrogen and Helium-4 experimental yields.

allow us to determine the $\dfrac{M_{1+}}{E_{0+}}$ ratio of the most influential
multipoles at threshold. Information about the multipoles
and the reaction mechanisms will be obtained in the general
analysis of all our data, which is still in progress.

REFERENCES

1. P. Argan et al., to be published.
   W. de Botton, previous paper in these Proceedings.

NEUTRAL PION PHOTOPRODUCTION FROM COMPLEX NUCLEI NEAR THRESHOLD*

F.L. Milder, E.C. Booth, and B.L. Roberts

Boston University

J. Comuzzi

Massachusetts Institute of Technology

H. Crannell

Catholic University of America

Preliminary data are presented for total cross section measurements of π° photoproduction from ^{6}Li, ^{7}Li, ^{12}C, ^{28}Si, and ^{40}Ca. The data were taken utilizing the bremsstrahlung spectrum from the Bates Linear Accelerator at electron energies of 140 and 145 MeV. Below threshold background measurements at an endpoint energy of 134 MeV were also obtained.

The experimental arrangement consisted of targets with a thickness of approximately 1 gm/cm^2 centered in the photon beam between eight detectors. Four detectors (15 x 15 cm) were on each side of the target at a distance of 15 cm, covering one-third of the total solid angle. Additional solid angle and angular information were obtained by placing the target upstream, midstream, and downstream of the detector array. Each detector consisted of a pair of wave-shifted plastic Cherenkov counters, 3 cm thick, a one radiation length lead photon converter, and 10 gm/cm^2 of carbon absorber to suppress charged particles. Twofold coincidence events from

* Work supported in part by the National Science Foundation.

all possible detector pairs were stored in a PDP-11 computer with pulse height and timing information retained. A selected set of detector pairs was also stored on visual scalers. Table I presents the total counts per equivalent quanta per nucleus from the visual scalers. This total is approximately proportional to the total yield per equivalent quanta given by

$$Y(E_o) = \int_{E_T}^{E_o} \Phi(E_o,E)\sigma(E)dE / \int_{o}^{E_o} \Phi(E_o,E)E/E_o \; dE.$$

$\Phi(E_o,E)$ is the bremsstrahlung flux, E_T and E_o are the threshold and beam endpoint energies, and $\sigma(E)$ is the total π^o production cross section.

We have not yet calubrated the detection efficiency so that Table I represents only relative measurements. A liquid hydrogen target is nearly ready for operation and will be used in subsequent measurements for the efficiency calibration. Retrieval of the additional information stored in the computer will improve our statistical accuracy and should provide a rough picture of the angular distribution.

TABLE I.   YIELDS PER EQUIVALENT QUANTA PER NUCLEUS (ARBITRARY UNITS) FOR BREMSSTRAHLUNG ENDPOINT ENERGY E_o.

E_o (MeV)	134	140	145
^{6}Li	.54 ± .32	1.7 ± .21	14.6 ± 1.3
^{7}Li	----	2.2 ± .42	----
^{12}C	-.40 ± .94	7.21± .73	76.8 ± 2.7
^{28}Si	----	23.0 ± 12.7	----
^{40}Ca	3.0 ± 6.0	48.5 ± 6.3	220.3 ± 14.3

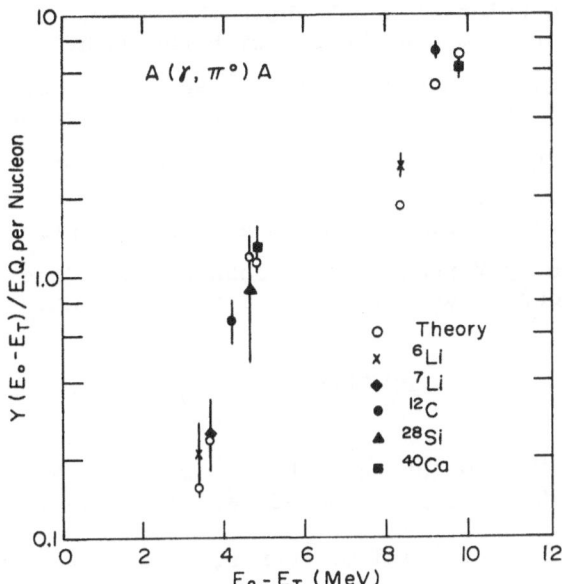

Figure 1. Yields per equivalent quanta per nucleon (arbitrary units) compared with theory.

Figure 1 presents the observed counts per equivalent quanta per nucleon for the various targets with the below threshold backgrounds subtracted. For comparison, we have calculated the p-wave coherent production cross section based on the theory of Chew, Goldberger, Low and Nambu[1] (CGLN). The cross section (for spin zero nuclei) is approximated by

$$\frac{d\sigma}{d\Omega} = KA^2Ep^3\sin^2\theta F^2(q),$$

where K is a constant at low q depending on the fundamental amplitude for π° production from the nucleon, E is the photon energy, p is the pion momentum, q is the momentum transfer to the nucleus, A is the target nucleus mass number, and F(q) is the nuclear form factor, taken to be identical to the electron scattering form factor. The open circle points in Fig. 1 (no error bars) are the relative yields obtained by folding the theoretical bremsstrahlung flux together with the integrated CGLN cross sections. The theory is arbitrarily normal-

ized to the ^{12}C point at 4 MeV above threshold. The overall
result is a fairly good agreement with the data. This indicates an $A^2F^2(q)$ dependence of the cross section and predominantly coherent production. Similar results have been reported at higher energies by Shrack, Leiss and Penner[2] (170 MeV)
and by Govorkov, Denisov and Minarik[3] (180 and 200 MeV). Our
results will also be compared with a more realistic theoretical calculation by Vergados and Woloshyn[4] which includes rescattering processes and final state interactions.

REFERENCES

1. G.F. Chew, M.L. Goldberger, F.E. Low and Y. Nambu, Phys.
    Rev. 106 (1957) 1345; L. Saunders, Ph.D. Thesis, M.I.T.
    (1967).
2. R. Shrack, J. Leiss and S. Penner, Phys. Rev. 127 (1962),
    1772.
3. B. Govorkov, S. Denisov and E. Minarik, Akad. Nauk, S.S.R.
    Lebedev Ins. Proc. 34 (1968) 1.
4. J.D. Vergados and R.M. Woloshyn, Phys. Rev. C16 (1977)
    292.

THE ^{6}Li$(\gamma,\pi°d)^4$He REACTION

R. Sridhar, K. Srinivasa Rao and S. Susila

MATSCIENCE, The Institute of Mathematical

Sciences, Madras, India

We propose two models for the study of the cluster knock-out reaction ^{6}Li$(\gamma,\pi°d)^4$He, with coincidence between the outgoing pion and the deuteron. One of these is based on the familiar cluster model wave function for ^{6}Li and the other is based on the pole-approximation model.

WAVE FUNCTION FORMULATION

We assume a cluster model wave function for ^{6}Li in the notation of Tang et al.[1]

$$\psi = N_o \exp[-\frac{\alpha}{2}\sum_{i=1}^{4}\rho_i^2 - \frac{\alpha}{2}\sum_{j=5}^{6}\rho_j^2 - \frac{2}{3}\beta R^2]\, Y_M^1(\hat{R})\xi_{\sigma\tau} \tag{1}$$

where N_o is the normalization factor, $\xi_{\sigma\tau}$ is the appropriate spin-isospin function. In the laboratory frame, the initial state is expressed as:

$$\Psi_i = e^{i\vec{v}\cdot\vec{r}_j}\,\Phi_\alpha\,\Psi(\vec{r})\,\Psi(\vec{R})\,\xi_{\sigma\tau} \tag{2}$$

where Φ_α and $\Psi(\vec{r})$ are the internal wave functions of the α and the two-nucleon system with $\vec{r} = 2\vec{\rho}_5$ and $\Psi(\vec{R})$ is the inter-cluster wave function. The final state is given by:

$$\Psi_f = e^{i\vec{\mu}\cdot\vec{r}_j} \Phi_\alpha e^{i\vec{k}_d\cdot\vec{R}_d} e^{i\vec{k}_\alpha\cdot\vec{R}_\alpha} \Psi_D(\vec{r}) \xi_{\sigma\tau} \tag{3}$$

where $\Psi_D(\vec{r})$ is the deuteron wave function.

The transition matrix element is given by:

$$T_{fi} = \langle \Psi_f | \sum_{j=5}^{6} t_j | \Psi_i \rangle \tag{4}$$

where the single-nucleon amplitude is as in Ref. 2. Transforming to the relative $(\vec{R})$ and c.m.$(\vec{Z})$ coordinates of the α and d cluster, (4) becomes:

$$T_{fi} = \pi^3 \langle \xi_{\sigma\tau} | \sum_j t_j | \xi_{\sigma\tau} \rangle \int e^{\frac{i}{2}\vec{k}\cdot\vec{r}} \Psi_D^*(\vec{r}) \, \psi(\vec{r}) d\vec{r}$$

$$x \; \delta(\vec{k}-\vec{k}_d-\vec{k}_\alpha) \int e^{-i(\vec{k}-\vec{k}_d)\cdot\vec{R}} \Psi(\vec{R}) \, d\vec{R} \tag{5}$$

where $\vec{k} - \vec{\vartheta} - \vec{\mu}$, is the momentum transfer, $\vec{\vartheta}$ and $\vec{\mu}$ being the photon and pion momenta, respectively.

For the free $\gamma d \rightarrow d\pi^\circ$ case, the transition matrix element is given by:

$$T'_{fi} = \langle \Psi'_f | \sum_{j=1}^{2} t_j | \Psi'_i \rangle \tag{6}$$

where $\Psi'_i = e^{i\vec{\vartheta}'\cdot\vec{r}_j} \Psi_D(\vec{r}) \xi'_{\sigma\tau}$, $\Psi'_f = e^{i\vec{\mu}'\cdot\vec{r}_j} e^{i\vec{k}'\cdot\vec{R}_d} \Psi_D(\vec{r})\xi'_{\sigma\tau}$

After integrating over the deuteron c.m. coordinate, we get:

$$T'_{fi} = (2\pi)^3 \delta(\vec{\vartheta}'-\vec{\mu}'-\vec{k}')\langle\xi'_{\sigma\tau}| \sum_j t_j |\xi'_{\sigma\tau}\rangle \int \Psi_D^2 (\vec{r})e^{\frac{i}{2}\vec{k}'\cdot\vec{r}}d\vec{r} \tag{7}$$

Replacing the pion photoproduction matrix element in (5), by the elastic neutral pion photoproduction matrix element for the free deuteron in (7), the cross section for the $^6Li(\gamma,\pi^\circ d)^4He$ reaction can be shown to be:

$$\frac{d^5\sigma}{d\Omega_d\ d\Omega_\pi\ d\mu_o} = \frac{m_d}{8}\ \frac{\mu k_d}{\mu'}\ (\frac{d\sigma}{d\Omega_\pi})_{free}\ |P_D\int e^{i(\vec{k}'-\vec{k}_d')\cdot\vec{R}}\ \psi(\vec{R})\ d\vec{R}|^2 \qquad (8)$$

where

$$P_D^2 = |\int e^{\frac{i}{2}\vec{k}\cdot\vec{\tau}}\ \psi_D^*(\vec{r})\psi(\vec{r})d\vec{\tau}/\ e^{\frac{i}{2}\vec{k}\cdot\vec{\tau}}\ \psi_D^2(\vec{r})d\vec{\tau}|^2 \qquad (9)$$

is the probability of finding the deuteron with ^{6}Li.

Thus, in this wave function formulation, the cross section for ^{6}Li(γ, π°d)^{4}He is related to the cross section for γd $\rightarrow$ dπ° reaction, the momentum distribution function of the two-nucleon system to be knocked out as the deuteron and the probability of finding the deuteron within ^{6}Li.

POLE MODEL

The amplitude associated with the simple deuteron pole diagram (Fig. 1) is

$$M_{m_d,m_6}^{(d)} = \sum_{m_d'} M_{m_d,m_d'}^{(\gamma)}\ M_{m_d',m_6}^{(6)} / (\frac{q^2}{2m_d} - \varepsilon - i0) \qquad (10)$$

where m_d m_d' and m_6 are the spin components of the outgoing deuteron, the deuteron pole and ^{6}Li, respectively. In Eq. (10), $M_{m_d',m_6}^{(6)}$ is the amplitude for the virtual decay ^{6}Li $\rightarrow$ α+d, and if it is assumed that the main contribution to the ^{6}Li-α-d vertex amplitude arises from the S-wave component of α-d relative motion, we have[3]:

$$M_{m_d',m_6}^{(6)} = \frac{g_6}{\sqrt{4\pi}}\ \delta_{m_d',m_6} \qquad (11)$$

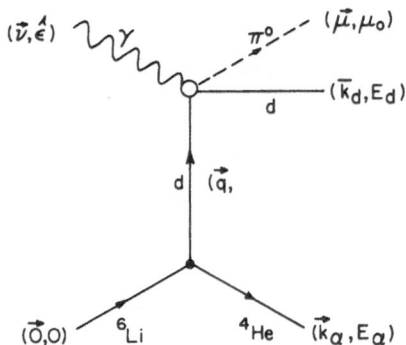

Figure 1. The deuteron pole amplitude. Momenta and kinetic energies of the particles are shown. $\hat{\epsilon}$ is the photon polarization and $\vec{\nu}$ its momentum.

where g_6 is the Li- -d vertex constant[4]. The amplitude $M^{(\gamma)}_{m_d,m'_d}$, for the deuteron on the mass shell and for a certain region of kinematic variables (e.g., as in the case of $\gamma d \to pn$, Ref. 5), is expected to reduce the free $\gamma d \to d\pi^0$ amplitude.

The cross section for the $^6Li(\gamma,\pi^0 d)^4He$ reaction, after summing over final spin states and averaging over the initial spin states and the incident photon polarization is:

$$\frac{d^6\sigma}{d^3k_d d^3\mu} = \frac{(2\pi)^{-5}}{6} \sum_{m_d,m_6,\hat{\epsilon}} |M^{(d)}_{m_d,m_6}|^2 \, \delta(\mu_0 + E_\alpha + E_d + B_2 - \nu_0) \quad (12)$$

where the delta function represents the total energy conservation and $B_2 = M_d + M_\alpha - M_6 = 1.47$ MeV, the binding energy of $\alpha + d$ into ^6Li. From Ref. 2, we have

$$\left(\frac{d\sigma}{d\Omega_\pi}\right)_{free} = (2\pi)^{-2} \frac{\mu\mu_0}{6} \sum_{m_d,m_\alpha,\,\hat{\epsilon}} |M^{(\gamma)}_{m_d,m_d'}|^2 \quad (13)$$

Substituting Eqs. (10) and (11) into (12) and using (13), the cross section for ^{6}Li(γ,π°d)^{4}He is:

$$\frac{d^5\sigma}{d\Omega_d \, d\Omega_\rho \, d\mu_o} = K \, P(\vec{q}) \left(\frac{d\sigma}{d\Omega_\pi} \right)_{free} \tag{14}$$

where $P(\vec{q}) = |g_6|^2/4\pi B_2^2 \, (1+q^2/\alpha^2)^2$, $\alpha^2 = 2\mu_{\alpha d}B_2$, $\mu_{\alpha d}$ being

the reduced mass of the α and d, and $K -k_d E_d/(2\pi)^3$ is the

kinematical factor.

Numerical results based on these models are to be obtained and these will be reported elsewhere.

ACKNOWLEDGEMENT

The authors thank Professor Alladi Ramakrishnan for his interest, and one of us (S.S.) acknowledges with thanks the award of a U.C.G. Teacher Fellowship.

REFERENCES

1. Y.C. Tan, K. Wildermuth and L.D. Pearlstein, Phys. Rev. 123 (1961) 548.
2. K. Ananthanarayanan and K. Srinivasa Rao, Nuo. Cim. 44 (1966) 31.
3. A. Ghovanlou and F. Prats, Phys. Rev. C10 (1974) 1309.
4. T.K. Lim, Phys. Letts. 56B (1975) 321.
5. E.G. Mititel and S.F. Timashev, JETP Letts. 7 (1968) 215.

CORRECTIONS TO THE QUASI-FREE PROCESS

J. M. Laget

DPh-N/HE, CEN Saclay, BP 2, 91190 Gif-sur-

Yvette, France[+]

Laboratory of Nuclear Science

Massachusetts Institute of Technology

Cambridge, Massachusetts   02139

ABSTRACT

At intermediate energy the photon-nucleus interaction is dominated by the quasi-free process: the photon interacts with one nucleon moving in a mean potential. The other nucleons are spectator in the sense that their overall effect is to create a mean optical potential in which the outgoing particles move (Distorted Wave Impulse Approximation).

The limits of the validity of this process will be discussed on the basis of the results of the analysis of the two following kinds of reactions:

1.   The $A(\gamma, \pi)B$ reactions induced by a monochromatic photon beam. Here the spectrum of pions emitted at a given angle is measured.

2.   The $A(\gamma, p\pi)B$ reactions induced by a bremsstrahlung beam. Here the proton and the pion are detected in coincidence.

The dominance of this mechanism decreases quickly when the momentum of the spectator system increases, and strong

deviations appear.   They are understood in terms of the
onset of two nucleon mechanisms.   Two of them are of dominating importance: the Δ-N interaction and the meson
exchange corrections to the γN → Nπ vertex.   We shall put
the emphasis on their consequences on the yield of pion
photoproduction reactions induced on few body systems.

        At intermediate energy the photon nucleus interaction
is dominated by the quasi-free process.   The photon interacts with one nucleon moving in a mean potential.   The
other nucleons are spectator, in the sense that their overall
effects are to create a mean optical potential in which the
outgoing particles move (Distorted Wave Impulse Approximation (DWIA).

        In the Δ(1236) region, this mechanism accounts for
almost all of the total photoabsorption cross section on
light nuclei.   When the Fermi motion and the Pauli
exclusion principle are taken into account, the measured
total cross section for absorption of a photon on the
Deuterium[1] or the Li and Be isotopes[2] are well reproduced[1,3].

        This mechanism is rather simple and we are interested
in going beyond this one nucleon background, from which
nothing new can be learned of the elementary amplitude and
the low momentum part of the nucleon wave function.

        The deviations from the quasi-free process appear when
the momentum transfer becomes high and it must be realized
that they represent only a few percent of the total photoabsorption cross section.   However, their study is very
important, because they can be understood as the onset of
the two nucleon mechanisms.   I shall devote the remainder
of my talk to their analysis.   The first reason is that the
Δ(1236) plays a capital role in the description of the
elementary amplitude of the reaction γN→πN, and the electromagnetic probe is the cleanest way to create it from a
nucleon, allowing us to study unambiguously its interactions
with the other nucleons inside nuclei.   Most of the observed
deviations from the quasi-free process can be explained in
terms of the final state ΔN interaction[4-7].   The second
reason is that the Δ(1236) also plays an important role in
the analysis of the γN→ Nππ reactions above Eγ=400 MeV.

The double pion photoproduction reactions can be described, with great degree of accuracy, as the quasi two-body reactions $\gamma N \to \pi \Delta$[8,9]. Above $E\gamma = 400$ MeV the influence of this double pion photoproduction channel on the single pion photoproduction reactions on nuclei has been recently observed[7,10]. One of the pions emitted at a nucleon is reabsorbed by another nucleon in the nucleus. This meson exchange current correction to the elementary $\gamma N \to N\pi$ amplitude accounts for the remaining deviations from the quasi-free process.

Let us now develop these ideas in more detail and choose as an example the simplest nucleus, the deuterium. This is the best laboratory to study the Δ-N interaction and to compare this simplified description of the three-body πNN system to the exact result of a more general theory, in which the practical calculation can be performed with a degree of accuracy which is difficult (or impossible) to reach when heavier nuclei are considered. In Figure 1, the different steps of the history of a pion, created by a photon at a nucleon in the deuterium, are depicted. It can escape freely from the target without suffering any interaction. This is the pion photoproduction reaction on quasi-free nucleons (diagram II). The two outgoing nucleons can interact after the elementary reaction. This is a correction coming from final-state interaction (diagram III). The pion can suffer one of several scatterings before escaping the nucleus (diagrams IV and V). Those mechanisms allow us to study the Δ-N interaction which can be split into an exchange part (diagram VI) and a direct term (diagram VII). The first term in the development of the exchange part corresponds precisely to the exchange of a real pion, and the study of the pion-nucleon rescattering graphs leads to an understanding of this part of the Δ-N interaction. The last possibility for the pion is to be reabsorbed by another nucleon and never escape (diagram VIII). This internal pion is now very far from its mass shell and cannot be distinguished from the virtual pions which mediate the nucleon-nucleon interaction; these diagrams correspond to the so-called meson-exchange current effects.

The Born terms play an important role in all these diagrams and are of dominating importance near the single pion photoproduction threshold. The physical problems are the

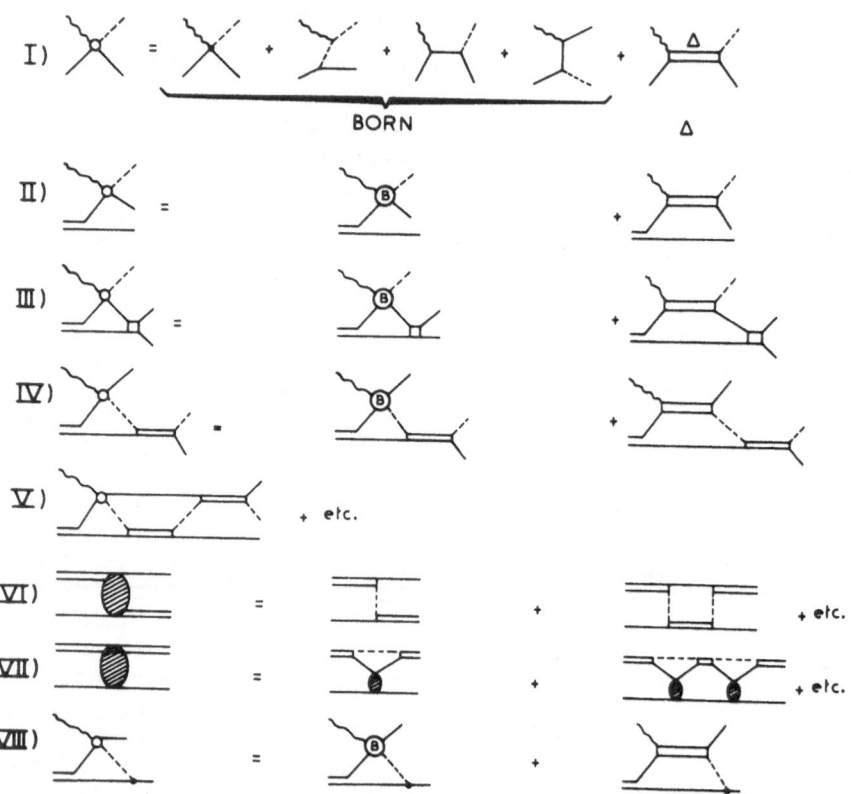

Figure 1: The different steps of the history of a pion in the deuterium. I: The pion photoproduction amplitude on a free nucleon is expanded in terms of the Born terms and the s-channel Δ formation amplitude. II: Pion photoproduction on a quasi-free nucleon. III: Final state nucleon-nucleon interaction. IV: Pion-nucleon single rescattering. V: Pion-nucleon multiple scattering. VI: Exchange part of the Δ-N interaction. VII: Direct part of the Δ-N interaction. VIII: Pion reabsorption mechanism. In all those diagrams the symbol B at the γN→Nπ vertex represents the Born terms of diagram I.

same in this kinematical region and can be handled with the
same methods as those we shall describe here (cf. Ref. 5,11).
If the Born amplitude is extrapolated in the unphysical
region (below one pion threshold) the corresponding operators
are those which permit us to treat the meson exchange current
effects on the cross-section of the reactions induced by a
low energy photon beam ($E\gamma \leq 140$ MeV). Therefore, strong
similarities appear in the description of all these mechanisms
in quite different energy ranges. The topology of the relevant diagrams and their methods of calculation are nearly the
same. The only change concerns the operators which appear at
the various vertices and the emphasis can be put on the
different aspects of the interaction by adequately choosing
the kinematics. In the following we shall always take into
account these Born terms, but we shall put the emphasis on
the study of the Δ-nucleus interaction.

Let us now briefly describe this method of calculation.
Clearly we start with the pion-nucleon multiple scattering
expansion. Since strong interactions occur in the final
state, this method would be impracticable if a small set of
relevant diagrams could not be selected. The best way to
single out a diagram is to look for its singularities and to
choose the kinematics of the reactions in such a way that
they come close to the physical region. The contribution of
the corresponding diagram exhibits a rapid variation above a
slowly varying background. This method presents two advantages. For an experimentalist it is easy to select the relevant kinematical variables to be measured in an experiment,
and for a theoretician it is possible to compute the amplitude
of the dominant mechanism and to check the model in detail.

Let us illustrate this method on the simplest mechanism
which occurs when an intermediate energy photon interacts with
the deuteron. The pion photoproduction on quasi-free nucleons
(diagrams II in Figure 1), which accounts for almost all the
total photo-nuclear cross section[1-3]. The cross section

$$\frac{d\sigma}{d\Omega_\pi dp_{\underset{\sim}{r}}} = (1 + \beta_r \cos \theta_r) \; \rho(p_r) \; \frac{d\sigma}{d\Omega_\pi} (Q,\omega) \qquad (1)$$

is nothing but the relation between the counting rate of the

$D(\gamma,p\pi^-)p$ reaction, the cross section $\frac{d\sigma}{d\Omega}$ (ω) of the elementary $\gamma n \to p\pi$ reaction (which depends only on the pion angle ω and the total energy Q available in the center of mass frame of the πN pair), the number $\rho(\underset{\sim}{p}_r)$ $d\underset{\sim}{p}_r$ of target nucleons per unit volume in the momentum space and the photon flux $(1+\beta_r\cos\theta_r)$ seen by the target nucleon moving with the velocity - $\underset{\sim}{\beta}_r$ and momentum - $\underset{\sim}{p}_r$. This is known as the spectator nucleon model and has been extensively used in the past to deduce the cross section of the $\gamma n \to p\pi^-$ reaction cross section from the $\gamma D \to pp\pi^-$ yield. It exhibits also the two kinds of singularities which appear in the diagrams in Figure 1. The singularities of the vertex, and the singularities of the matrix element itself. For instance when the amplitude of the elementary reaction $\gamma N \to N\pi$ is expanded (diagram I in Figure 1) the $\Delta(1236)$ pole in the s-channel leads to a strong variation of the cross section when the energy at the corresponding vertex comes close to the mass of the $\Delta(1236)$ resonance. The other singularity is associated with the intermediate nucleon propagator and is contained in the deuteron wave function. For low values of the spectator nucleon momentum the momentum distribution reduces to the nucleon pole

$$\rho(p_r) \underset{p_r \to 0}{\thicksim} \left[\frac{1}{p_r^2+\alpha^2}\right]^2 \tag{2}$$

The singularity $p_r^2 = -\alpha^2$ is very near the physical region and makes dominant the quasi-free process when the spectator nucleon momentum approaches zero. This behaviour is clearly apparent in Figure 2 where the momentum distribution of the nucleons emitted in the $\gamma D \to pp\pi^-$ reaction, studied in a bubble chamber experiment[12], is shown. When the spectator nucleon momentum is low the model in which only one nucleon is active accounts well for the data. Its contribution decreases quickly when the momentum increases and becomes small enough to be overwhelmed by the mechanism involving two nucleons. The excess of events observed for high values of the recoiling nucleon momentum is entirely due to the pion-nucleon rescattering mechanism[4], which will be discussed in more detail later. However, the statistical

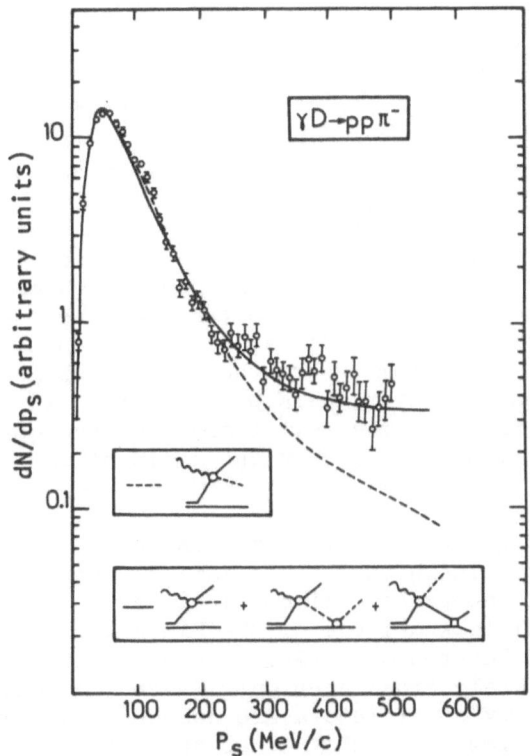

Figure 2: The momentum distribution of a nucleon emitted in the γD→ppπ⁻ reaction. The experimental points are taken from Ref. 12. The dashed line curve is obtained when only one active nucleon is considered (spectator nucleon model). The full line curve is obtained when pion-nucleon single scattering and nucleon-nucleon rescattering are included (reprinted from Ref. 4).

accuracy of these data does not allow us to extract differential cross sections and the momentum distribution was obtained by selecting the events with a given value of the momentum p_r and integrating over all the remaining independent kinematical variables.

Another example is provided by the spectrum of the pions emitted at a given forward angle, when a monochromatic beam is used. Such an experiment has been recently carried on at Saclay with the new monochromatic photon beam ($N_\gamma \simeq 5.10^7$ par-

ticles/s in $\Delta E_\gamma \sim 3$ MeV at $E_\gamma = 300$ MeV). The preliminary
data[13] confirm and reproduce the theoretical predictions[5]
depicted in Figure 3 where the π^+ and π^- spectrum are drawn.
The enhancement near the maximum pion momentum is due to
the strong NN interaction in the $^1S_0$ state. (Note the different effects of the pp and the nn interaction coming from
the Coulomb force). Their relative velocity is vanishing
here. When the pion momentum decreases the momentum of one
of the emitted nucleon goes through zero and the broad bump
is entirely due to the quasi-free mechanism. When the pion
momentum decreases again, the importance of the quasi-free
process decreases and small deviations might appear in the
part of the pion spectrum, where the emitted nucleon momenta
begin to be high. However, this pion spectrum is still an
integrated quantity and the deviations from the quasi-free
process are still overwhelmed by the dominant one nucleon
mechanism.

To go beyond and to learn something more, experiments
must be performed in which the statistical accuracy is good
enough when all independent kinematical variables are measured. The complete knowledge of the kinematics implies the
detection of at least two particles in coincidence, and the
measurement of small cross sections. Hence high duty-cycle
and high intensity accelerators are needed. The performances
of the 600 MeV electron Saclay Linac have made possible the
achievement of the experiments which are now described.

We have seen that the quasi-free mechanism is dominant
when the spectator nucleon momentum is low. In that region
the nucleon momentum distribution in a nucleus is well known.
The different deuterium wave-functions differ by about $\pm 5\%$
below $p_r \sim 150$ MeV/c. Therefore, this quasi-free region allows
us to determine the cross-section of the elementary reaction
$\gamma n \rightarrow p \pi^-$ with the help of eq. 1. In Figure 4 the value of this
cross section, deduced from the yield of the $\gamma D \rightarrow p p \pi^-$ reaction[6]
or the $^4He(\gamma, p\pi^-)$ reaction[14] analyzed in the frame work of
DWIA[15], are compared. The agreement between the two sets of
values is excellent and they are well reproduced by the theoretical model[16].

To study in detail the mechanisms involving more than one

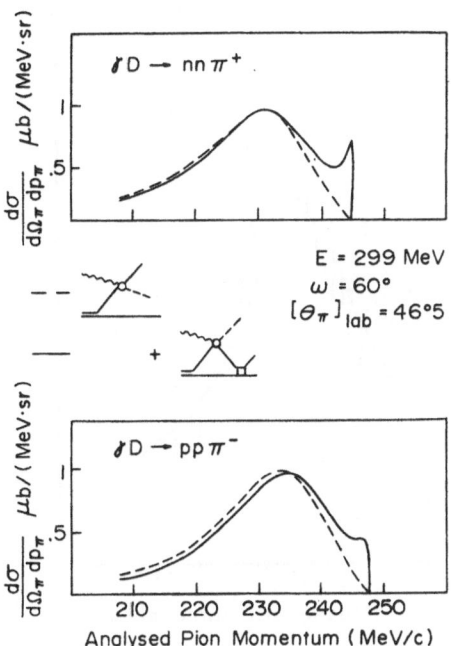

Figure 3:   The spectra of the pion, emitted in the $\gamma D \to nn\pi^+$ and the $\gamma D \to pp\pi^-$ reactions at $(\theta_\pi)_{Lab} = 46.5°$, when $E\gamma = 299$ MeV. The broken line curves include only the quasi-free contribution, whereas the full line curves include also the final state N-N interaction.

nucleon, which appear when the momentum of the emitted nucleons increases, it is convenient to perform an experiment in which the one nucleon contribution is kept constant and minimized. By inspection of eq. 1, it is easy to be convinced that the absolute value p_r of the recoiling system, the energy Q and the angle ω must be kept constant. The remaining independent kinematical variable is the angle θ_r made by the recoiling system and the incoming photon direction (or equivalently the energy of the photon). Hence, we have measured[6,7] the angular distribution of a proton emitted with a constant momentum $p_r = 400$ MeV/c in the reaction $\gamma D \to pp\pi^-$, when Q = 1200 MeV and $\omega = 90°$. It is depicted in Figure 5

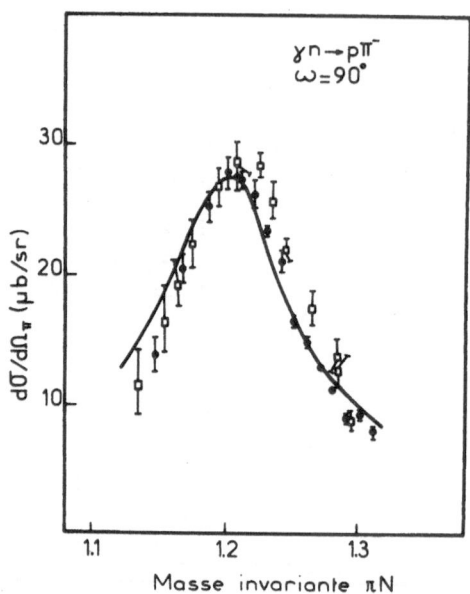

Figure 4: Comparison of the γn→pπ⁻ reaction cross section
extracted (in the framework of the pion photoproduction on
a quasi-free nucleon model) from the yield of the D(γ,pπ⁻)
reaction (full circles) or the ⁴He (γ,pπ⁻) reaction (open
squares). The curve is the prediction of Ref. 16.

and exhibits a strong anisotropy which contrasts strongly
with the spectator nucleon model predictions. A strong peak
clearly appears near θ_r~45°. This violent structure is almost entirely due to the pion-nucleon single scattering
mechanism (diagram IV in Figure 2) and is the signature of
the corresponding singularity. The matrix element exhibits
a moving logarithmic singularity (associated with the on-shell
propagation of the internal pion), which comes close to the
physical region when the angle θ_r approaches 45° (see Ref. 5
for a detailed discussion of this phenomenon). This result
is very important, the exchange part of the Δ-N interaction
is very simple. The dominance of the one-pion exchange term
and the smallness of the pion multiple scattering are strongly
supported by this result. However, the measured cross section
decreases less rapidly, when θ_r increases, than the pion-

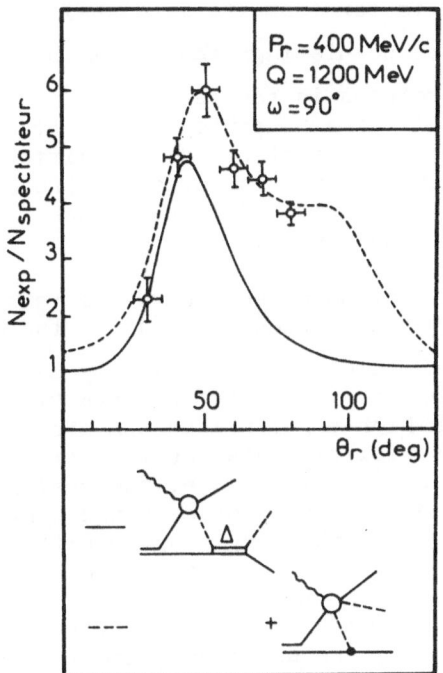

Figure 5: The angular distribution of a nucleon emitted in the $D(\gamma,pp)\pi^-$ reaction with a constant momentum p_r=400 MeV/c when Q=1200 MeV and ω=90°. The ratio between the experimental yield[6] and the yield which should have been obtained if only one nucleon were active (diagrams II, in Figure 1) is plotted. The full line curve is obtained when pion-nucleon single rescattering mechanisms are considered (diagrams IV in Figure 1) whereas the dashed line curve takes also into account the meson exchange corrections to the single pion photo-production amplitude.

nucleon rescattering contribution. This difference is well accounted for by the following mechanism. Two pions are emitted at one nucleon and one of them is reabsorbed by the other nucleon. The cross section of the reaction $\gamma N \rightarrow N\pi\pi$ increases very quickly between 400 MeV and 600 MeV[8,9] and this is precisely in this energy range that the influence of this elementary reaction on the yield of the $\gamma D \rightarrow pp\pi^-$ reaction is the strongest. This mechanism can also be viewed as a meson exchange correction to the amplitude of the $\gamma N \rightarrow N\pi$.

The internal pion is far from its mass shell (-16 $m_\pi^2 \le q^2$ ≤ -9 m_π^2) and cannot be distinguished from the virtual pions surrounding the nucleons in the deuterium. It is impossible to know whether this pion is photoproduced at one nucleon and then reabsorbed by the other, or is emitted at one nucleon and reaches the other when it absorbs the photon; the two descriptions are equivalent. A detailed description of this phenomenon can be found in Ref. 10.

In the experiment described above the relative kinetic energy $T_{\Delta N}$ between the emitted $\Delta(1236)$ and the other nucleon is high and the Δ-N interactions is mainly mediated by the peripheral one pion exchange term (diagram VI in Figure 1). Partial waves with high angular momentum are involved and the centrifugal barrier prevents the two particles from coming close together. Furthermore, the real nature of the exchanged pion makes the one pion exchange contribution very important in describing the long ranged part of the Δ-N interaction. The question now is to learn something more about the ΔN interaction at low energy, where the low angular momenta are involved and where the centrifugal barrier does not exist ($L_{\Delta N}=0$) or is so small that the two particles can interact at short distances. In these circumstances the $\Delta(1236)$ and the nucleon can interact strongly and clearly a single diagram cannot reproduce reality: a multiple scattering series must be summed, an example of which is drawn in diagram VII of Figure 1 (see also Ref. 5). This is a very difficult task and the problem can be simplified, as in the N-N scattering case, in the following way. The partial waves with high momenta ($L_{\Delta N} \ge 1$) are parametrized by the one pion exchange term (diagram VI in Figure 1) which has been shown to account fairly well for the experimental data when the relative kinematic energy $T_{\Delta N}$ of the ΔN pair is high (see Figure 5). The s-wave is parametrized in terms of an effective range expansion near the Δ-N threshold. This kind of analysis permits us to reproduce the angular distribution[6] of the proton emitted in the $\gamma D \rightarrow pp\pi^-$ reaction with a constant momentum ($p_r=150$ MeV/c) when the mass of πN pair is Q=1220 MeV. The quasi-free contribution is still dominant but significant deviations (about 20%) appear clearly in Figure 6. The one loop diagrams (III and IV in Figure 1) cannot reproduce the entire angular distribution. The rise near the forward angles is due to the nucleon-nucleon scat-

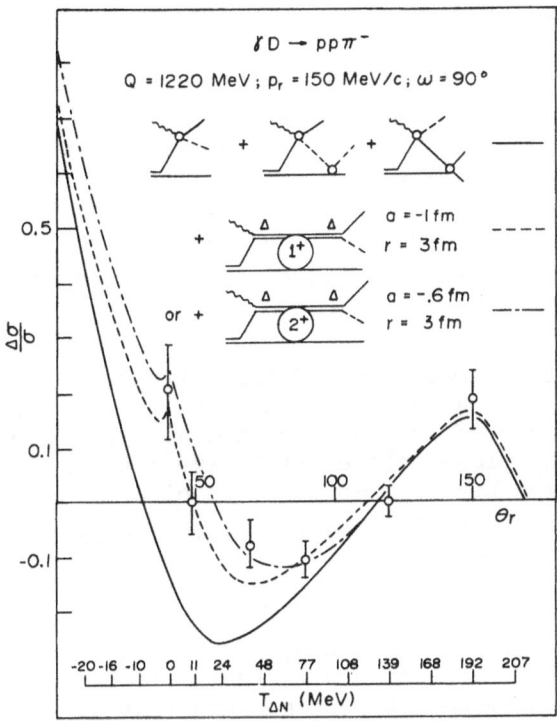

Figure 6: The angular distribution of a nucleon emitted in the D(γ,pπ⁻) reaction, with a constant momentum p_r=150 MeV/c when Q=1220 MeV and ω=90°. The relative deviations from the quasi-free process contribution are plotted. The full line curve includes the contribution of the one loop diagrams III and IV in Figure 1. The dashed (dash-dotted) line curve is obtained when the Δ-N interaction in the $J=1^+$ $(J=2^+)$ state is taken into account. The relative kinetic energy between the Δ and the nucleon is plotted on abscissa. Note the cusp at the Δ-N threshold.

tering diagram III (their relative kinetic energy decreases as θ_r decreases) and the enhancement near θ_r=120° is due to the pion nucleon scattering diagram IV (e.g. the peripheral one pion exchange ΔN interaction). The remaining discrepancy is accounted for by parametrizing the s-wave ΔN amplitude by its effective range expansion and adjusting the values of the scattering length a and the effective range r to reproduce the experimental data. When $L_{ΔN}$=0 (s-wave) two states are possible: $J^{\pi}=1^+$ ($^3S_1$wave) or $J^{\pi}=2^+$ ($^5S_2$wave). The data can be fitted equally well either by the $J^{\pi}=1^+$ state alone with a=-1.fm and r=3fm, or the $J=2^+$ state also with a=-.5fm and r=3fm (note that the sign convention is the same as in the N-N scattering case: q cotg $\delta \sim -1/a$). It is worthwhile noting the relative kinetic energy scale. The s-wave interaction reproduces the data near $T_{\Delta N}=0$ and the peripheral one pion exchange ΔN interaction reproduces the data for high values of $T_{\Delta N}$.

The relative importance of these two waves cannot be decided on the basis on this experiment alone. However, they can be distinguished by measuring the high energy part of the spectrum of the pions emitted at backward angle when a mono-chromatic photon beam is used. It is clearly apparent in Figure 7 that the $J^{\pi}=2^+$ state ($^5S_2$ wave) contribution is strongly suppressed when compared to the $J^{\pi}=1^+$ state ($^3S_1$ wave) contribution (the same parameters as in Figure 6 are used). This suppression is easily understood if we realize that, after its interaction with one nucleon, the $\Delta(1236)$ decays by emitting another nucleon and that all the nucleon-nucleon states are not allowed. In the high energy part of the pion spectrum, the relative kinetic energy of these two outgoing nucleons is small and they must emerge in the $^1S_0$ state. When the pion wave is coupled to this state it is possible to obtain the $J=1^+$ state [$^1S_0$(NN), $\ell_{\pi}=1$], but not the $J^{\pi}=2^+$ state (the lowest momentum required are [$^3P_2$(NN), $\ell_{\pi}=0$] or [$^3P_0$(NN), $\ell_{\pi}=2$] or [$^1D_2$(NN), $\ell_{\pi}=1$]), due to momentum and parity conservation.

This high energy part of the pion spectrum was recently studied at Saclay[13]. To improve the statistical accuracy of the data, it was integrated over 20 MeV/c below the maximum pion momentum, and the excitation function of this quantity was measured. The data seems to favor the $J=2^+$ state near the ΔN threshold, where a cusp effect is clearly apparent. However, these data are still preliminary. This experiment is very difficult because the cross section is rather small

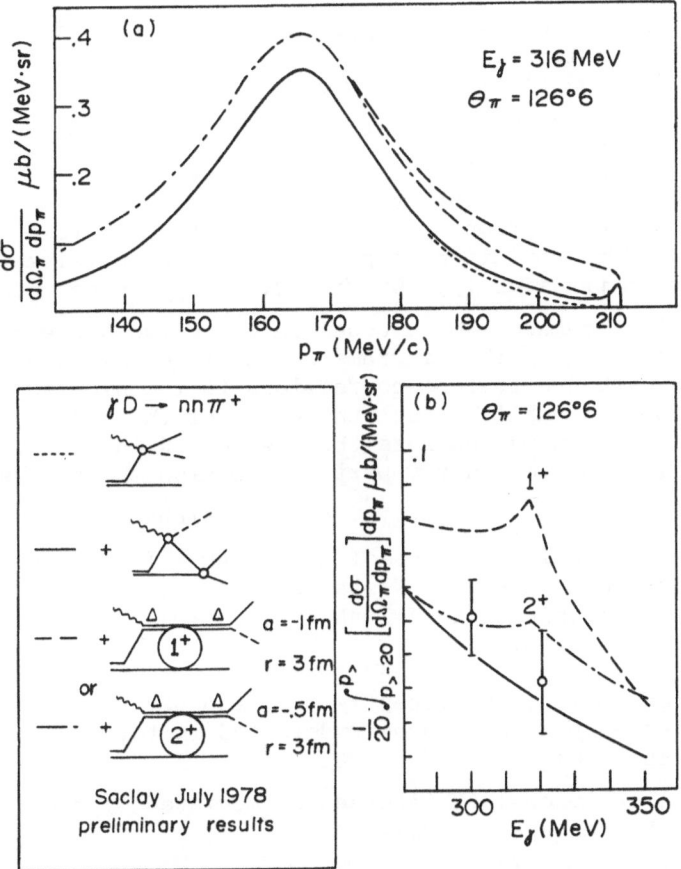

Figure 7: a) The spectrum of the pion emitted in the $\gamma D \rightarrow nn\pi^+$
reaction, at $(\theta_\pi)_{Lab}$=126.6° when $E\gamma$=316 MeV. The dotted line
curve includes only the quasi-free contribution, whereas the
full line curve includes also the final state nn interaction.
The dash-dotted (resp. broken) line curve includes also the
ΔN interaction in the $J=2^+$ (resp $J=1^+$) state. b) The excitation function of the high energy part of the pion spectrum
integrated over 20 MeV/c. The experimental points are very
preliminary, and show the tendency of the experimental data.

and because the monochromatic photon flux is not high enough.
More experimental work is needed to improve the accuracy
before a definite conclusion can be drawn.

Let us now describe an application of these ideas to photopion reactions involved on a heavier nucleus. In Figure 8 we have plotted the excess of events found when the measured yield of the ^{4}He $(\gamma, p\pi^-)$ reaction[14] is compared to the prediction of a DWIA Model[15]. For low values of the momentum of the recoiling system, this model provides us with a good description of the experimental data (see also Figure 4). The deviations appear only for high values of the momentum transfer $(p_{\tilde{r}} \approx 200 \text{ MeV/c})$. They can be understood in the framework of a model in which two nucleons are active, as shown by the diagrams in the inset of Figure 8. The meson exchange current corrections to the single pion photoproduction amplitude account well for the high energy part of the spectrum, whereas the s-wave Δ-N interaction accounts for its low energy part. Here also the data seem to prefer the J=2$^+$ $(^5S_2)$ state, but the situation is more complicated than in the case of deuterium. In ^{4}He it is also possible to reach the T=2 ΔN states. However, their contribution is small since the T=1 ΔN state seems to account for almost all the data.

In conclusion, the recently measured deviations from the quasi-free process can be understood in terms of three simple mechanisms:

- The peripheral ΔN interaction mediated by the real pion exchange diagram.

- The central ΔN interaction which can be parameterized at low energy by an effective range expansion. The J=2$^+$ state $(^5S_2$ wave), with a=-.5fm and r=3fm, seems to be preferred.

- The meson exchange corrections to the $\gamma N \to N\pi$ amplitude, which are connected to the $\gamma N \to N\pi\pi$ reaction and are of rising importance above $E_\gamma \sim 400$ MeV.

These three mechanisms provide us with an elegant way of analyzing the deviations measured in different kinematical conditions, but it must be realized that they represent the minimal framework for describing a small set of experiments. For instance, the description of the $^5S_2(\Delta N)$ wave near the ΔN

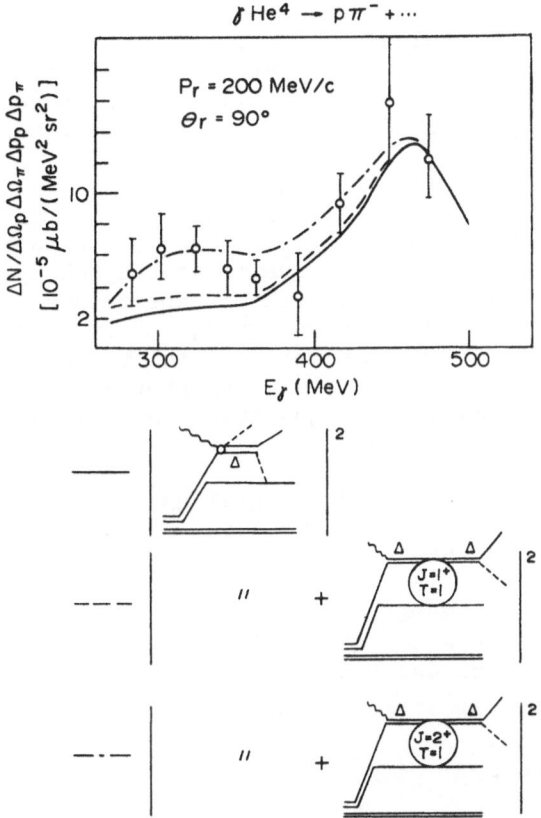

Figure 8: The excess of events found in the analysis of ^{4}He $(\gamma, p\pi^-)$ reaction, when the experimental yield is compared to the predictions of a DWIA model. The full line curve takes into account the meson exchange correction as indicated in the inset. The dot-dashed (resp. broken) line curve takes also into account the effects of the ΔN rescattering in the J=2$^+$ (resp. J=1$^+$) state. The parameters are the same as in Figures 5 and 6.

threshold by a real scattering length is a very strong limi-tation of the model. Since this state can be coupled to the $^1D_2$(NN) wave and since the ΔN states must also be coupled to the πNN non-resonant states a complex scattering length should have been used, or the fit should have been made in the framework of a multi-channel effective range expansion. However, the scarceness of the data prevents, for the moment, such sophistication. Clearly more experimental results are needed to check these ideas in more detail and under other kinematical conditions.

On the theoretical side it is necessary to understand why such a simple scheme works. Clearly we have parametrized the three-body πNN systems by the quasi-two-body ΔN system. The compatibility of the present bulk of experimental data with such a parametrization is encouraging, but it is necessary to understand it at a more fundamental level by solving the corresponding relativistic three-body problem, or by taking into account explicitly the constraints of the three-body unitarity.

The Δ-N system is as fundamental as the nucleon-nucleon system. It is also the starting point of most of the studies of the behavior of the Δ(1236), or more generally of the pion inside the nucleus. Its importance justifies more efforts in its understanding. Moreover, the study of photopion reactions on few body targets are a good test case where the approximation needed by the complex nature of heavy nuclei must be checked.

+
Permanent Address

REFERENCES

1. T. A. Armstrong et al., Nucl. Phys. B41, 445 (1972).
2. J. Ahrens et al. This conference. B. Ziegler, in "Few-Body Systems and Electromagnetic Interaction". Frascati, March 7-10, 1978. To be published in lecture notes in Physics Springer-Verlag.
3. J. S. O'Connel et al. This conference.
4. J. M. Laget. Phys. Lett. 68B, 58 (1977).
5. J. M. Laget. Nucl. Phys. A296, 388 (1978).
6. P. E. Argan et al. Nucl. Phys. A296, 373 (1978).
7. P. E. Argan et al. Phys. Rev. Lett. 41, 86 (1978).
8. D. Luke et al. Nuo. Cim. 53B, 235 (1968).
9. D. Luke et al. Springer Tracts in Modern Physics 59, 39 (1971).
10. J. M. Laget. Phys. Rev. Lett. 41, 89 (1978).
11. P. Bosted et al. Nucl. Phys. A296, 413 (1978).

12. P. Benz et al.   Nucl. Phys. B65, 158 (1973).
13. P. E. Argan et al.   Private communication.
14. P. E. Argan et al.   To be published and communications
     D1 and L5 in the Abstract Volume of the Zurich Confer-
     ence (September 1977).
15. J. M. Laget.   Nucl. Phys. A194, 81 (1972).
16. I. Blomqvist et al.   Nucl. Phys. A280, 405 (1977).

PHOTOPRODUCTION OF CHARGED PIONS FROM C^{12} USING A TAGGED PHOTON BEAM

J. Arends, J. Eyink, H. Hartmann, A. Hegerath,

B. Mecking, G. Nöldeke, and H. Rost

Physikalisches Institut der Universität Bonn

Nussallee 12, 5300 Bonn, West Germany

The first experiment using the tagged photon beam facility at the Bonn 500 MeV synchrotron is described. An internal tagging system determines the photon energy with an accuracy of 9 MeV in the photon energy range from 200 to 390 MeV. Charged photoreaction products are analyzed by a large solid angle magnetic spectrometer. Data are presented for the energy distributions of charged pions produced by the tagged photons from a C^{12} target. The data are compared with the results of an intranuclear cascade Monte Carlo calculation.

INTRODUCTION

The investigation of the photoproduction of pions from complex nuclei can give information on the nuclear structure and on the interaction of pions with nuclear matter. The elastic production, leaving the final nucleus in a state with low excitation energy, is sensitive to the wave functions of the initial and the final nucleus and to the pion nucleus optical potential. Inelastic production of pions is considered to take place on quasifree nucleons. Therefore the cross sections are determined by the momentum distribution of the nucleons inside the nucleus and by the interaction of the outgoing pion and the recoil nucleon with the rest system. Compared to quasielastic electron scattering the quasifree pion production process offers the advantage that the interaction

used to probe the momentum of the nucleon has approximately
the same strength for protons and neutrons. In charged pion
photoproduction the charge of the pion already determines the
charge state of the target nucleon.

In the photon energy range around 300 MeV the prime
interaction mechanism of photons with nucleons is the excitation of a Δ-resonance. The investigation of photoprocesses
in this energy range seems to be especially interesting because effects due to the Δ-nucleon interaction may influence
the cross sections.

Up to now the full potential of these reactions has not
been exploited. The main reason is that no suitable high intensity source of monochromatic photons is available. In the
photon energy region of interest three different methods have
been used to produce quasimonochromatic photons:
   a) the photon difference method (using the difference be-
      tween the yields obtained with electron bremsstrahlung
      spectra of different endpoint energies),
   b) positron annihilation in flight (using the enhance-
      ment at the endpoint energy of the bremsstrahlung
      spectrum due to the $(e^+e^- \rightarrow 2\gamma)$-reaction,
   c) the tagging of electron bremsstrahlung photons (by
      measuring the energy of the recoil electron).
Due to various drawbacks none of these methods can be called
a general purpose source of monochromatic photons.

Most experiments on nuclear pion photoproduction were
performed with bremsstrahlung beams. The results are difficult to interpret because only the pion yield integrated over
the kinematically allowed photon spectrum is measured. In a
few experiments this difficulty has been avoided by using the
photon difference technique. Due to statistical problems the
accuracy of this method decreases with increasing excitation
energy of the undetected rest system. Up to now tagged bremsstrahlung or positron annihilation photon beams have not been
used to study nuclear pion photoproduction.

To study photonuclear reactions in the energy range from
200 to 400 MeV a tagged photon beam was set up at the Bonn 500
MeV synchrotron. In the first experiment the facility was
used to measure pion (and proton) energy spectra from a C^{12}
target. Carbon was chosen as a target because a considerable

amount of information on charged pion photoproduction has [1] already been obtained either by using bremsstrahlung beams or the photon difference technique[2-5]

1. EXPERIMENTAL SETUP

The tagged photon beam facility consists of an internal tagging system, a magnetic spectrometer and a set of scintillation counters. The layout is shown in Fig. 1.

2.1 The Tagging System

The bremsstrahlung beam is produced by the circulating electrons hitting a gold foil placed into the synchrotron magnet 5. The recoil electrons are deflected by the guiding field of magnet 5 and leave the synchrotron vaccuum system through a thin foil in front of the hodoscope counters.

The recoil electrons are detected in a 20 scintillation counter hodoscope placed between magnets 5 and 6. The width of the scintillators has been adjusted to give equal photon energy resolution over the whole tagging range and thus a linear relationship between the tagging counter number and the photon energy.

Independent of the synchrotron endpoint energy E_o the tagging system covers the photon energy range from 0.45 to 0.85 of E_o with a resolution $\Delta k = 0.02 \cdot E_o$.

Since the determination of the photon energy relies on a coincidence between the tagging counters and the reaction product detector the method is subject to intensity limitations due to random coincidences. The ratio of random to true coincidences η depends on the process under investigation and on the limits of the tagging range relative to the endpoint energy of the bremsstrahlung spectrum. η is related to the width τ of the coincidence between the product detector and the tagging counters, to the duty cycle of the photon beam d and to the counting rate of the tagging hodoscope N

$\eta \sim N \cdot \tau / d$

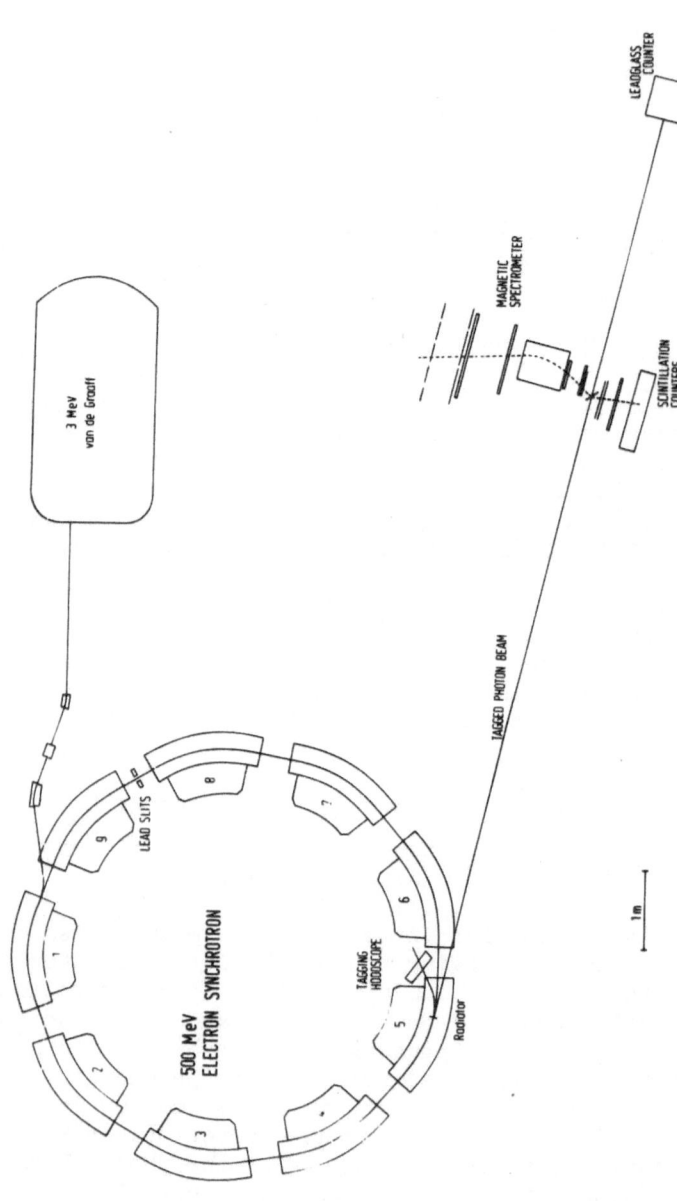

Figure 1. Schematic diagram of the tagged photon beam facility. (Not shown is the experimental equipment associated with the normal bremsstrahlung beam which is produced in the fieldfree region between the magnets 5 and 6.)

If a mean correction of $\eta = 0.1$ is tolerated this gives in our case ($\tau/d = 2 \cdot 10^{-7}$ sec) a maximum rate of $N^{max} = 2.5 \cdot 10^5$ tagged photons/sec.[+]

In the normal operating mode the 500 MeV synchrotron proauces $1.3 \cdot 10^9$ photons/sec in the tagging range. Therefore, the photon beam intensity has to be reduced by a factor 5000. This is achieved by running the synchrotron in a dedicated mode with the tagged beam as the prime user.

The parameters of the tagged photon beam are summarized in Table 1.

2.2   Detection Apparatus

To compensate for the low intensity of the tagged photon beam the detection apparatus has to cover a large solid angle and momentum range simultaneously. This has been achieved by combining a wide acceptance magnetic spectrometer for the detection of charged particles with a set of scintillation counters to register coincident particles. The complete set-up is shown in Fig. 2.

2.2.1 Magnetic Spectrometer. The magnetic spectrometer consists of a C-type magnet, a scintillation counter trigger system and the four drift chambers D1-D4. An event is defined by coincident signals from the start counter ST in front of the magnet and the two hodoscopes H1 and H2 in the rear. The time difference between ST and H2 determines the velocity of the detected particle. Its momentum and angle are calculated from the coordinates given by D1-D4. Combining momentum and velocity yields a clean mass information for the separation of pions, protons and deuterons. This is demonstrated in Fig. 3 for positively charged particles. The properties of the spectrometer are summarized in Table 2.

[+]Using a DC electron accelerator with $\tau/d = 10^{-9}$ sec the tagging rate could be increased to $N = 5 \cdot 10^7$/sec.

TABLE 1

Parameters of the Tagged Photon Beam

Radiator thickness	5 µm gold
Tagging range (k/E_o)	0.45 - 0.85
Energy resolution $(\Delta k/E_o)$	0.02
Duty factor	0.03
Bunch separation	6 nsec
Mean tagging rate	$3 \cdot 10^5$/sec
Photon definition probability	0.98
Angular divergence (fwhm)	5.5 mrad

TABLE 2

Properties of the Magnetic Spectrometer

Pole face area	50×60 cm^2
Gap height	20 cm^2
Maximum field (B_{max})	0.45 T
Momentum acceptance at B_{max}	$(80 - 1000)$MeV/c
Mean solid angle	80 msterad
Momentum resolution at B_{max} (fwhm)	
for pions	2.4 %
for protons	4.6 %
Minimum range required	1.3 g/cm^2

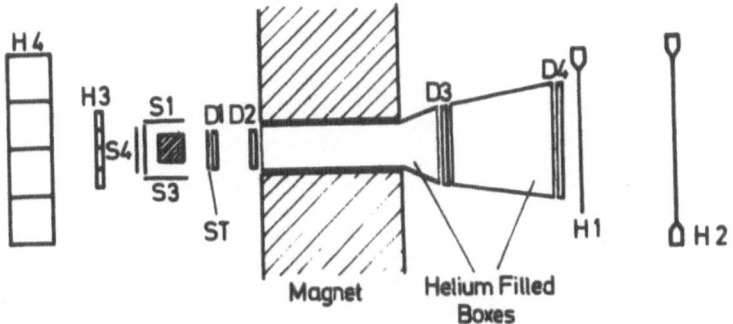

Figure 2(a). Setup of the magnetic spectrometer - vertical cut. Details are given in the text.

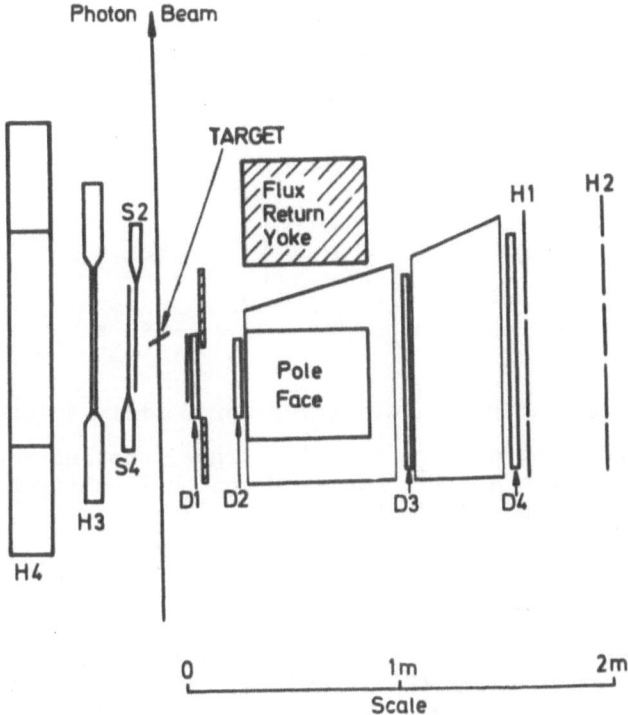

Figure 2(b). Setup of the magnetic spectrometer - horizontal cut. Details are given in the text.

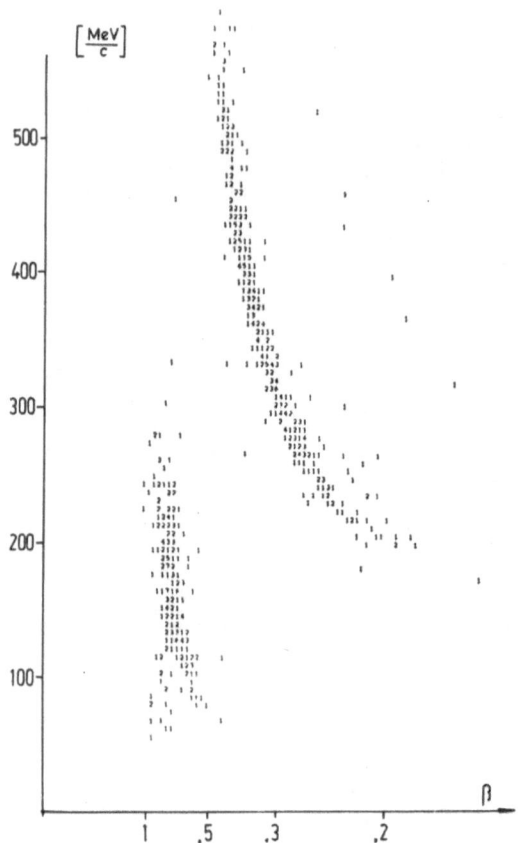

Figure 3. Momentum p vs. velocity β for positively charged
particles photoejected from the C^{12} target, showing the separation between d, p and π^+. Some e^+ can be identified at the
lower left corner of the distribution.

   2.2.2 Scintillation counters. Opposite to the magnetic
spectrometer a set of scintillation counters was used to detect coincident charged or neutral particles in a solid angle
of 2 sr. It consists of the scintillation counters S1-S4
surrounding the target and the hodoscopes H3 and H4. The mass
and energy of charged particles are determined by combining
the measured pulse heights of at least 3 counters in a fit
procedure. Neutral particles are characterized by a signal
in H4 only.

3. DATA TAKING AND REDUCTION

In the first experiment the tagging facility was used in a four weeks run to investigate the emission of charged particles from C^{12} by monochromatic photons in the energy range from 200 to 390 MeV. The tagged beam was operated at an intensity of $(2-3) \cdot 10^5$ tagged photons/sec. Four different spectrometer positions were necessary to cover the angular range from 38° to 136°. At each run, data were taken for positively charged particles (π^+, p, d) and, with the magnetic field reversed, for π^-.

3.1 Electronics

Following a trigger in the magnetic spectrometer the characteristic data of the event are recorded by a CAMAC system. The position information of the drift chambers is digitized by an independent system and then read by a CAMAC interface. For each event the complete data set is sent to an on-line computer system and then stored on magnetic tape for the off-line analysis.

3.2 Data Analysis

In the off-line analysis the drift chamber data are used to calculate the momentum of the charged particle in the spectrometer. By combining the momentum with the time-of-flight, electrons, pions, protons and deuterons can be distinguished. Muons cannot be separated from pions. However, since the track of the charged particle is overdetermined by the four drift chambers, muons coming from the π-μ-decay are strongly suppressed, thus avoiding a distortion of the pion momentum spectrum.

After the reconstruction of angle and kinetic energy the events are binned into pion energy and angular intervals. Taking into account the acceptance of the spectrometer and the number of photons the double differential cross section for the inclusive pion photoproduction as a function of pion and photon energy is determined.

3.3   Corrections

The data were corrected for random coincidences, pion decay and non-target contributions.

If a random coincidence between a tagging counter and the magnetic spectrometer occurs, a false photon energy will be assigned to the event. The corresponding correction is determined by analyzing delayed coincidences. In general the correction rises with decreasing pion momentum (because the momentum spectrum of pions produced by bremsstrahlung rises with decreasing pion momentum). The mean correction is 7%.

Since the path length of pions in the spectrometer is relatively short, the loss of pion events due to the π-μ-decay is small even for low momentum pions. The corresponding correction has been calculated by a Monte Carlo program taking the geometry of the spectrometer and the track finding procedure into account. The correction amounts to 25% for $T_\pi = 40$ MeV and decreases with increasing pion energy to 3% for $T_\pi = 200$ MeV.

Non-target contributions are kept small by using the position information of the front drift chambers D1 and D2. The correction amounts to 3%.

3.4   Errors

Uncertainties in the absolute normalization and in the energy determination have to be distinguished.

The uncertainty in the photon flux, the spectrometer acceptance and the pion decay correction is ± 4% for high energy pions. It rises to ± 8% for the low energy end of the pion spectrum.

The error in the energy calibration of the tagged photon beam is estimated to be $\Delta k/k = \pm 2\%$, in the momentum calibration of the magnetic spectrometer $\Delta p/p = \pm 2\%$.

4. RESULTS AND DISCUSSION

The basic results of the measurements are double differential cross sections for the inclusive photoproduction of π^+ and π^- as a function of the photon energy and the pion energy and emission angle. The cross sections are integrated to obtain the differential and total cross section. The results are compared to an intranuclear cascade calculation.

4.1 The Cascade Calculation

A Monte Carlo program to describe photonuclear reactions has been developed at the Oak Ridge Laboratory by Gabriel[6].

The prime interaction between the photon and the nucleon is assumed to be the pion production and the quasideuteron process. The final state interaction of the outgoing pions and nucleons is described by taking pion-nucleon scattering, pion absorption by two nucleons and nucleon-nucleon scattering into account.

4.2 The Double Differential Cross Sections

As an example, Fig. 4 shows the results for the π^- energy distributions in 10 different photon energy bins. In the forward angular range the data exhibit a broad peak. The position of the peak is approximately 25 MeV below the energy of the free pion production process. At low photon energies and in the backward angular interval the peak is no longer visible. Additionally to the peak there is a broad distribution extending to the lowest pion energies measured.[+] There is no essential difference between the π^- and the π^+ data.

[+]Contributions from the double pion production have been estimated and were found to be negligible in this energy range.

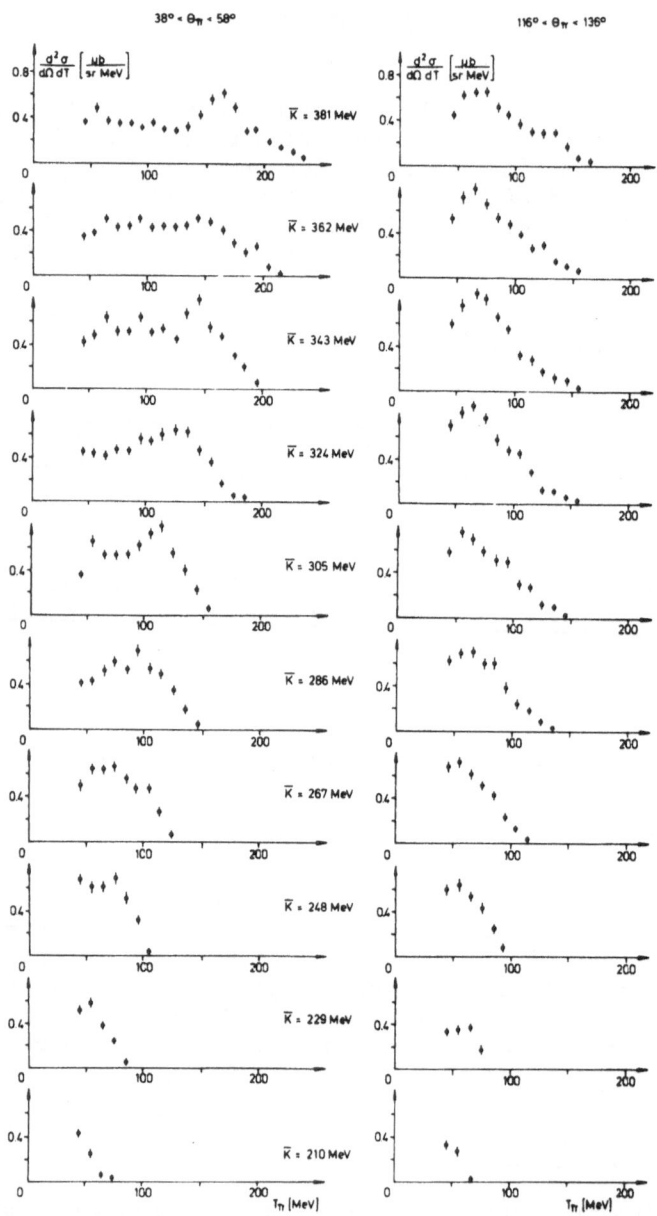

Figure 4. Double differential cross sections for π^- emission from C^{12} in 10 different photon energy and two angular bins. (Some information exists for $T_\pi < 40$ MeV, but no cross sections are given due to large acceptance corrections.)

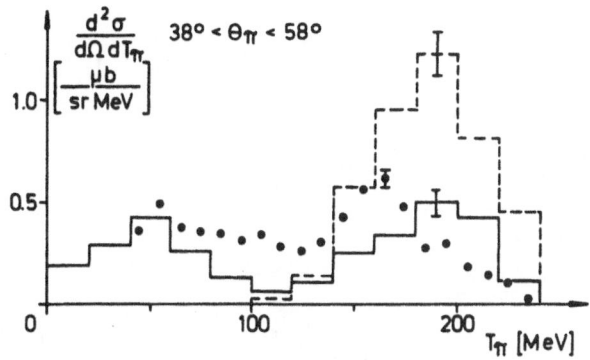

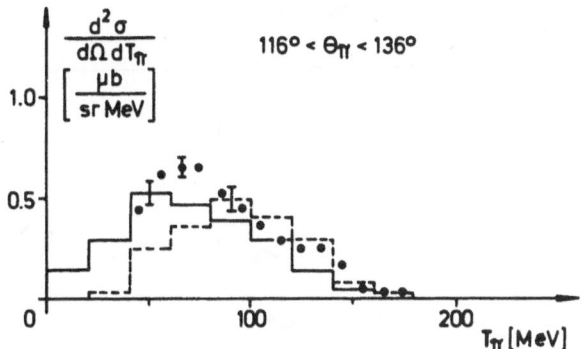

Figure 5. Comparison between two measured π⁻ energy distributions and the corresponding prediction of the cascade calculation for k̄ = 385 MeV. Dashed histogram: final state
interactions are not included; solid histogram: final state
interactions are taken into account. ⌐-measured data points.

    A comparison with the predictions of the cascade model is
shown in Fig. 5 for two selected energy spectra. If the final
state interactions are included in the Monte Carlo calculation
predicts shape and magnitude of the measured distributions
quite well. Especially it explains the broad distribution at
low pion energies as being due to pions that were originally
produced quasielastically, but then shifted down in energy by
pion-nucleon scattering.

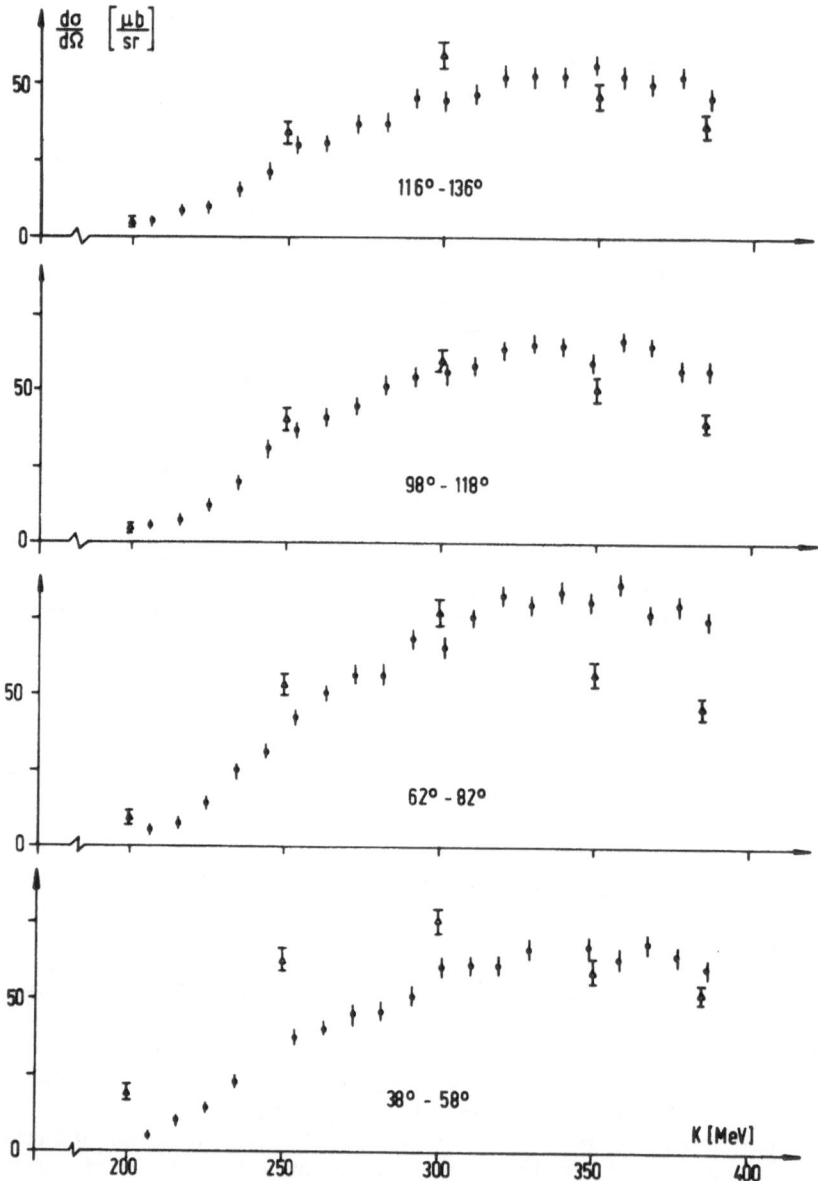

Figure 6. Differential cross sections for the emission of a
π with $T_\pi >$ 40 MeV as a function of the photon energy k in
four different angular intervals. $\triangle$-measured data points,
predictions of the cascade program using the same pion energy
threshold.

4.3   The Differential Cross Sections

Integration of the measured pion energy distributions yields the differential cross sections for the emission of pions with T_π > 40 MeV. Figure 6 gives the results together with the predictions of the Monte Carlo calculation using the same pion energy threshold. Again, no significant difference exists between π^+ and π^- data.

While the Monte Carlo calculation predicts a peak near 300 MeV due to the Δ-excitation, the data shows a rise up to 320 MeV and then stay essentially flat.

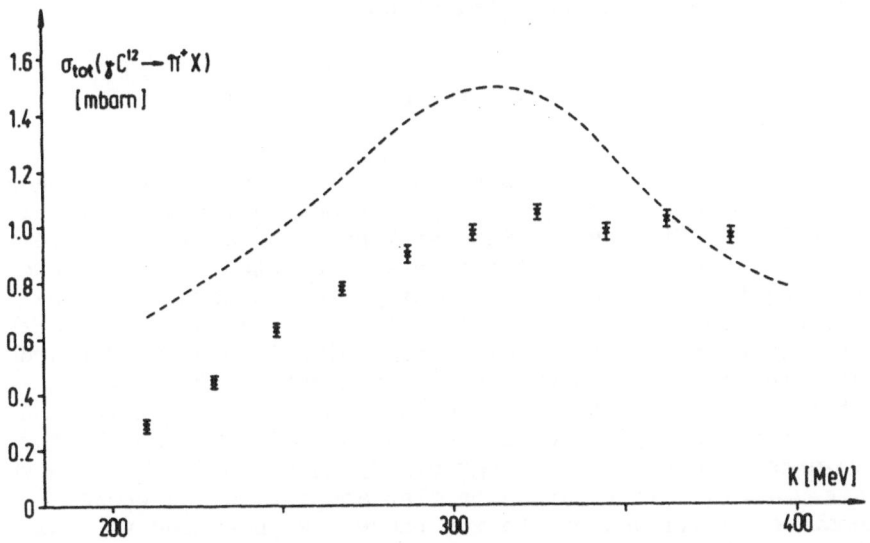

Figure 7. Total cross section for the emission of a π^+ from C^{12} as a function of the photon energy k. The cross section for the elementary process $\gamma p \rightarrow \pi^- n$ (multiplied by 6) is shown for comparison (dashed line).

4.4   The Total Cross Section

The total cross section for pion photoproduction was obtained by extrapolating the measured pion distributions into
the energy and angular range not covered by the experimental
setup. The shape of the distributions was taken from the
cascade calculation, while the absolute height was adjusted
for π^+ is shown in Fig. 7 together with the total cross section for the elementary reaction $\gamma p \to \pi^+ n$ multiplied by the
factor 6. Again, no peak due to the Δ-excitation is visible
in the data. However, it cannot be excluded that the resonance is broadened and shifted to higher proton energies.

The Monte Carlo predictions are approximately equal to
the elementary cross section multiplied by the factor 5. This
might indicate that the Pauli principle has not been properly
taken into account in the calculation.

4.5   The Excitation Energy Distribution

The pion energy distributions (Fig. 4) have shown already
that very often the energy of a pion produced quasielastically
is degraded by final state interactions. For a small sample
of the π^- events the recoil proton was detected in coincidence
in the scintillation counters opposite to the magnetic spectrometer. (Due to the kinematics of the π^- production process
the recoil protons are predominantly emitted into the forward
angular interval not covered by the present scintillation
counter setup.) For the coincident π^--p pairs the excitation
energy of the final system can be calculated. The result for
all events in the photon energy bin $350 \leq k \leq 390$ MeV is shown
in Fig. 8. The distribution is flat and covers the whole
kinematically allowed excitation range. No peak due to a direct production process is visible. However, it should be
noted that with the present coincidence setup events with
large final state interactions are favored.

4.6   Comparison with Previous Measurements

A few measurements of the pion production from C^{12}, taken
with the photon difference technique, are available for comparison. Our data agree very well with the results of Kabe et

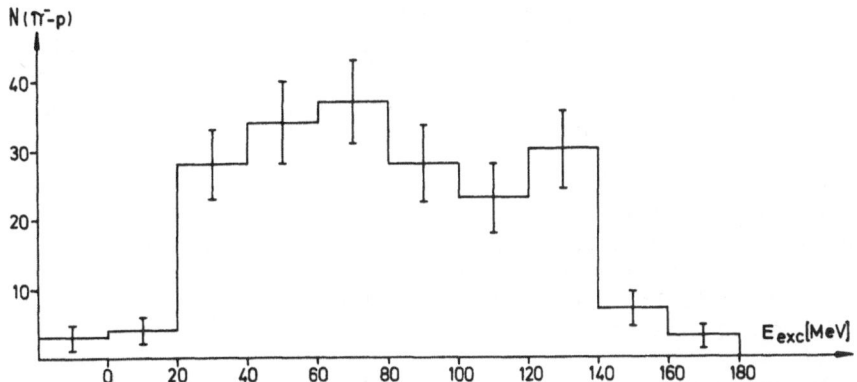

Figure 8. Excitation energy distribution of the final nuclear system for coincident π⁻-p pairs. All events produced by photons in the energy bin 350 MeV < k < 390 MeV are used. The numbers of events in the bin -20 MeV < E_{exc} < 0 corresponds to the level of accidental coincidences (between the spectrometer and the tagging counters) which have not been subtracted.

al.[3] who have measured π⁺ energy spectra at the Tokyo synchrotron for photon energies between 180 and 550 MeV.

A second group[4] at the Tokyo synchrotron has investigated π⁺ and π⁻ production at 28° and 44° for photon energies between 300 and 850 MeV. The agreement between their pion energy distribution and ours is reasonable near and above the quasielastic peak.

For highly inelastic events the accuracy of the photon difference technique decreases rapidly, and a comparison is no longer meaningful. Especially, our data show that the extraction of the width of the quasielastic peak is a somewhat arbitrary procedure unless final state interactions can be handled properly.

5. POSSIBLE IMPROVEMENTS AND FUTURE EXPERIMENTS

At present the properties of our tagged photon beam are adequate for the investigation of inclusive single particle

energy spectra. However, the energy resolution is not
sufficient to investigate narrow structures in the excitation
energy of the final nuclear system, as expected in direct reactions. Also the low intensity makes it difficult to study
processes with low counting rates (e.g. coincidence reactions).

Both problems are presently worked on. A new tagging
system is expected to give an energy resolution of the order
of one MeV. Improvements on the effective duty cycle of the
synchrotron will hopefully increase the intensity of the
tagged beam by a factor of 10.

Future experiments with the tagging system will include
the quasielastic production of charged pions from various
nuclei to study the influence of the final state interactions.
Additionally to the pion the recoiling nucleon will be detected in coincidence by using a time-of-flight spectrometer.

6. SUMMARY AND OUTLOOK

The performance of a tagged photon beam facility for
medium energy photonuclear physics has been described. First
measurements of the photoproduction of charged pions from
C^{12} have demonstrated that the equipment is working well. Compared to competing monochromatization techniques it is especially useful for the investigation of highly inelastic processes. The experimental results can be basically explained
by combining the elementary cross sections for pion photoproduction and pion-nucleon scattering with a simple model of
the nucleus. However, the Δ-resonance does not appear at the
expected position.

Future accelerator developments will make high energy
DC electron beams available. The corresponding increase in
the intensity of tagged photon beams will facilitate the
measurement of low cross sections. Since high energy resolution can be obtained without a further sacrifice in beam intensity the tagged bremsstrahlung photon beam will probably be
the most suitable general purpose source of monochromatic
photons for medium energy physics.

REFERENCES AND NOTES

This work was supported by the BMFT (Federal Ministry for Research and Technology).
1. J. Steinberger, A.S. Bishop, Phys. Rev. 78(1950), 494.
    J. Steinberger, A.S. Bishop, Phys. Rev. 86(1951), 171.
    R.M. Littauer, D. Walker, Phys. Rev. 86(1952), 838.
    W.S.C. Williams et al., Phys. Rev. 105(1957), 1840.
    W. Imhof et al., Phys. Rev. 108(1957) 1040.
    J.R. Waters, Phys. Rev. 113(1959), 1133.
    W. Melville McClelland, Phys. Rev. 123(1961), 1423.
    N.V. Goncharov et al., Sov. Jour. Nucl. Phys. 14(1972), 18.
    L.A. Grishaev et al., Sov. Jour. Nucl. Phys. 14(1927), 20.
2. T.R. Palfrey et al., Phys. Rev. 122(1961), 1323.
3. S. Kabe et al., Jour. Phys. Soc. Japan 19(1964), 1800.
    S. Kabe et al., Phys. Lett. 10(1964), 354.
4. I. Endo et al., Phys. Lett. 47B(1973), 469.
    S. Noguchi et al., Hiroshima University Report HUPD-7513 (1975).
    K. Baba et al., Hiroshima University Report HUPD-7804 (1978).
5. S. Arai et al., Nagoya University Report DPNU-9(1972).
6. T.A. Gabriel, Phys. Rev. C13(1976), 240.

QUASI-FREE PION PHOTOPRODUCTION FROM COPPER AND LEAD

K. Baba, I. Endo, M. Fujisaki*, S. Kadota and Y.
Sumi, Department of Physics, Hiroshima University,
Hiroshima, Japan; H. Fujii** and Y. Murata, Insti-
tute for Nuclear Study, University of Tokyo, Tana-
shi, Japan; S. Noguchi***, Department of Physics,
University of Tokyo, Tokyo, Japan; and A. Mura-
kami, Department of Physics, University of Saga,
Saga, Japan

INTRODUCTION

We have recently reported the experimental results of
charged pion photoproduction from carbon in an energy range
between 300 and 850 MeV of the incident photons[1]. Momentum
spectra at lab angles 28.4° and 44.2° have been measured and
in these spectra clearly revealed are the striking features
characteristic of the quasi-free production (QFP) of the pions.
The data have in fact been well reproduced by a plane-wave impulse approximation (PWIA) calculation in which pion absorption due to the final state interactions is properly taken into account.

The present experiment has been undertaken with the purpose of extending the above measurement to two other elements,
copper and lead, as typical examples of medium-heavy and heavy
nuclei, respectively. It is interesting to see whether the
QFP features can also be observed in these nuclei, and if

there is a certain kinematical region where the QFP is dominant. Even for such heavier nuclei, the processes $(\gamma, \pi^{\pm})$ are
very adequate to study the single particle properties of
various nuclei just as in the reactions (e,e'), (e,e'p) and
$(p,2p)$.[2]

RESULTS

The 1.3-GeV electron synchrotron at the Institute for
Nuclear Study, University of Tokyo, was used to obtain the
bremsstrahlung photon beam.   The charged pions produced by
the beam were detected by a 700-MeV/c magnetic spectrometer.
Details of the apparatus and experimental procedure employed
were the same as in 1. The experimental conditions are listed
in the table below.

element and thickness (g/cm^2)	detected particle	lab angle	range of photon energy (MeV)
natural copper	π^-	28.4°	400 − 750
(.450 ± .002)	π^+	28.4°	550 − 750
	$\pi^{\pm}$	44.2°	650 − 700
natural lead (1.21 ± .02)	$\pi^{\pm}$	44.2°	650 − 700

The photon energy was defined to ± 25 MeV by the subtraction
technique.

Typical examples of the data are shown in Fig. 1. The
spectra have a single-peaked structure similar to that for carbon observed in I. In the present case, however, the structure
is less prominent than that for carbon particularly at high
energies. Even at lower energies where the peak structure is
clearly seen, the spectral shape is far from a symmetric form,
whereas for carbon we have been able to fit the data very
well with a Gaussian distribution.

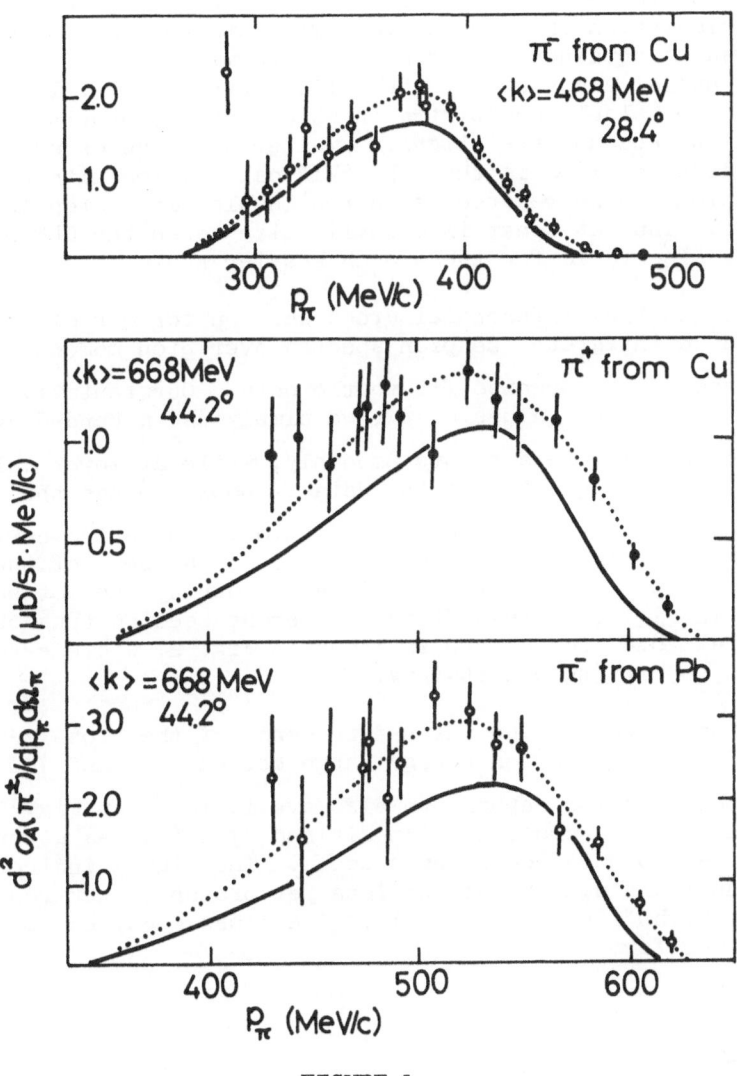

FIGURE 1

DISCUSSION

First, to test the validity of the QFP model, we make a PWIA calculation in a way quite similar to that given in 1. The results obtained by assuming the Fermi-gas model for the target nuclei are shown by the solid curves in Fig. 1. Although the calculated spectra in which pion absorption is taken into account are slightly smaller than the observed, our general feeling is that the PWIA calculation with pion absorption can be regarded as a reasonable first step approximation. Thus at least in a qualitative sense the QFP picture is also valid for these much heavier nuclei.

To get the differential cross section for QFP of pions we need to integrate the pion spectra over pion momentum p_π, and hence a more precise fit to the data. Our tentative one is that at higher values of p_π, we simply use a hand-drawn curve connecting data points smoothly, while at lower values of p_π where the quality of the data is poor, we use the spectral shape given by the PWIA calculation. These two curves are connected smoothly in the vicinity of the peak of the spectra. The results of this fit are shown by the dotted curves in Fig. 1. The differential cross section thus obtained at $28.4°$ for copper is shown in Fig. 2, where errors are due to fitting uncertainty.

Finally we examine the A-dependence of the cross section at $44.2°$ and the photon energy range between 650 and 700 MeV, where the data for carbon are also available[1]. In Fig. 3, we plot the π^+ (π^-) cross section divided by Z (N), as a function of A, both in logarithmic scale. The figure indicates that the individual cross sections per proton or neutron lie on a straight line quite nicely. The cross sections are fitted by the form

$$\frac{d\sigma_A}{d\Omega_\rho}(\pi^\pm) = C^{(\pm)} N^{(\pm)} A^{-\alpha^{(\pm)}},$$

where $N^{(+)} = Z$ and $N^{(-)} = N$. The fitted values of these paramters are $C^{(+)} = 24.5\pm2.5$ ($\mu b/sr$), $C^{(-)} = 1.3\pm2.5$ ($\mu b/$

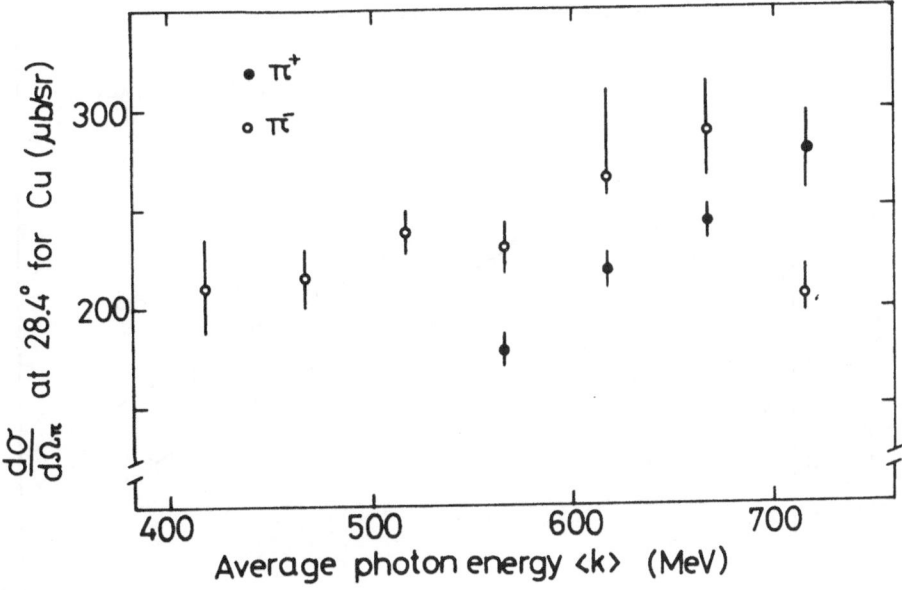

FIGURE 2

$\alpha^{(+)}$ = .30±.03 and $\alpha^{(-)}$ = .32±.04. Note that the values of α's are closely equal to 1/3. In Fig. 3 also plotted are the $\pi^{\pm}$ cross sections for deuterium taken from the data by Fujii et al.,[3] and the above fit shows an accidentally good agreement with the data.

CONCLUSION

The present data clearly exhibit that the main features of QFP of pions are observed even for sufficiently heavy nuclei provided that kinematical conditions are set up appropriately. The PWIA calculation performed in a way analogous to the carbon case shows a reasonable agreement with the data.

As for the A-dependence of the cross sections, this experiment gives a very transparent result that in the QFP, pions are produced to be proportional to the number of rele-

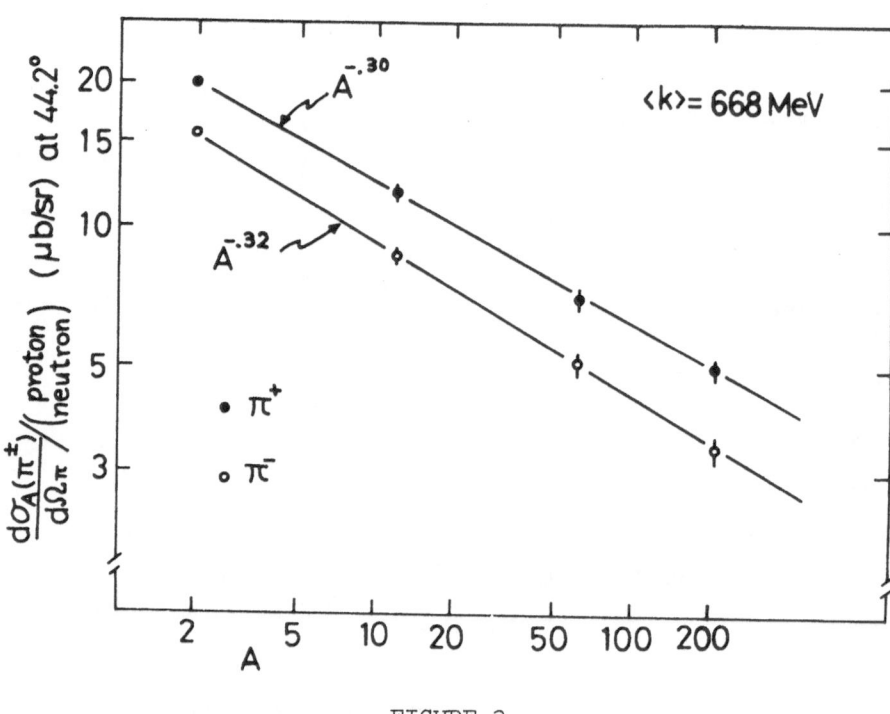

FIGURE 3

vant nucleons and the produced pions are subject to the attenuation effect approximately proportional to $A^{-1/3}$ which is inversely proportional to the path length of the pions.

REFERENCES

   * Now at CEN-Saclay, B.P. No. 2, F-91190, Gif-sur-Yvette, France.
  ** Now at Department of Physics, University of Tokyo, Tokyo.
 *** Now at Institute for Nuclear Study, University of Tokyo, Tanashi
1. K. Baba et al., Nucl. Phys., to be published.
2. See for example, E.J. Moniz et al., Phys. Rev. Letters 26, 445 (1971).
3. T. Fujii et al., Nucl. Phys. B120, 365 (1977).

PION-NUCLEUS INTERACTIONS AND PHOTOPRODUCTION*

F. Tabakin

Department of Physics and Astronomy

University of Pittsburgh

Pittsburgh, Pennsylvania   15260

I.   INTRODUCTION

In the photoproduction process the interaction of the meson with the residual nucleus plays a significant role. The basic process consists of the photon coupling to the electromagnetic currents of the constituent nucleons, the nucleons are thereby moved to act as meson radiators. The distribution of the nucleons in the nucleus and the associated currents are ascertained by using electron scattering and nuclear spectra information to provide nuclear wave functions.

The radiated pion not only scatters repeatedly from the nucleons in the nucleus, but can actually disappear. Thus flux is lost from the meson-nucleus system both by quasi-elastic pion-nucleon scatterings, wherein the pion either excites the nucleus or knocks out nucleons, or by physical annihilation of the pion. Knowledge of the strong interaction dynamics, including the above two absorption processes, is stored in the pion optical potential.

Fortunately, information about pion-nucleus scattering and associated absorption is available, and is of rapidly improving quality[1]. Experiments on pionic atoms of increased precision tell us the extent of physical absorption at low energies that is required to explain the level widths[2]. Recent

accurate pion-nucleus elastic and inelastic scattering experiments have been reported at a range of energies and for
a variety of target nuclei[3]. Between zero and 28 MeV,
however, it is very difficult, due to pion decay, to perform
such experiments. At this stage one must rely on physically-
motivated interpolations to specify the low-energy pion-nucleus interaction. At higher energies the 3-3 resonance
dominates the situation and provides considerable absorption
via the quasi-elastic scattering. Other processes, such as
(p,π), (π^+,π^-) and (γ,p) shed light on the pion-nucleus interaction and on the isobar dynamics.

In this paper, aspects of the pion-nucleus interaction
will be discussed in the context of meson photoproduction.
To discuss the role of the pion-nucleus final state interaction in the (γ,π) reaction, it is necessary to confront the
interplay of the nuclear structure dynamics, the production
operator, and the pion wave functions. It is therefore useful to separate the nuclear information into appropriately
defined transition densities and to deal with the photon and
pion systems by suitable tensor operator couplings. First I
will present some salient features of the photoproduction
operator developed for use in the nuclear context by Blomqvist and Laget[4]. Then I will indicate the tensor operator
structure of a γ,π calculation[5], with emphasis on the extraction of the requisite nuclear transition densities, including
those arising from nonstatic terms due to nucleon motion, recoil and from the dependence of the photoproduction process
on pion momenta. Hence we will see clearly how to use recently developed optical potentials, which incorporate the
strong interaction information[6], to obtain the distorted pion
waves needed in the γ,π work. Finally, I will comment quantitatively on the sensitivity of photoproduction to the pion
waves. Several topics for future research are also delinea-
ted.

II.   THE PHOTOPRODUCTION OPERATOR

Blomqvist and Laget (BL)[4] have developed a photoproduction operator that can be conveniently used in a nuclear context if one assumes that the amplitude determined by produc-

tion from a single nucleon applies to the emission from a nucleon moving in a nuclear medium. There are, of course, several basic problems that one overlooks by making the above impulse approximation. The effect of Pauli excusion on intermediate state nucleons, of the nucleon binding, and of nonlocal aspects due to finite ranged propagators are being set aside by assumign that a zero-ranged, nonrelativistic form of production operator is appropriate for the nuclear problem. It is a reasonable beginning and certainly goes beyond the often used $\vec{\sigma} \cdot \hat{\epsilon}$ form. Although BL make a non-relativistic reduction, they keep terms to order p^2/m_N^2 and, at least approximately, provide an operator valid in a general frame of reference. Hence, their operator is particularly convenient for studying photoproduction from moving nucleons in the nucleus; it includes the nonstatic nucleon recoil terms and dependence on pion momenta. Their form has been fitted to photoproduction data from a nucleon, but with our subsequent approximations needed for simple application in coordinate space studies, we have restricted its use to pions of ≤ 50 MeV.

For π^- production, their (pseudoscalar) production operator is

$$G_1 \, \vec{\sigma} \cdot \hat{\epsilon}_\lambda + G_2(\vec{\sigma} \cdot \vec{q})(\vec{\sigma} \cdot \vec{k} \times \hat{\epsilon}_\lambda) + G_2(\vec{\sigma} \cdot \vec{k} \times \hat{\epsilon}_\lambda)(\vec{\sigma} \cdot \vec{q}) +$$

$$+ G_3[\vec{\sigma} \cdot (\vec{P}_f - \vec{P}_i)](\vec{q} \cdot \hat{\epsilon}_\lambda) + G_4(\vec{P}_f \cdot \hat{\epsilon}_\lambda)(\vec{\sigma} \cdot \vec{q}) + H_{\gamma\pi-}^\Delta$$

$$\equiv J^5 \cdot \hat{\epsilon}_\lambda = H_{\gamma\pi-} \tag{2.1}$$

where the isobar (3-3 resonance) term $H_{\gamma\pi-}^\Delta$ will be discussed later. One can readily extract the pseudovector current $\vec{J}^5$ from the above equation, but the first form permits ready identification with the basic diagrams describing the process, as shown in Fig. 1. Three of these diagrams describe post emission (D1) and pre-emission (D2) of pions and also the "photoelectric" term (D3) which involves the pion's electric current. With the production of pions one associates the usual pseudoscalar $\vec{\sigma} \cdot \vec{q}_\pi$ operator, while the absorption of the photon by a neutron (D1) involves a $\vec{\mu}_N \cdot \vec{B} \rightarrow \mu_N \vec{\sigma} \cdot (\vec{k} \times \hat{\epsilon}_\lambda)$

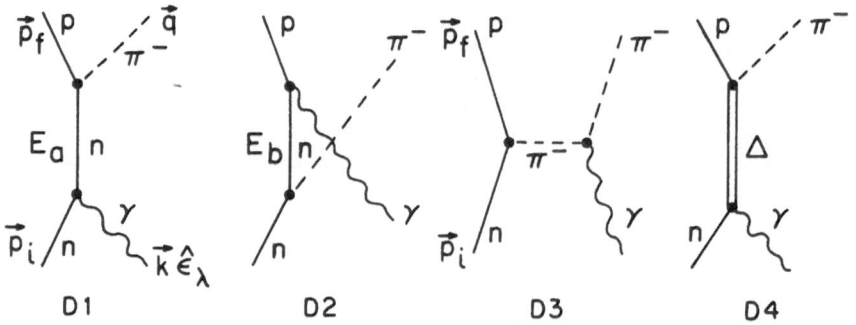

Figure 1.   The basic photoproduction diagrams relating to
Eq. 2.1.

term (G_2). The proton can absorb a photon (D2) by its magnetic moment $\vec{u}_p \cdot \vec{B} \to \mu_p \vec{\sigma} \cdot (\vec{k} \times \hat{\epsilon}_\lambda)$, or by its convective current
which involves a $\vec{j} \cdot \vec{A} \to \vec{p}_f \cdot \vec{\epsilon}_\lambda$ in D2 (see terms G_2'). The full
convective current of the pion, along with the virtual pion
emission via $\vec{\sigma} \cdot (\vec{p}_f - \vec{p}_i)$, accounts for the G_3 term in Eq.
(2.1). The $\vec{\sigma} \cdot \hat{\epsilon}_\lambda$ term is the usual static result for a γ_5
pseudo-scalar coupling which involves the antinucleon terms
buried in D2 and D1. Blomqvist and Laget also provide the
results for a p.v. theory and for π^+, π^0 production.

The above physical description of the BL operator correlates with their result for the amplitudes:

$$\frac{G_1}{i} = g \left\{ \frac{2m_N^2 \left(1 + \frac{\kappa_p k}{2m_N}\right)}{E_b(E_b + P_b^{\,0})} + \frac{2m_N^2 \kappa_n k}{2m_N E_a (E_a + P_a^{\,0})} \right\}$$

(2.2a)

$$G_2 = \frac{\mu_N g}{2E_a(P_a^{\,0} - E_a)} ; \quad G_2' = \frac{\mu_p g}{2E_b(P_b^{\,0} - E_b)}$$

(2.2b)

$$\frac{G_3}{i} = \frac{-2g}{(q - k)^2 - m_\pi^2} ; \quad \frac{G_4}{i} = \frac{-g}{E_b(P_b^{\,0} - E_b)}$$

(2.2c)

where $\sqrt{2}\ g = 4\pi g_\pi e/m_N$ (the electromagnetic · strong coupling), and the magnetic moments $\mu_N = \kappa_K = -1.91$ $\mu_p = 1 + \kappa_p = 2.78$ in usual notation. The energies $E_a^{} = \sqrt{p_b^2 + m_N^2} \simeq m_N$, $P_b^{\ o} = E_i - k \simeq m_N - k,$ $P_a^{\ o} = E_\pi + E_f \simeq m_N + m_\pi$ with $\vec{P}_a = \vec{P}_f + \vec{q}, \vec{P}_b = \vec{P}_i - \vec{k}.$ Under the above approximation the amplitudes (G) are shown in Fig. 2a as a function of pion energy in the laboratory frame for a $^{12}C(\gamma,\pi^-)\ N^{12}$ reaction. In general the G's involve operators, but are greatly simplified by the above approximations; they do, however, limit our use of the BL amplitude to ≤ 50 MeV. From Fig. 2a, it is clear that the $\sigma \cdot \epsilon$ dominates, that the magnetic moment terms are smaller (with the larger term being $f \approx (|\mu_p| + |\mu_N|)$ and the proton convective current G_4 term is quite small. The pion propagator has been roughly approximated by $(q - k)^2 - m_\pi^{\ 2} \simeq -2E_\pi E_\gamma$, which accounts for the fall-off of G_3 with E_π.

The isobar term (D4) has been described by BL as

$$H_{\gamma\pi-} = \frac{(C_\pi G_3)(C_\gamma G_1)}{W^2 - M^2 - i\Gamma M} \ (\vec{S}\cdot\vec{Q})\ [S^+\cdot\vec{K} \times \hat{E}_\lambda], \tag{2.3}$$

where $\vec{Q} = \vec{q} - \frac{q_o}{M} \vec{P}_\Delta \simeq \mu(\vec{v}_\pi - \vec{v}_\Delta)$ and $\vec{K} = \vec{k} - \frac{(M - m)}{m} \vec{P}_i \simeq$ $M(\vec{v}_\Delta - \vec{v}_i)$ include a recoil effect at the pion production $(C_\pi G_3)$ and photon absorption $(C_\gamma G_1)$ vertices. The operator S couples the spin 1/2 nucleon to the spin 3/2 isobar; the bilinear form can be reduced using $(\vec{S}\cdot\vec{A})(\vec{S}^+\cdot\vec{B}) = \frac{2}{3} (\vec{A}\cdot\vec{B})$ $- \frac{1}{3} i\ \vec{\sigma}\cdot(\vec{A} \times \vec{B})$ to an operator in the nucleon space. The leading terms are expected to be operators of type G_2 and G_2'; $\frac{2}{3} \vec{q} \cdot (\hat{k} \times \hat{\epsilon}_\lambda)$ and $- \frac{1}{3} i\ \vec{\sigma} \cdot (\vec{q} \times (\vec{k} \times \hat{\epsilon}_\lambda))$, times the resonant form $(W^2 - M^2 - i\Gamma M)^{-1}$ determined by the isobar's mass (M = 1232 MeV) width Γ, and total c.m. energy W. Of course,

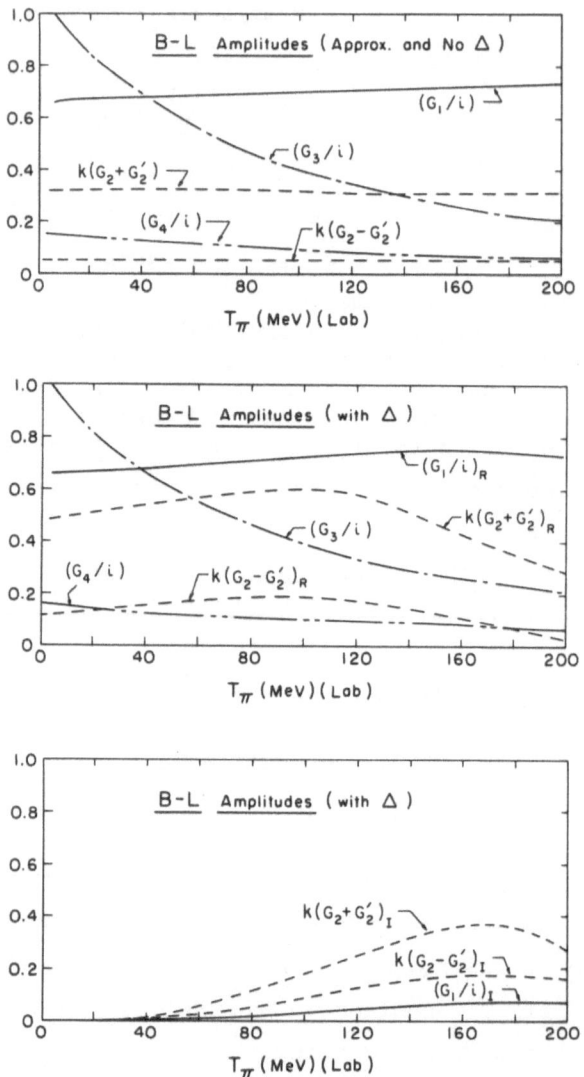

Figure 2. The BL amplitudes, G, versus pion lab. energy for
$C^{12}(\gamma,\pi^-)N^{12}$. Approximations made are: nucleon energies
$E_{a,b}$ are replaced by the nucleon mass and the pion propagator
term G_3 is replaced by $G_3/i = g/E_\pi k$. (a) no isobar term,
(b), (c) isobar of BL turned on.

these parameters must describe the isobar dynamics in the nuclear medium and, in the 3-3 resonance region, might be extractable from the associated isobar dynamics in pion-nucleus and γ,p reactions[8].

The effect of the isobar on the terms given in Eq. (2.1) is shown in Figs. 2b and 2c. Additional operators that are generated are included in M. Singham's contribution to this symposium. The $\vec{\sigma} \cdot \hat{\epsilon}_\lambda$ and the magnetic moment terms are enhanced; the essential effect is due to the $\mathrm{Re}(W^2 - M^2 - i\Gamma M)^{-1}$ which peaks just before its zero at $\sim$ M. The Im $(W^2 - M^2 - i\Gamma M)^{-1}$ is positive and has the usual resonant behavior, which accounts for the effect shown in Fig. 2c; here the G_2 and G_2' magnetic terms are indeed the ones that are mostly influenced by the isobar.

There are several advantages to the above BL operator. It does give an approximate description in a moving frame. It fits the basic $\gamma\pi$ data and it incorporates an isobar effect. Nevertheless, there are several basic problems. One is that the Born expressions viewed as driving terms generate the resonance itself - so one might have a serious double counting problem. Secondly, the imbedding of (2.1) in a nuclear system is subject to remormalization and a multitude of medium effects.

Despite these difficulties, it is a form that is convenient and provides insight into the production dynamics. Finally the coordinate space version of (2.1) is readily obtained[7] once the G's are simplified, by the substitution $iq \rightarrow \nabla_\pi$ and $ip \rightarrow \nabla_N$ where π and N denote pion and nucleon derivatives. This is in contrast to the CGLN versions in which awkward F_2/qk forms appear.

For the approximate forms of G's used here the photoproduction from a nucleon is modified as shown in Fig. 3; clearly the approximations agree with the full BL result only for $\leq$ 50 MeV. Above that the individual terms, for example, from the delta, deviate from the BL fit. Hence one must include the energy and momentum dependence of the G's above 50 MeV, which will require the development of momentum space techniques.

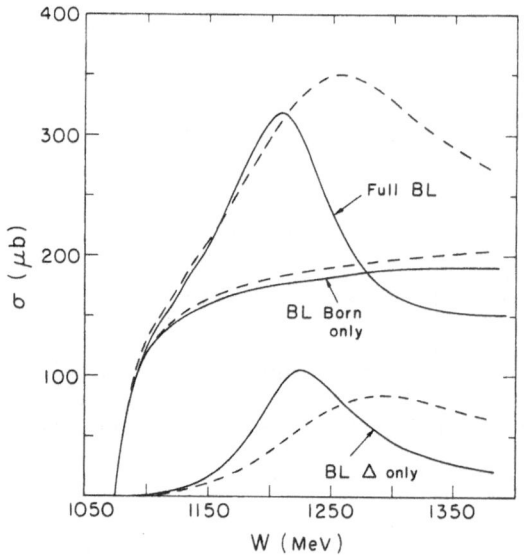

Figure 3. The basic meson photoproduction cross-section from a nucleon in microbarns versus the total c.m. energy W. Solid curves are for the full BL amplitude, with the Born (D_1, D_2, D_3) and isobar (D_4) contributions indicated. The dashed curves result when the approximations described in Figure caption 2 are made. This result limits the π-nucleus coordinate space calculation to $T_\pi \leq 50$ MeV, unless nonstatic aspects of the amplitudes G are treated.

III. OPERATOR STRUCTURE OF PHOTOPRODUCTION

The task of calculating photoproduction cross-sections from nuclear targets involves the many-body matrix element of the one-body operator generated by the BL operator $H_{\gamma\pi}$

$$M_{fi} = < n'|F|n> = \sum_{\alpha\beta} < n'|a_\alpha^+ a_\beta|n> < \alpha|\tau^- \vec{J}\cdot\hat{\varepsilon}_\lambda^5 e^{ik\cdot r} \phi_{-q}^{(+)}|\beta>,$$

$$(3.1)$$

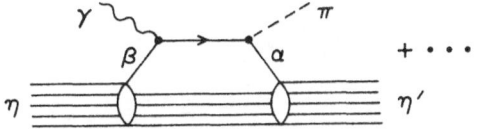

Figure 4. Schematic picture of the one body operator
applied to a nucleus.

along with the pion distorted wave $\phi^{(+)}_{-q}$ and the photon field
$\hat{\epsilon}_k e^{ik \cdot r}$. Here η and η' denote $JMTT_3$ for the initial and
final nuclei and α labels the single particle orbitals by
$nljmm_\tau$. The basic piture of the one body operator is shown
in Fig. 4, where the single nucleon β is extracted, acted
on by $H_{\gamma\pi-}$ (diagrams D1-D4) and then returned to level α of
the residual nucleus η', which for the π^- case has one less
neutron and one more proton than the target nucleus η. The
one-body nature is evidently a consequence of ignoring the
interactions of the propagating nucleon with the rest of the
nucleus via either meson exchanges (exchange currents) or by
the N-nucleus interactions (binding and Pauli effects).

It is understood that $H_{\gamma\pi-}$ includes derivatives ∇_N and
∇_π acting on the nucleons and pions, respectively. Our intent
is to separate the above matrix element into a purely nuclear
part times a tensor combination of the photon-pion system.
In addition to a momentum exchange of $\vec{q} - \vec{k}$, the photon-pion
system has an angular momentum exchange with the nuclear
system. This defines the tensor operator structure of the
dynamics illustrated in Fig. 5, where KQ is the total angular
momentum transferred from the $\gamma\pi$ system to the $\eta\eta'$ system,
while L' refers to the associated orbital angular momentum
transfer. The label S indicates if a spin operator $\vec{\sigma}$ occurs
(S=1) or not (S=0), which permits a spin angular momentum exchange. The orbital angular momentum L refers the result of
coupling of the pion to the photon orbital angular momentum.
The coupling scheme is shown in Fig. 6, where the photon spin
enters as shown.

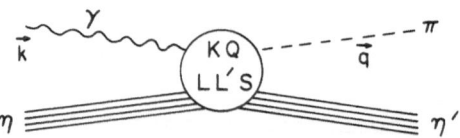

Figure 5. Illustration of the dynamics relevent to the tensor operator structure of the photon-pion system as discussed
in the text.

To carry out the above general scheme for separating
the nuclear dynamics from the $\gamma\pi$ system, the following dyadic
is useful.

$$\overleftrightarrow{1}_\Omega = \sum_{KQL} \int \overleftarrow{Y}^{KQ}_{L1}(\hat{r})\ \vec{Y}^{KQ*}_{L1}(\hat{r}')\ d\Omega_{\hat{r}'}, \qquad (3.2)$$

where $\vec{Y}$ is a vector spherical harmonic and an angle-space
integration is performed. For example, the one particle
matrix element $\langle\alpha|\vec{\sigma}\cdot\hat{\epsilon}|\beta\rangle$ can be separated using $\overleftrightarrow{1}_\Omega$ into

$$\int R_\alpha(\vec{r})\ d^3r(\vec{\sigma}\cdot\overleftarrow{Y}^{KQ}_{L1}(\hat{r}))\ R_\beta(\vec{r})\ \chi^{KQ}_{L1\lambda}(r) \qquad (3.3)$$

where the last quantity involves only a projection in angle

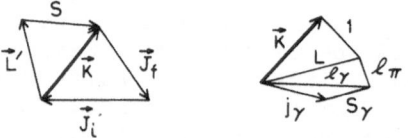

NUCLEAR $(\eta\eta')$ ×   PION-PHOTON (γ,π)

Figure 6. Coupling scheme of the various angular momenta
relevent in pion-nucleus photoproduction. The meaning of
the various symbols is discussed in the text.

space:

$$\chi_{L1\lambda}^{KQ}\ (r) = \int d\Omega_{\hat{r}'}\ \vec{Y}_{Li}^{KQ*}\ (\hat{r}')\ \cdot\ \vec{\epsilon}_\lambda\ e^{i\vec{k}\cdot\vec{r}'}\ \phi_{-q}^{(+)}\ (\vec{r}')\{\begin{matrix} \vec{r}':\ r\theta'\phi' \\ \vec{r}:\ r\theta\phi \end{matrix}$$

$$(3.4)$$

Following this procedure for all terms in $H_{\gamma\pi-}$, there results only three basic types of transition densities which are defined as <u>angle-space</u> reduced matrix elements

$$\rho_{LLS}^{K}(r) \equiv [L]^{-1}\ <n'||(\vec{\sigma}^S\cdot\vec{Y}_{L1}^K)||n> = [L]^{-1}\ <n'||(Y_L \times \sigma^S)_K||n>$$

$$(3.5a)$$

$$\rho_{LL'S}^{'K}(r) \equiv [L]^{-1}\ [L']^{-1}\ <n'||((Y_L \times \vec{\nabla})^{L'} \times \sigma^S)_K||n> \qquad (3.5b)$$

$$\tilde{\rho}_{LL'S}^{'K}(r) \equiv [L]^{-1}\ [L']^{-1}<\ \ ||((Y_L \times \overleftarrow{\nabla})^{L'} \times \sigma^S)_K||n>. \qquad (3.5c)$$

$[x] = \sqrt{2x+1}$, and $\vec{\nabla}$ acts on the final nucleon variables (i.e., it refers to $\vec{P}_f$). For a pure ℓ-shell ρ' and ρ' are identical. To evaluate the above transition densities we can express them in terms of single particle matrix elements (labelled by a, b: nlj); for example:

$$\rho_{LLS}^{K}(r) = \sum_{ab}<J_fT_f||(a_a^+ \times \tilde{a}_b)_{k;1}||J_iT_i>(-1)^{\ell_a+1}[S][j_a][j_b]$$

$$\begin{Bmatrix} j_a j_b K \\ \ell_a \ell_b L' \\ .5.5\ S \end{Bmatrix} \begin{bmatrix} \ell_a \ell_b\ L \\ o\ o\ o \end{bmatrix} R_a(r)R_b(r) \qquad (3.6)$$

Similar expressions hold for $\rho\acute{}$ and $\tilde{\rho}\acute{}$ except they involve the usual radial derivatives of the single particle orbitals and 6j symbols. Tensor operators of the ρ type have been discussed earlier by Saunders[9]. Also the above forms appear in studies of electron scattering[10], except for the intermediate coupling of the pion plus photon to a value L', which is a special feature due to needing $\ell_\pi > o$ above threshold.

In Figs. 7a-c, examples of these transition densities are shown for the Cohen-Kurath (CK), O'Connell, Donnelly and Walecka (DW), and Haxton (H)[11] values of $<J_f T_f || (a^+ \times \tilde{a})_{K1} || J_i T_i>$ which were determined by fits to M1 inelastic scattering to analogue states of those occurring in photoproduction. The ρ' densities result from nucleon derivatives appearing in $H_{\gamma\pi-}$. The three cases have similar ρ's except for the ρ_2 case of Haxton, which is quite small due to his density matrix for the 1/2 1/2 case (which involves L' = 2). Similarly the other L' = 2 transition densities $\rho\acute{}$ seen in Figs. 7b,c are small for Haxton's fit compared to CK and ODW values. The use of a Helm model corresponds to a simple choice of ρ only, ignoring the $\rho\acute{}$ and $\tilde{\rho}\acute{}$ nuclear derivative transition densities. It turns out to be a reasonable assumption at low energies, but might not suffice at higher energies. Clearly $\rho\acute{}$ terms are small (see the scale in Fig. 7b,c) and the nuclear derivative amplitudes (G_3 and G_4) are also smaller than the other terms.

An important aspect of these transition densities is that they peak in the region of the nucleons active in the photoproduction process, i.e., in the surface for $\ell = 1$ orbitals of C^{12}. The derivative terms, aside from the L' = 0 case, have a zero at the nuclear edge and behave as shown in Figs. 7b,c . It appears that volume information might depend on the derivative or nucleon motion in $H_{\gamma\pi-}$ playing a role.

To obtain the photoproduction cross-section one must combine the amplitude of Fig. 2 with the transition densities of Fig. 7, and the photon-pion "wave function" $\chi(r)$ in the following scheme:

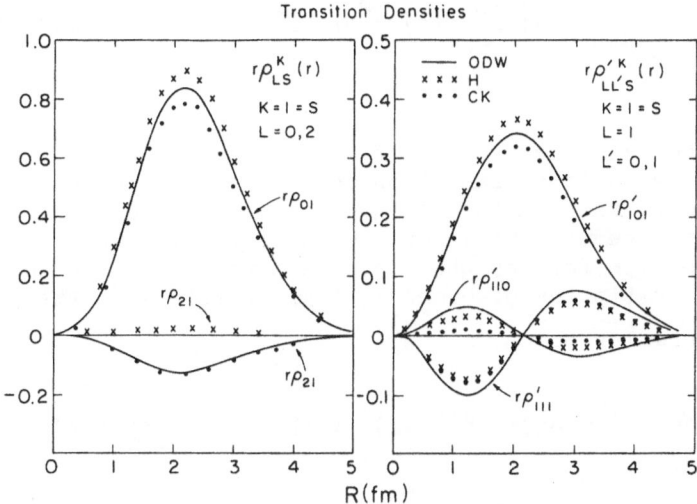

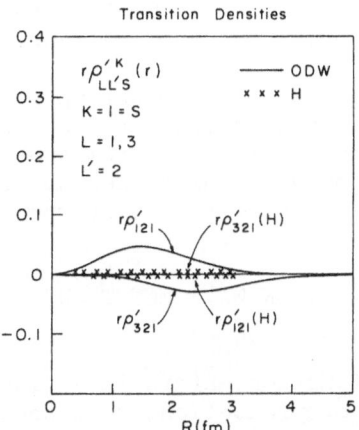

Figure 7.   The transition densities as defined by Eq. (3.5) for the O'Connell, Donnelly, Walecka (ODW), the Cohen-Kurath (CK) and the Haxton (H) values of $\langle \eta' || (a_a^+ \times \tilde{a}_b)_{K;1} || \eta \rangle$. The ρ' and $\tilde{\rho}'$ cases include the nucleon momentum oerators as they appear in $H_{\gamma\pi}$ (Eq. 2.1). Note for $L' = 2$, the Haxton case yields different transition densities.

$$M_{fi} = \sum_{KQLL'S} G \int \rho_{LL'S}^{KQ}(r) \, r^2 dr \, \chi_{L'}^{KQ} \, (r) \; + \; \rho' \text{ and } \chi'$$

terms. $\hspace{12cm}$ (3.7)

This symbolizes the process of combining the BL amplitudes, G, with the various transition densities and pion-photon functions χ, wherein the ρ' (nucleon derivative) χ' (pion derivative) terms also enter. Using Eq. (3.4) and a multipole expansion for the photons, one obtains for the case of magnetic multipoles of order j_γ (here ℓ refers to the pion orbital angular momentum)

$$\chi_{L1\lambda}^{KQ}(r) = \sum_{\substack{\ell m \\ j_\gamma > 1}} Y_{\ell m}^*(-q) i^{\ell+j_\gamma} (-1)^m [\hat{\ell}][\hat{L}] \phi_\ell^{(+)}(qr) j_{j_\gamma}(kr)(2j_\gamma$$

$$\times \begin{pmatrix} j\gamma\ell L \\ o o o \end{pmatrix} \begin{pmatrix} \ell L j\gamma \\ m m o \end{pmatrix} \hspace{4cm} (3.8)$$

Similar terms occur for electric multipoles. For χ'-type terms one tends to invoke pion derivatives.

In summary, we have found tensor operations for the nuclear transition densities and for the photon-pion system which account for the angular momentum transfer of the process. Now the pion waves have been separated and we are ready to discuss the role of pion distorted waves.

III. PION WAVE FUNCTIONS

The pion wave function enters in a crucial way in the evaluation of the photoproduction cross-section. It contains the physics of the final state collisions and pion annihilation near the residual nucleus. Much theoretical and experimental work has been devoted to determining that physics, which is customarily stored in an optical potential. It is important to realize that for $\gamma\pi$ and other reactions the wave function itself is needed, not just its asymptotic form which is directly related to the pion elastic scattering properties. Indeed, once the $\gamma\pi$ dynamics is understood it is possible that this reaction might be used to study the behavior of a pion in the vicinity of a nucleus even at low energies. Realization of that hope depends on the sensitivity of the process to the details of the pion waves $\Psi_{\pi\eta}$.

The optical potential which is most often used to determine $\Psi_{\pi\eta}$ is of the form[12]

$$2E_\pi V_{\pi-} = -4\pi[P_1 b_0 + P_1 b_1(\rho_n - \rho_p) + b_2 \nabla^2 \rho + P_2 B_0 \rho^2 - \vec{\nabla} \alpha(r)\vec{\nabla}]$$

where $\alpha(r) = \beta(1 + \frac{4\pi}{3} \xi \beta)^{-1}$ has the Ericson-Ericson form which arises from a polarization of the nuclear medium due to the p-wave pion field (or from higher order terms in multiple scattering). The function $\beta(r)$ is

$$\beta(r) = P_1^{-1} c_0 \rho(r) + P_1^{-1} c_1 (\rho_n - \rho_p) + P_2^{-1} C_0 \rho^2(r)$$

Here $\rho \equiv \rho_n + \rho_p$ and ρ_n, ρ_p are the neutron and proton densities normalized to N and Z, respectively. The above form applies for π^-; for π^+ one switches $\rho_n \leftrightarrow \rho_p$. The parameters $P_1 = 1 + (m_\pi/m_N)$ and $P_2 = 1 + (m_\pi/2m_N)$ account roughly for the transformation of the basic pion-nucleon interaction to the c.m. frame of the pion-nucleus system. The $\nabla^2\rho$ term also originates, in part, from that transformation.

The energy dependent paramters (b's and c's) can be related to the basic pion-nucleon amplitudes, when it is assumed that simple multiple scattering theory holds. Under that assumption b_0 originates mainly from the πN s-waves, b_1 from the associated isovector terms, while c_0 and c_1 arise mainly from the π-N, p-wave amplitudes, with c_1 carrying the isospin dependence. The $\rho^2(r)$ terms are a simple, indeed an overly simple, way of introducing annihilation of the meson; the quadratic form being suggested by the process of a pion depositing its energy and momentum onto two nucleons. The parameters B_0 and C_0 are consequently complex with the energy denominators associated with the $\pi + 2N \rightarrow 2N \rightarrow \pi + 2N$ process suggesting a form $|B_0|(-1+i)$. Such forms have been studied in a recent analysis of accurate pionic atom data by Batty et al.[2] The annihilation process and the associated optical potential representation will surely be studied further; it is of great importance for the $\gamma\pi$ process.

The complex nature of $V_{\pi\eta}$ originates from two sources: the above $\rho^2(r)$ terms for actual annihilation and the possibility of using complex b's and c's. The latter case represents quasi-elastic absorption wherein an elastic pion-nucleon collision leads to a recoil nucleon which causes the nucleus as a whole to leave the pion-nucleus elastic channel. At low energies one expects the above recoil effect to vanish and hence to have real b's and c's. From pionic atom information we know that the physical annihilation process (i.e., the $\rho^2(r)$ terms) are still on at low energies; consequently, the pion-nucleus cross sections display exothermic behavior at low energies, e.g., σ_{TOT} and σ_{abs} ~ v_π^{-1} at low energies (in the absence of a Coulomb potential). As the energy increases the physical absorption is probably fairly constant, but the quasi-elastic terms turn on.

This behavior is reflected in the parameters shown in Table I. The 0-10 MeV parameters are taken directly from the pionic atom studies of Hüfner, Tauscher and Schneider (HST)[13]. As expected from medium effects, these parameters differ from those obtained from πN scattering. The process of extracting these parameters from pionic atom experiments is subject to further study. For example, it is quite possible that neglected V_N^2 terms (leading to $b_0^2\rho^2$ contributions) might alter the parameters requisite for $\gamma\pi$ calculations[13].

As we proceed to higher energies in Table I, the c_0, c_1 parameters are made to acquire complex parts, corresponding to a turning on of quasi-elastic absorption which is a major feature of the recent Stricker et al. study of $V_{\pi\eta}$.[6]

TABLE I

$$V_{\pi\eta} \quad \text{Parameters}^{6,13}$$

	0-10 MeV	30 MeV	40 MeV	50 MeV
b_0	-0.0161	-0.03	-0.03	-0.03
b_1	-0.0873	-0.08	-0.08	-0.08
b_2	0	0.023 + 0.001i	0.024 + 0.001i	0.026 + 0.001i
B_0	-0.06 + 0.052i	-0.04 + 0.04i	-0.04 + 0.04i	-0.04 + 0.04i
c_0	0.23	0.23 + 0.014i	0.023 + 0.021i	0.23 + 0.030i
c_1	0.22	0.22	0.22	0.22 + 0.015i
C_0	0.076i	-0.076i + 0.076i	-0.076 + 0.076i	-0.076 + 0.076i
ξ_{LL}	1	1	1	1

Instead of discussing the additional aspects of their paper and of the Colorado potential[15] in which Pauli blocking and alternate absorptive terms are used, let us now examine the general features of the optical potential in the 10-50 MeV region. In Fig. 8, the fit we find for π^+C^{12} elastic scattering using the Stricker parameters of Table I is shown compared to the (revised) CMU data[3]. The corresponding reflection coefficients are shown in Fig. 9, where the CMU values are displayed along with those found at 50 MeV using Stricker's parameters. Also shown are the HST potential results at 1 MeV. For comparison a typical η_L is shown for a 200 MeV pion, where the greatly increased absorption is due to the quasi-elastic mechanism through the 3-3 resonance. At low energies the annihilation process dominates, but as

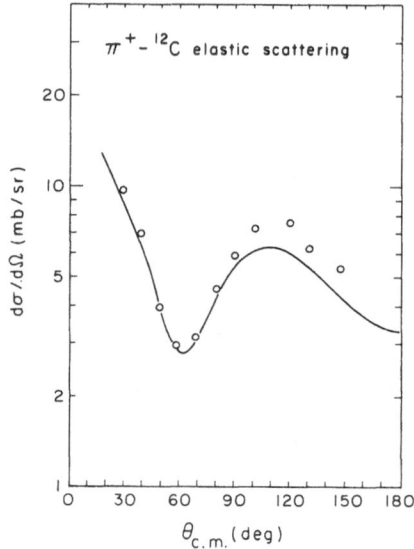

Figure 8. The $\pi^+ - C^{12}$ elastic scattering differential cross section obtained using Stricker et al. parameters[6] compared to the recent CMU measurements.[3]

one proceeds from 1 to 50 MeV a significant quasi-elastic effect is seen in Fig. 9. The isospin dependence is seen by comparing the C^{12} to N^{12} results. Another significant aspect of Fig. 9 is that the increased absorption of p relative to s-waves is used to provide the interference needed to fit the CMU data (Fig. 8).

As a further indication of the role of absorption on the pion-nucleus dynamics, the wave functions of Figs. 10 and 11 should be viewed. First consider the overall behavior displayed in Fig. 10 for $\pi^- - Ca^{48}$ scattering as calculated by Phatak from 100 to 250 MeV pion energy. Here one sees that absorption due to the 3-3 resonance damps the interior function making the surface region likely to be dominant in the $\gamma\pi$ process at these energies. Interesting quarter-wave plate effects can be invoked to explain some of the bumps in the η_L vs. L plots. These wave functions are normalized as $|qr\phi^{(+)}| \to \frac{1}{2} e^{-i\theta}L - S_L e^{+i\theta}L$, which for

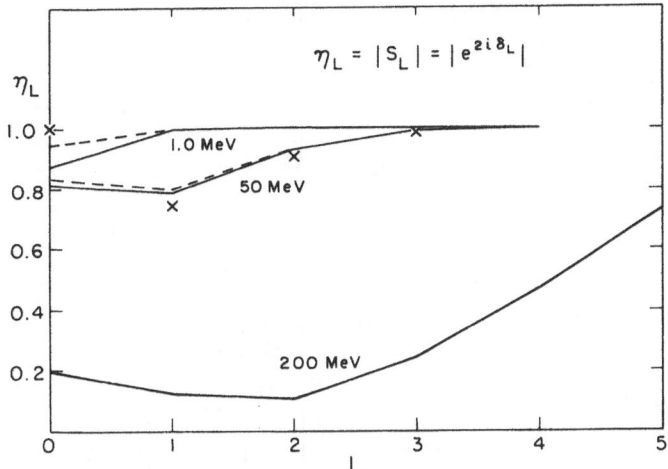

$$\eta_L = |S_L| = |e^{2i\delta_L}|$$

Figure 9. The reflection coefficients η_L vs. pion angular momentum L. The solid curves are for a N^{12}, dashed curves for a C^{12} target; wherein, the effect of the isospin term in the optical potential is seen. For comparison, η_L for a 200 MeV pion on N^{12} is shown, where one sees the large absorption arising from quasi-elastic scattering. The crosses indicate the CMU[3] values for $\pi^+ - C^{12}$.

large r goes to 1 for no absorption or to 0.5 for complete absorption. The corresponding pion wave functions at 1 and 50 MeV for N^{12} are shown in Fig. 11. Note the 50 MeV waves are really smaller than the 1 MeV case because of the pion momentum in $|qr\phi^{(+)}|$. The approximation of factoring out a constant $\phi_o^{(+)}$ at low energies seem reasonable from Fig. 11, but for $\ell>0$ and at higher energies caution is required to include the curvature of $\phi_{\pi\eta}$. The curvature of $\phi_{\pi\eta}$, of course, determines the contribution of q_π-dependent terms of the photopion operator. The terms have been studied, along with $\ell > 0$ and isobar contributions and are presented in M. Singham's contribution.

Here the sensitivity of the photopion process to changes in the pion wavefunction is studied by a phase shift equiva-

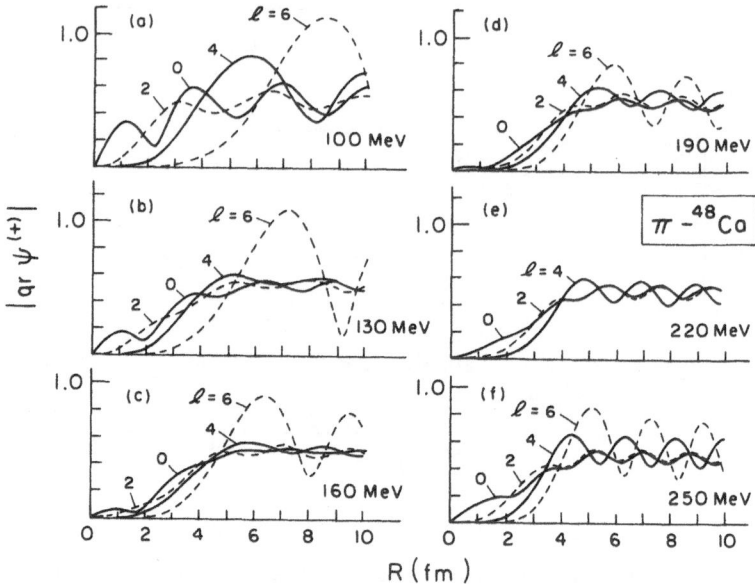

Figure 10. The π-Ca^{48} wave functions from 100 MeV to 250 MeV pion energy. Here the onset of quasi-elastic absorption via the 3-3 resonance is clearly seen to yield an internally damped wave function. Note also that $|qr\phi^{(+)}|\to 1/2$ when $\eta_L\to 0$, i.e., for complete absorption. Some wave function penetration occurs, but clearly one has from this behavior a predominance of surface production at these higher energies.

lent potential approach.

IV. EQUIVALENT OPTICAL POTENTIALS

The basic idea of a phase shift equivalent (PSE) potential is to introduce a short-ranged unitary transformation to modify the scattering wave function without changing its asymptotic value. Elastic scattering information in $\phi_{\pi\eta}$ can thereby be held fixed and short-distance effects can be neatly probed. B. Keister has recently used that approach

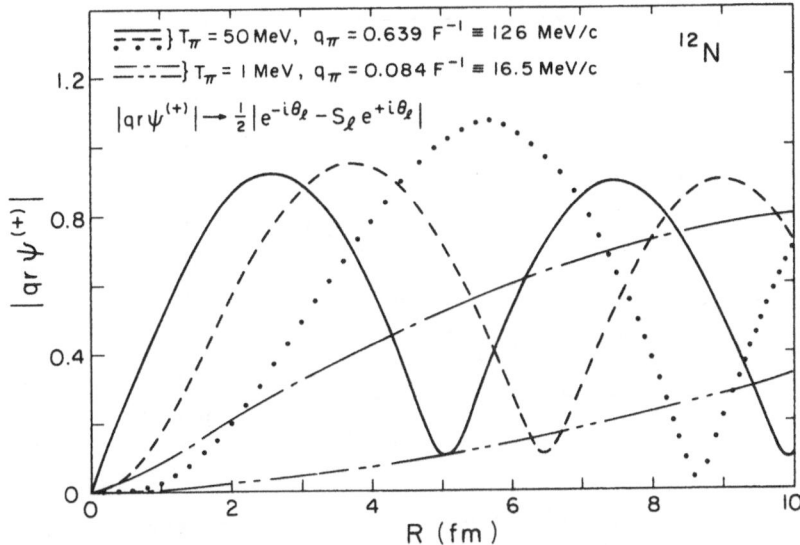

Figure 11. The $\pi^- - N^{12}$ wave function for the optical potentials of $(HTS)^{13}$ at 1 MeV, and Stricker et al.[6] at 50 MeV.
The absorption is small compared to that seen in Fig. 10,
as is clear from Fig. 9. At 1 MeV absorption is from annihilation; some quasi-elastic absorptions is turned by 50 MeV
and develops at 200 MeV to be the dominant source of absorption (Fig. 9).

for pion-nucleus optical potentials and has studied the influence of the pion wave function on inelastic scattering.[16]
One of his transformations consists of a modification of the
pion's phase velocity in the nuclear surface region. He
makes the transformation

$$\tilde{\phi}^{(+)}(r) = e^{i\lambda\rho(r)} \phi^{(+)}(r) \xrightarrow[r \gg R]{} \phi^{(+)}(r), \qquad (4.1)$$

where since $\rho(r) \to 0$ beyond the nuclear surface R, the asymptotic form of ϕ is preserved. The norm of the wave function
is unchanged, but the phase is modified by $\lambda\rho(r)$. Conse-
quently the phase velocity is affected as $\tilde{v}_\phi = v_\phi + \lambda \dfrac{d\rho}{dr}$;
i.e., in the surface.

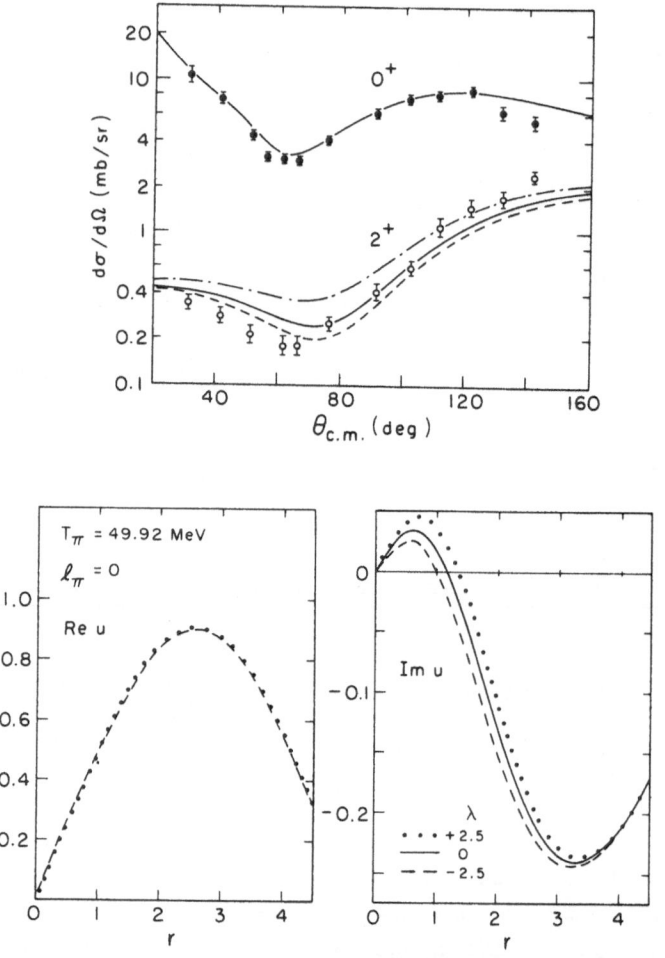

Figure 12. (a) Keister's[16] results for a phase shift equivalent potential of gauge transformation type (Eq. 4.1) applied to the $C^{12}(\pi,\pi')\,C^{12}*$ inelastic scattering to the 2^+ final state and for $\lambda = 0,\,\pm\,2.5\,F^3$. The wave function given in Fig. 7 for 50 MeV still applies but changes in the Real (u) and Imag (u), with our normalization, that are generated are shown in (b). These changes correspond to a modified phase velocity $\tilde{v}_\phi - v_\phi = \lambda\,\dfrac{d\rho}{dr}$ in the nuclear surface region.

The inelastic scattering $C^{12}(\pi^+,\pi^{+'})C^{12}$ is significantly affected by the above change in the pion's phase velocity as is seen in the semi-log plot of Fig. 12 taken from a preprint by B. Keister. Using our wave function normalization the changes in $\phi_{\pi\eta}$ due to (4.1) with $\lambda = \pm 2.5$ F^3 is mainly in the imaginary part (Fig. 12). The effect on the $\gamma\pi^-$ cross-sections for $C^{12}(\gamma\pi^-)N^{12}_{g.s.}$ proved to be slight. For example, at 50 MeV we get:

λ	$\sigma_T(ub)$	$\sigma(o^o)$
0	7.716	4.2175
+2.5	7.707	4.2166
-2.5	7.724	4.2182

This result is understood by noting that $\phi_{\pi\eta}$ appears only once in the $\gamma\pi$ amplitude in contrast to the double entry ϕ^*_π, ϕ_π in inelastic scattering. That obvious fact, along with the surface peaking of the dominant transition densities, Fig 7a, explains the relative insensitivity of $\gamma\pi$ for C^{12} to changes in the pion's phase velocity near the nuclear surface.

Recall that we have dealt with a special PSE case, other cases might reveal some sensitivity and might serve to guide us to the extraction of pion-nucleus information from the $\gamma\pi$ process.

V.   CONCLUSION

The three ingredients of a $\gamma\pi$ study have been discussed along with some results. The Blomqvist-Laget operator has permitted study of photoproduction from moving nucleons in what is essentially a pseudopotential approach. Improved description of the production process is clearly required, especially now that the pion-momentum dependent terms and the isobar dynamics, in conjunction with the early onset of pion p-waves, have been shown to be significant at

relatively low energies (T_π = 50 MeV)[5]. Nucleon motion
effects in the BL operator proved to be small at 50 MeV,
but might increase at higher energies as suggested by

Figures 2 and 3. The replacement of $E = \sqrt{p^2 + m_N^2}$ by m_N,
or equivalently the use of approximate nucleon propagators,
as well as a simple approximation for the pion propagator
are steps that need to be improved before proceeding to
reliable $\gamma\pi$ calculations at higher energies. That is, the
energy and momentum dependence of the amplitudes, G, must
be accounted for in the future. Also needed is an improved
treatment of the isobar dynamics, consistent with information from $\pi\eta$ elastic, inelastic scattering, and γ,p
reactions.

    The isolation of nuclear structure effects has been
made by defining transition densities (Figure 7), some of
which include nuclear derivatives. That step permits useful
contact with information gathered from inelastic electron
scattering concerning one-body nuclear operators. The role
of two-body operators from exchange currents in $\gamma\pi$ is
essentially unexplored. In the resonance region, we will
need to extend the process of isolating the nuclear structure
by introducing nuclear form factors, which, because of
nonstatic effects, will not be simple functions of the
momentum transfer.

    Finally, recent pionic and low energy scattering analy-
ses store the quasi-elastic and real absorption (annihilation)
into reasonable optical potentials, which have provided a way
to calculate total and differential cross-sections. It will
be interesting to watch the confrontation of such calculations with data! There is some hope in that way we might
learn something about the behavior of a pion near a nucleus.
The present information is contained in calculated pion waves,
some of which are displayed here. Can we make our $\gamma\pi$ work
sufficiently precise to extract pion wave function information? That question is open. One small insight is that the
$\gamma\pi$ cross-sections are insensitive in DWIA calculations to
changes in the pion's phase velocity near the nucleus.

ACKNOWLEDGEMENTS

It is a pleasure to express my appreciation to my collaborators G.N. Epstein and M.K. Singham. Thanks are also due to Prof. C.M. Vincent for his help with aspects of the nuclear information, and Prof. R.A. Eisenstein for information about the 50 MeV data.

REFERENCES AND NOTES

* Supported in part by NSF

1. International Conference on High-Energy Physics and Nuclear Structure (proceedings edited by M.P. Locher); See surveys presented by E. Boschitz, B.M. Preedom and R.A. Eistenstein.
2. C.J. Batty et al., Phys. Rev. Letts. $\underline{40}$, 931 (1978), and E. Friedman (private communication).
3. S.A. Dytman et al., Phys. Rev. Lett. $\underline{38}$, 1059 (1977) and preprint (1978) R.R. Johnson et al. Nucl. Phys. $\underline{A296}$, 444 (1978); M.A. Moinester et al., (LAMPF preprint, 1978).
4. I. Blomqvist and J.M. Laget, Nucl. Phys. $\underline{A280}$, 405 (1977).
5. G.N. Epstein, M.K. Singham, and F. Tabakin, Phys. Rev. $\underline{C17}$, 702 (1978) and papers presented to this conference by: M. Singham et al., A. Nagl and H. Uberall, and S. Furui (reference given there).
6. K. Stricker, H. McManus and J.A. Carr (preprint, 1978).
7. J. Delorme, M. Ericson and G. Fäldt, Nucl. Phys. $\underline{A240}$, 493 (1975).
8. J.T. Londergan and G.D. Nixon, "Isobars and the Medium-Energy (γ,p) Reaction" preprint (1978) and J.L. Mathews et al., Phys. Rev. Lett. $\underline{38}$ 8 (1977).
9. L.M. Saunders, Nucl. Phys. $\underline{B7}$, 299 (1968).
10. T. de Forest, Jr., and J.D. Walecka, Advances in Physics $\underline{15}$, Nos. 56, 1(1966).
11. W.C. Haxton, Phys. Lett. $\underline{76B}$ 165 (1978) and references therein.
12. M. Ericson and T.E.O. Ericson Ann. Phys. (N.Y.) $\underline{36}$, 323 (1966).

13. J. Hüfner, Phys. Rev. 21C, 1(1975). L. Tauscher and W. Schneider, Z. Physik <u>271</u>, 409 (1974).

14. R. Kwon and F. Tabakin (to be published).

15. N.J. DiGiacomo, A.S. Rosenthal, E. Rost and D.A. Sparrow, Phys. Lett. <u>B66</u>, 421 (1977) and Phys. Lett. <u>B71</u>, 237 (1977).

16. Bradley, D. Keister (to be published) "Phase-Shift-Equiv-alent Potentials and Pion-Nucleus Reactions."

$^{12}C(\gamma,\pi^-)$ ^{12}Ng.s. for (0-50) MeV PIONS

G. N. Epstein, M. K. Singham, F. Tabakin

Department of Physics and Astronomy

University of Pittsburgh

Pittsburgh, Pennsylvania 15260

We calculate both total and differential cross-sections for the reaction $^{12}C(\gamma,\pi^-)^{12}N$g.s. for pion laboratory kinetic energies ranging from 0-50 MeV. There are essentially three elements in the analysis: nuclear structure; pion distortions; the photoproduction operator ($H_{\gamma\pi}$).

Nuclear Structure

We use Haxton's[1] wave-function configurations which are fitted to Ml form factors obtained from recent electron-scattering data from ^{12}Cg.s. to ^{12}C (1^+, 15.11 MeV) state, where the latter is customarily taken to form an isospin triplet with ^{12}Ng.s. and ^{12}Bg.s. This calculation was done for the momentum transfer range 0 --- 300 MeV/c. For (γ,π) reactions, these wave functions suffice since the momentum transfer is about 150 MeV/c at threshold and varies from 100 MeV/c to 300 MeV/c for a 50 MeV pion. We compare the results with those obtained using the configurations of O'Connell, Donnelly, and Walecka[2] which were obtained in a similar fashion but with earlier electron scattering data.

Pion Distortions

For pions with K. E. 0 to 10 MeV, we use one set of pionic atom optical potential parameters given by

Backenstoss[3] and another set given by Hüfner[4] and
Tauscher and Schneider[5] (quoted in ref. (1)). For 30, 40,
and 50 MeV pions we use the optical potential parameters of
Stricker et al.[6], which are consistent with π-nucleus
elastic scattering data at those energies.

Photoproduction operator $H_{\gamma\pi}$
_____

    We use the Blomqvist-Laget[7] (BL) model for the
$\gamma N \rightarrow \pi N$ amplitude (N $\equiv$ nucleon) and convert this into a
co-ordinate space operator $H_{\gamma\pi}$. This is done by replacing
the pion and nucleon momenta in $H_{\gamma\pi}$ by derivatives operating
on the appropriate wave function. This amplitude gives a
good description of the $\gamma N \rightarrow \pi N$ data up to 250 MeV pions
and it enables us to handle the full $H_{\gamma\pi}$ operator up to
$O(p^2/m_N^2)$ and not just the leading $\underline{\sigma}\cdot\underline{\epsilon}$ term. To be useful
as a co-ordinate space operator in a nuclear context however,
we had to set the pion and nucleon three-momenta to zero in
the coefficients of the operators and this limited the range
of validity of our calculation to 50 MeV pions.

    The BL amplitude enables us to estimate the importance
of the higher order corrections to the amplitude and to
study the role of the $\nabla(1232)$ in the photoproduction process.
It also provides a physically motivated and convenient means
of coping with problems that exist in transforming the CGLN[8]
amplitude from πN center-of-mass to π-nucleus center-of-mass.

Results
_____

    From Figure 1 we see that there is not much difference
in the curves for total cross-sections obtained with the
Haxton and ODW configurations, which suggests that this
particular reaction is not too sensitive to nuclear structure.
We also see that our earlier calculation[9] at threshold
energies with $H_{\gamma\pi} = C \underline{\sigma}\cdot\underline{\epsilon}$ (where C was a constant adjusted
to reproduce reduced cross-sections at threshold) is consis-
tently about 10% higher than that obtained with the full $H_{\gamma\pi}$.

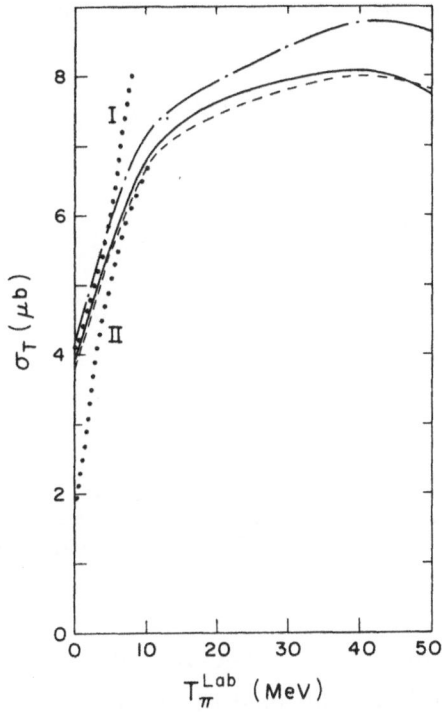

Figure 1: Total cross-section (σ_T) for $^{12}C(\gamma,\pi^-)^{12}$N g.s. T_π^{Lab} is the pion laboratory kinetic energy. The dash-dot curve is obtained using $H_{\gamma\pi}$ of ref. 9 and ODW wave functions; the dashed curve using BL $H_{\gamma\pi}$ and ODW wave functions; the solid curve using BL $H_{\gamma\pi}$ and Haxton's wave functions. The dotted lines marked I and II represent the experimental data range of ref. 10.

We also found that for $\sigma(\theta)$ at 50 MeV, the Haxton configurations yield slightly more forward peaking than the ODW set. The forward peaking also increased at higher energies (Figure 2) with a minimum at around 90°. At 90°, $\sigma(\theta)$ was 0.4 µb/sr at 5 MeV; 0.1 µb/sr at 30 MeV; 0.015 µb/sr at 50 MeV.

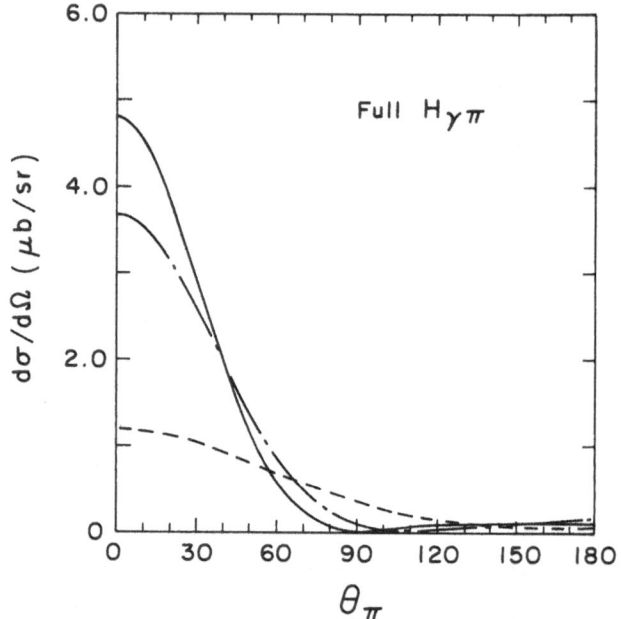

Figure 2: Differential cross-sections for $^{12}C(\gamma,\pi^-)^{12}N$g.s.
as a function of T_π^{Lab}. The dashed curve is at 5 MeV; the
dash-dot curve is at 30 MeV; the solid curve is at 50 MeV.
θ_π is in the center-of-mass of the pion-nucleus system.

     The pion partial wave contribution to the data is shown
in Figure 3, where we see that the S-wave contribution increases slightly from 0 - 10 MeV and then decreases sharply
with energy. We also see that the higher partial waves start
contributing at much lower energies than for π-nuclear
scattering. The p-waves contribute almost from threshold,
the d-waves at about 20 MeV and f-waves at 50 MeV. The effect
of the higher partial waves on $\sigma(\theta)$ was to enhance the production at forward angles and suppress it at backward angles.

     The effect of higher order terms in $H_{\gamma\pi}$ is shown in
Figure 4. It was found that setting the <u>nucleon</u> momenta to
zero in $H_{\gamma\pi}$ had almost no effect on σ_T or $\sigma(\theta)$ at 50 MeV
which suggests that it is <u>pion</u> momentum dependent terms that
contribute significantly. The effect of the pion momentum

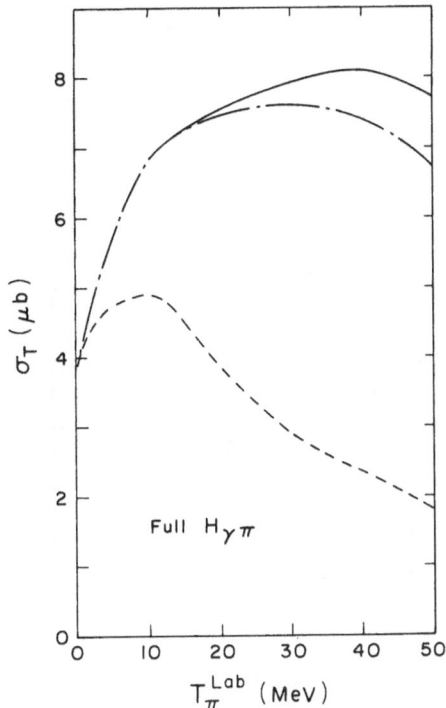

Figure 3: Partial wave dependence of σ_T using BL $H_{\gamma\pi}$ and Haxton's wave functions. The dashed curve is obtained using only s-wave pions; the dash-dot curve using s and p wave pions; the solid curve using s,p and d wave pions.

dependent terms is to increase σ_T very slightly close to threshold, but decrease it by about 25% at 50 MeV. If we keep the pion momentum terms, but turn off the terms coming from the isobar, we see that σ_T increases by 15%. Both the isobar and the pion-momentum terms in $H_{\gamma\pi}$ start contributing signi-ficantly at about 20 MeV. These terms do not drastically change the angular distribution of the cross-section.

Conclusions

It is clear that these results demonstrate the inadequacy of a simple $\underline{\sigma}\cdot\underline{\epsilon}$ model for $H_{\gamma\pi}$ and illustrate the need to de-

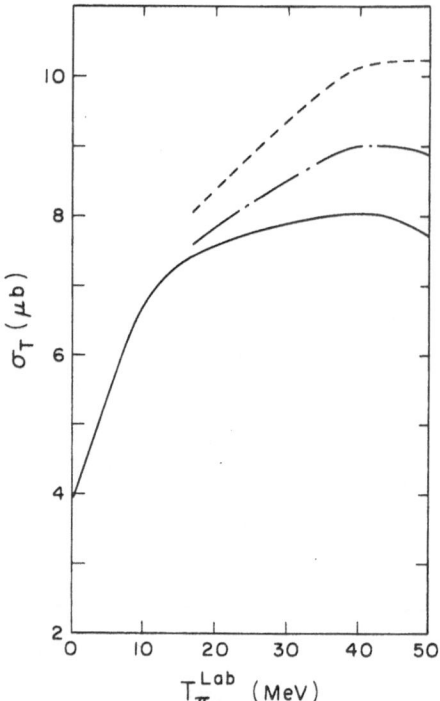

Figure 4:   Operator dependence of σ_T using Haxton's wave functions.   The solid curve is obtained using full BL $H_{\gamma\pi}$; the dash-dot curve is with the isobar terms turned off; the dashed curve is with the pion-momentum dependent terms turned off.

velop a good understanding of the role of the Δ.   It also illustrates the need to include higher pion partial waves at energies lower than one would expect from analogy with π-nucleus scattering.   In this study, we have included all photon partial waves.   The effect of particular photon multi-poles on the cross-sections and the role of $V_{\pi\eta}$ will be presented later.

<u>References</u>

1.   W. C. Haxton, Phys. Lett. <u>76B</u>, 165 (1978).
2.   J. S. O'Connell, et. al. Phys. Rev. <u>C6</u>, 719 (1972).

3. G. Backenstoss, Ann. Rev. Nucl. Sci. <u>20</u>, 467 (1970).

4. J. Hüfner, Phys. Re. <u>21C</u>, 1 (1975).

5. L. Tauscher and W. Schneider, Z. Physik <u>271</u>, 409 (1974).

6. K. Stricker, H. McManus, J. Carr, (Preprint).

7. I. Blomqvist and J. M. Laget, Nucl. Phys. <u>A280</u>, 405 (1977).

8. G. F. Chew, et. al., Phys. Rev. <u>106</u>, 1345 (1957).

9. G. N. Epstein, M. K. Singham, F. Tabakin, Phys. Rev. <u>C17</u>, 702 (1978).

10. A. M. Bernstein, et. al., Phys. Rev. Lett. <u>37</u>, 819 (1976).

PION-NUCLEUS INTERACTIONS IN THE RESONANCE REGION*

E.J. Moniz

Center for Theoretical Physics

Laboratory for Nuclear Science and Department of

Physics

Massachusetts Institute of Technology

Cambridge, Massachusetts   02139

I.   INTRODUCTION

Two aspects of pion-nucleus interactions and photopion reactions in the resonance region will be discussed. The first is essentailly the theory of the pion optical potential. Obviously, we need the optical potential in order to calculate any coherent process (one in which the residual nucleus is in a definite quantum state) such as elastic pion scattering, coherent π^0 photoproduction, analog state charge exchange, etc. Most theoretical work up to now has focussed on developing the optical potential. The second aspect involves an understanding of pion reaction mechanisms. Important questions here involve the nuclear response function for pion scattering, the means of energy transfer between the pion and nucleus, and the role of pion absorption. Typically these questions are studied experimentally in various types of inclusive reactions and theoretically are mostly ignored. Yet our understanding of these reaction mechanisms is crucial to our construction of the optical potential.

This last point is clear if we recall the projection operator form of the optical potential. Let P project onto the states consisting of a pion plus the nucleus in the

ground state and Q onto all other states. Using the identi-
ty P + Q = 1, we can solve formally the Lippmann-Schwinger
equation for pion-nucleus elastic scattering[1]:

$$T(E) = V + V\,G(E^+)T(E), \quad G(E) = (E - k_\pi - H_N)^{-1}$$

$$T_{PP}(E) = \mathcal{V}_{PP}(E) + \mathcal{V}_{PP}(E)G_P(E)T_{PP}(E) \tag{1}$$

where $T_{PP} \equiv PTP$ and the optical potential is

$$\mathcal{V}_{PP}(E) = V_{PP} + V_{PQ}\,G_Q(E)\,V_{QP}$$

$$G_Q^{-1} = G_Q^{-1} - V_{QQ} \tag{2}$$

The imaginary part of the optical potential is then given by

$$Im\ \mathcal{V}_{PP} = \mathcal{V}_{PQ}(Im\ G_Q)\,V_{QP} \tag{3}$$

making it clear that absorption out of the elastic channel
determines directly the optical potential. Therefore, it is
crucial to understand at least the "early" states of pion
reactions in order to construct a sensible theory of the
optical potential. The usual first order optical potential
$V = \rho t_{\pi N}$ corresponds, in this language, to the assumption
that flux is lost from the elastic channel only through
quasifree nucleon knockout. Given the experimental fact that
almost half the $\pi - {}^{12}C$ reaction cross section in the resonance region (and even more for lower energies or heavier
targets) corresponds to true pion absorption, complete dominance of the quasifree reaction mechanism is rather suspect.

In Section II, we shall describe briefly the isobar-hole
formalism for pion scattering and show some results of an
application to $\pi - {}^{16}O$ scattering[2,3]. The isobar-hole approach is an alternate to the optical potential approach and
will deal with the more complicated reaction mechanisms such
as pion absorption through a Δ-nucleus spreading interaction.
This spreading interaction, derived from elastic scattering,
will then be used to predict total absorption cross sections.
In Section III, we discuss briefly some results for (π,π')

reactions with the intent of seeing whether or not the simple quasifree knockout response function is in evidence.
In both sections we shall stress the role of photopion reactions in the resonance region as a complement to pion-induced reactions for learning about pion and isobar dynamics.
Coherent π° photoproduction in the isobar-hole formalism will be discussed in Section II and incoherent photopion production in Section III. Section IV consists of a few concluding remarks.

II. PION SCATTERING AND COHERENT PHOTOPRODUCTION IN THE Δ-HOLE FORMALISM

The Δ-hole formalism is described in great detail in References 2 - 5, so we shall give a fairly brief review here. This work has been done in collaboration with M. Hirata, J. Koch and F. Lenz. It is important to realize at the beginning that the formalism is, in principle, completely equivalent to the usual multiple scattering formalism. This equivalence is removed only when approximations are introduced, as of course they must be in addressing the many-body problem. The physical basis for the Δ-hole approach is the experimental fact that both πN scattering and $\pi NN \rightarrow NN$ (i.e., absorption) are dominated in the relevant energy region by intermediate Δ formation (Fig. 1). These are the most important microscopic processes and so we choose to focus upon the Δ-nucleus dynamics.

The basic idea underlying the isobar-hole approach can be seen clearly in the projection operator formalism. Again let P project onto the pion plus ground state nucleus. In the model represented by Fig. 1, it is clear that, upon striking the nucleus, the pion forms a doorway state consisting of a Δ-isobar and a nucleon hole; let D project onto this Δ-hole space. As before, Q projects onto "the rest". Using the same procedure employed above, we have for the pion-nucleus elastic transition matrix

$$T_{PP} = H_{PD} \, G_{\Delta h} \, H_{DP} = H_{PD} \, \frac{1}{E - E_R + i\Gamma/2 - H_{\Delta h}} \, H_{DP}$$

$$= H_{PD} \, \frac{1}{E - E_R + i\Gamma/2 - (H_{DD} + H_{DD}^{\uparrow} + H_{DD}^{\downarrow})} \, H_{DP} \qquad (3)$$

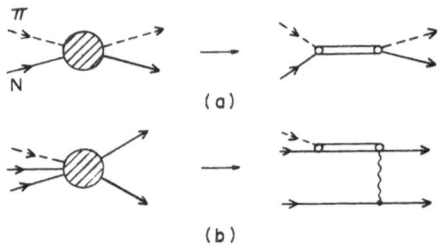

Figure 1.   Isobar dominance of (a) πN scattering and (b) pion absorption (the wavy line may also be a pion).

$$H_{DD}^{\uparrow} = H_{DP}\, G_P\, H_{PD}$$

$$H_{DD} = H_{DQ}\, G_Q\, H_{QD}.$$

In deriving this, we have used the conditions $H_{PP} = 0$ and $H_{PQ} = 0$, the latter being the usual doorway hypothesis. These conditions correspond to complete isobar dominance of the πN interaction, and we shall use this for convenience in the formal development. However, we point out now that the calculated results compared with data will include the nonresonant πN and Coulomb interactions as well. Equation 3 is represented diagrammatically in Fig. 2. The πNΔ vertex operator H_{DP} provides the entrance into the D-space, and the Breit-Wigner form represents the free-space Δ-isobar propagator. The escape interaction $H_{DD}^{\uparrow}$ sums intermediate coupling back to the P-space and is represnted in Fig. 2b; iteration of T_{PP} in powers of $H_{DD}^{\uparrow}$ corresponds to a multiple scattering expansion in the coherent approximation. The diagonal interaction H_{DD} includes terms such as isobar propagation, binding and Pauli blocking effects. The Pauli term provides a correction to the free width Γ due to the fact that nucleon states below the Fermi level are already occupied (Fig. 2c).

Isobar-hole residual interactions (e.g. ρ-exchange, as indicated in Fig. 2c, could also be included in H_{DD}, but these are not expected to play an important role in the resonance region (we use πNΔ vertex functions with a range of ~ .6 fm). The spreading interaction $H_{DD}^{\downarrow}$ incorporates inter-

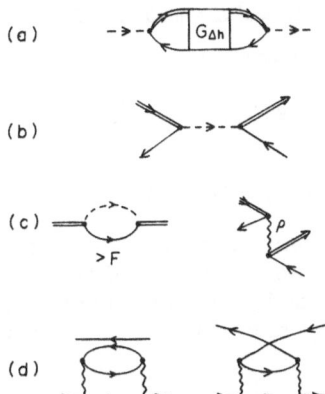

Figure 2. Diagrammatic representation of the isobar-hole formalism. (a) Pion elastic scattering via excitation of and propagation of Δ-hole states. (b) $H_{DD}^{\uparrow}$: pion "exchange" between -hole states. (c) Contributions to H_{DD}: Pauli blocking and ρ-exchange. (d) Simplest pion absorption contributions to $H_{DD}^{\uparrow}$.

mediate propagation in the Q-space. This is obviously very complicated and includes terms such as the pion absorption contributions indicated in Fig. 2d. Consequently, we treat the spreading interacting phenomenologically by replacing it with a Δ spreading potential

$$V_{sp}(r) \equiv V_{sp}(o)\, \frac{\rho(r)}{\rho(o)} \qquad (4)$$

where $\rho(r)$ is the nuclear ground state density. Note that this would correspond to including in the pion optical potential terms of second and higher order in the density. The spreading potential strength $V_{sp}(0)$ is treated as a complex parameter. A test of the formalism comes from relating this parameter, determined from elastic scattering, to other reactions. Eventually, one would hope to calculate it microscopically. To give an idea of the scale involved, we find $V_{sp}(0) \simeq -50i$ MeV in the resonance region.

It is worth going into a few details on the calculational procedure. As remarked earlier, Δ-hole states form the doorway for the reaction. More precisely, we define a normalized doorway state for each angular momentum (i.e., pion partial wave in $\pi - {}^{16}O$ scattering)

$$|D_o^L> \equiv N_L H_{DP}|0;(\vec{k})_L>$$

$$N_L = <0;(\vec{k})_L|H_{PD}H_{DP}|0;(\vec{k})_L>^{-\frac{1}{2}}. \tag{5}$$

This state is the coherent superposition of Δ-hole states formed by linear response of the nucleus to a spherical pion wave of angular momentum L and wave number k. The elastic partial wave scattering amplitude is the expectation value of $G_{\Delta h}$ in this state:

$$T_L(k) = N_L^{-2} <D_o^L|G_{\Delta h}|D_o^L> . \tag{6}$$

Of course, $|D_o^L>$ is generally not an eigenstate of $H_{\Delta h}$ so that we need a convenient basis for evaluating Eq. 6. We do this with the Lancsoz construction[2]

$$|D_i^L> \equiv N_i\{H_{\Delta h}|D_{i-1}^L> - \sum_{j=o}^{i-1} |D_j^L>H_{j,i-1}\}$$

$$H_{i,j} \equiv <D_i^L|H_{\Delta h}|D_j^L> \tag{7}$$

This "doorway basis" is built up from the doorway state by repeated application of the Hamiltonian and orthonormalization at each step. This basis is convenient because $H_{ij} = 0$ for $|i-j| > 1$, leading to a tridiagonal Hamiltonian matrix and thus the continued fraction expansion.[2,3]

$$T_L = \cfrac{N_L^{-2}}{E-E_R+i/2 - H_{oo}-\cfrac{H_{01}^2}{E-E_R+i/2 - H_{11}-\cfrac{H_{12}^2}{E-E_R+i/2-H_{22} \cdots}}} \tag{8}$$

This is a <u>systematic</u> expansion which can be truncated at any desired level of accuracy. We find that the one-doorway approximation (i.e., truncating the doorway basis at $|D_o^L>$) is already accurate to about 30%; with a truncation after three states, the calculation is accurate to much less than 1%.

This "collectivity" is understood easily for the central partial waves by analogy with the Brown-Bolsterli schematic model. The on-shell part of the pion rescattering term $H_{DD}^{\uparrow}$ is a separable Hamiltonian

$$H_{DD} \sim iH_{DP}H_{PD} \qquad (9)$$

with the doorway state as an eigenstate. Therefore, to the extent that the escape width dominates the Δ-hole self-energy, and neglecting binding energy differences for the nucleons, the doorway state would specify completely the scattering amplitude. More formally, the matrix element H_{01} represents the fluctuation of the Hamiltonian in the doorway state

$$H_{01} = [<D_o^L|H_{\Delta h}^2|D_o^L> - <D_o^L|H_{\Delta h}|D_o^L>^2]^{\frac{1}{2}} \qquad (10)$$

and this would vanish if $|D_o^L>$ were an eigenstate. The calculations give $|H_{01}| \simeq |H_{00}|/3 \simeq \Gamma/2$ for the $J^{\pi} - 0^-$ doorway state. The decomposition of the imaginary part of H_{00} is shown in Fig. 3.[2] The escape width clearly dominates the other self energies.

This is no longer true for the more peripheral partial waves[3,6]. Nevertheless, it turns out that the convergence of Eq. 8 is equally rapid because now $|H_{01}| \simeq |H_{00}| \simeq \Gamma/6$. Therefore, the Lancsoz construction leads to a rapid convergence in all cases.

Before leaving Fig. 3, we point out that the Pauli and spreading widths are by no means negligible. Note that the sum of these is positive so that the ratio of reaction to elastic cross sections is increased; this is especially pronounced at the lower energies. Further, the Pauli width "quenches" the free width substantially, implying that the quasifree reaction mechanism is reduced in importance compared to the spreading width. We conjecture that the latter arises primarily from pion absorption.

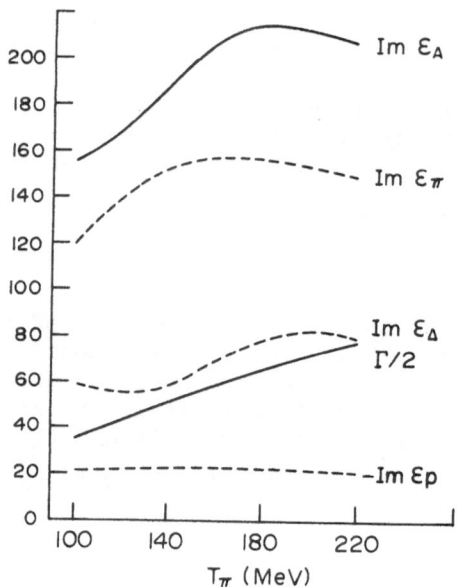

Figure 3.   Decomposition of Im $H_{00} \equiv \mathrm{Im}\varepsilon_A$.   The free width of
the isobar is denoted by Γ, while $\mathrm{Im}\varepsilon_\pi$, $\mathrm{Im}\varepsilon_\Delta$ and $\mathrm{Im}\varepsilon_p$ are the
escape, spreading and Pauli half-widths, respectively.   The
scale is in MeV.

        Figure 4 shows our calculated results[3] for $\pi^+ - {}^{16}O$ elas-
tic scattering at 163 MeV.   The data are from SIN[7].   We   show
for comparison the calculated result using the first order
optical code PIPIT[8].   Figure 5 shows the partial wave decomposition of the theoretical amplitudes and the results of a partial wave fit to the data.[3]   There are several interesting
points evident in the comparison.   One of the major improvements in the Δ-hole approach is a good description of the peripheral partial waves.   These are very important for reproducing
the diffraction-structure of the data.   The spreading potential, being proportional to the local density, is comparatively unimportant for these partial waves.   The crucial difference to the PIPIT calculation is the proper treatment (within
the shell model context) of the isobar propagation and nucleon
binding effects.   The spreading potential does play an impor-

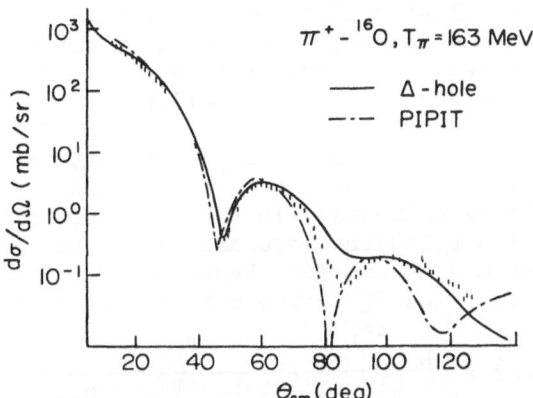

Figure 4.   Elastic differential cross section for π^+-^{16}O
scattering at 163 MeV.   Solid and dot-dashed curves result
from the Δ-hole and PIPIT calculations, respectively.
Data from Reference 7.

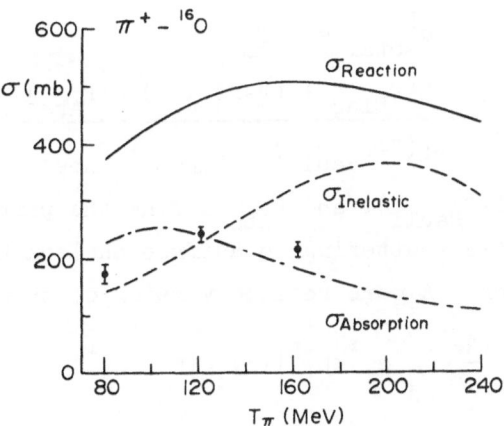

Figure 5.   Decomposition of the reaction cross section into
absorption and inelastic scattering cross sections.   The "data"
are interpolations of preliminary absorption cross section
measurements for other nuclei.[9]

tant role in the central partial waves. We remind the reader
that this potential has been fit to the total cross section
and forward angle elastic differential cross section. The
interesting result here is that a central strength of order
50i MeV was required. Our next task is to test this result
by examining the absorption cross section.

We shall assume that the spreading interaction represents
primarily the effects of pion absorption. This assumption is
motivated by the experimental finding that most of the reaction cross section below the resonance is found in the single
nucleon knockout and absorption channels. Our basic approach
is understood most clearly in the one-doorway approximation

$$
T_L(E) \equiv \frac{\eta_L e^{2i\delta_L} - 1}{2i} = \frac{-\Gamma_{Elas}/2}{E-E_R+i\Gamma/2 - H_{00}}
$$

$$
\equiv \frac{-\Gamma_{Elas}/2}{E-\tilde{E}_R + \frac{i}{2}[(\Gamma-\Gamma_{Pauli}) + \Gamma_{Elas} + \Gamma_{Abs}]} \tag{11}
$$

where Γ_{Abs} is identified as the spreading width. The partial
wave total cross section can then be written as

$$
\sigma_{Total}^L \propto ImT_L
$$

$$
= \frac{\frac{1}{2}\Gamma_{Elas} \cdot \frac{1}{2}[\Gamma_{Elas} + (\Gamma-\Gamma_{Pauli}) + \Gamma_{Abs}]}{(E-\tilde{E}_R)^2 + \frac{1}{4}[(\Gamma-\Gamma_{Pauli}) + \Gamma_{Elas} + \Gamma_{Abs}]^2} \tag{12}
$$

Clearly, Γ_{Elas}, $(\Gamma-\Gamma_{Pauli})$, and Γ_{Abs} define the partial cross
sections for elastic scattering, quasifree nucleon knockout,
and pion absorption. A more refined version of this is[3]

$$
\sigma_L^{Abs} = \frac{4\pi}{k} \sum_{ij} \alpha_i^* <D_i^L|ImV_{sp}|D_j^L>\alpha_j
$$

$$
\alpha_j = N_L<D_j^L|G_{\Delta h}|D_o^L> . \tag{13}
$$

The results are shown in Figure 5. Note that the inelastic
cross section now peaks slightly above the free resonance
position, as expected for knockout of a bound nucleon. The
absorption cross section "data" are interpolation of prelim-

inary results from other nuclei.[9] The calculation is in rough
agreement, perhaps falling too rapidly at the higher energies.
Recall that our calculation actually represents the loss of
flux into the Q-space out of the elastic channel, so that our
calculation is expected to be too low when multistep reactions
become important. Overall, this is a success for the model.
We note in passing that the isobar spreading width, if extrapolated to threshold, is also consistent with the measured
widths of pionic atom p-states.

We turn now to coherent π^0 photoproduction. This work
has done in collaboration with J. Koch and is described in
greater detail elsewhere in these proceedings.[6] The Δ-hole
formalism presents a powerful way for unifying the description of the (γ,π^0) and elastic scattering cross sections.[*]
This is evident from Fig. 6: the same Δ-hole propagator
enters in both cases. The difference is. of course, that
the photon excites a different linear combination of Δ-hole
states, so that we define a "photon doorway state" $|D_\gamma^L\rangle$.
The production operator is then

$$T_{\gamma\pi^0} = H_{PD} G_{\Delta h} H_{D\gamma} \tag{14}$$

giving a partial wave amplitude

$$T_{\gamma\pi0}^L = N_L^{-1} N_L^{-1} \langle D_o^L | G_{\Delta h} | D_\gamma^L \rangle$$

$$= N_L^{-1} N_L^{\gamma-1} \sum_{i=0}^{\infty} \langle D_o^L | G_{\Delta h} | D_i^L \rangle \langle D_i^L | D_\gamma^L \rangle \quad . \tag{15}$$

Note that the off-diagonal matrix elements of $G_{\Delta h}$ enter, with
weight proportional to the overlap between the pion doorway
basis states and the photon doorway state. These off-diagonal
matrix elements are the new information which may be obtained,

Figure 6.   Coherent π^0 photoproduction in the Δ-hole formalism.

_____

[*]A related discussion for the deuteron appears in the following
contribution by Dillig et al.

corresponding to the half-off-shell pion-nucleus scattering amplitude. A quantitative discussion is given in Reference 6.

It is interesting to note that the production operator in Eq. 14 can be rearranged to read[6]

$$T_{\gamma\pi^0} = [T_{pp}G_p+1]\{H_{PD}[E-E_R+i\Gamma/2-H_{DD}-H_{D\overset{\downarrow}{D}}]^{-1}H_{D\gamma}\} . \qquad (16)$$

The production operator in curly brackets is modified by the spreading interaction $H_{D\overset{\downarrow}{D}}$. Therefore, the same spreading potential fit to elastic pion scattering predicts here the magnitude of multistep contributions to the cross section.

Another coherent photoreaction which would be especially useful as a complement to pion measurements is analog state $(\gamma,\pi^{\pm})$ production in the resonance region. The pion charge exchange measurements, such as the activation measurements of $^{13}C(\pi^+,\pi^0)^{13}N_{gs}$, have eluded successful theoretical interpretation. The corresponding (γ,π^-) measurement would help to determine whether the problem lies with the pion optical potential or with an overly simplified reaction mechanism. Such a comparison would take advantage of the fact that the spatial distribution in the nucleus of the "charge exchange" may be quite different in the two cases because of the long photon mean free path.

III. INELASTIC PION SCATTERING AND INCOHERENT PHOTOPRODUCTION

We saw that the reaction mechanisms responsible for attenuating the elastic channel pion flux seemed to be basically understood in the resonance region in terms of Δ-nucleus dynamics. The same Pauli-quenching and spreading interaction which produced reasonable agreement with the elastic data also gave a reasonable description of the total inelastic and pion absorption cross sections. Of course, a very important assumption was made in identifying the spreading interaction as arising from pion absorption. The resulting picture is at least internally consistent. Nevertheless, it is important to test this picture of the reaction mechanism by examining the inclusive data.

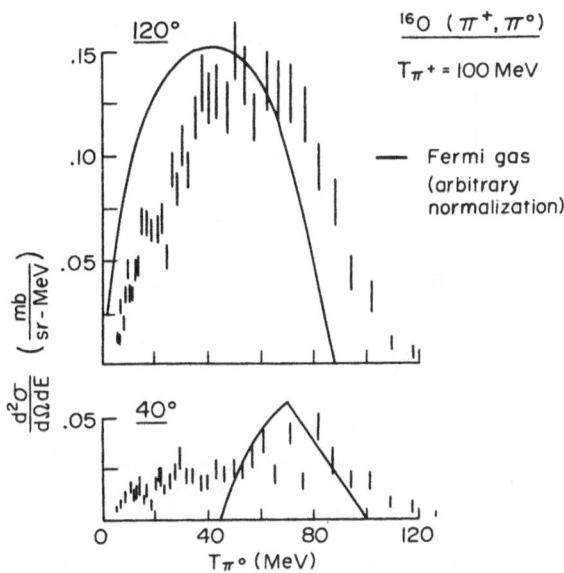

Figure 7. Inclusive single charge exchange[10] on ^{16}O for 100 MeV incident pion energy. Curves are simple Fermi gas results.

Figure 7 shows the inclusive pion spectrum[10] for the $^{16}O(\pi^+,\pi^0)$ reaction with incident energy T = 100 MeV. To the extent that quasifree scattering dominates the response function, the shape should be the same for $(\pi^+,\pi^{+\prime})$. The curves are simply the (arbitrarily normalized) Fermi gas response function. For a large scattering angle $\theta = 120°$, the data more or less follows the Fermi gas description (apart from a shift by $\sim$ 20 MeV). For the more forward angle $\theta = 40°$, the cross section is smaller in magnitude and has an appreciable low energy "tail" with respect to the Fermi gas response.** The only detailed calculations which can be compared reasonably with the data are those given in the classical intranuclear cascade model.[11] These include much the same physics as do the Δ-hole calculations for elastic scattering (i.e., πN scattering through the Δ, isobar propagation, and pion absorption through the Δ) and do reasonably

_____

**This is related to the results shown by B. Mecking previously in this volume.

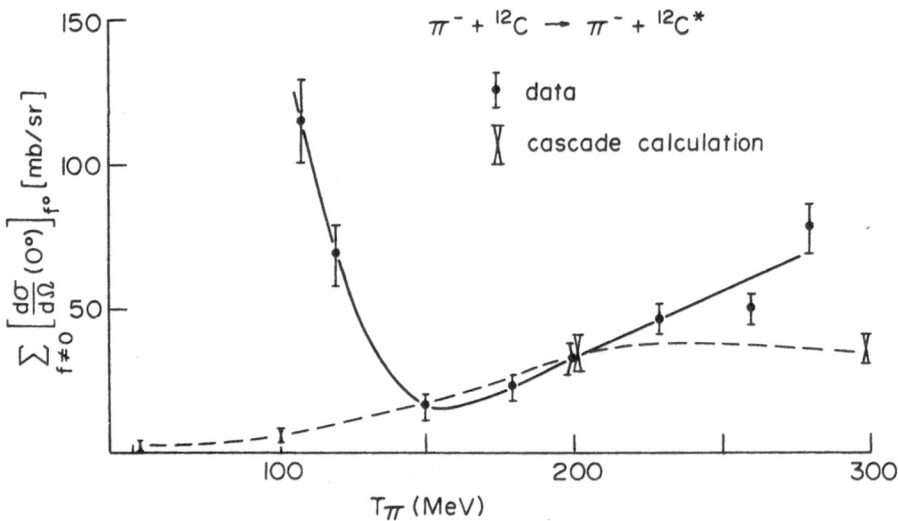

Figure 8. Summed forward inelastic cross section for π^--^{12}C scattering.[13] Crosses are the results of the intranuclear cascade calculations.[12] The solid and dashed lines are intended only to guide the eye through the data and cascade predictions, respectively.

well in reproducing both the shape and magnitude of the cross section.[11] These calculations also do reasonably well for the pion absorption cross sections,[12] so the basic reaction mechanism seems fairly well in hand. On the other hand, there have been measurements of the measured inelastic cross section in the very forward direction (θ = 0°).[13] The result for ^{12}C is shown in Figure 8, displaying a significant dip in the resonance region. The experiment would seem to be quite difficult but, if correct, would present a real puzzle in our understanding the reaction mechanism. We show the results of Ginocchio's calculations[12] and they disagree considerably. One can also perform a simpler closure calculation (note that closure should not be a very good approximation, but we are interested here only in the qualitative behavior). Using the DWIA framework with Glauber distorted waves, neglecting nuclear correlations, taking a uniform nuclear density, and keeping only the imaginary part of the πN scattering amplitude, we have[14]

$$\sum_{f \neq 0} \left| \frac{d\sigma}{d\Omega} (0°) \right|_{f_0} = A \left| \frac{d\sigma}{d\Omega} (0°) \right|_{\pi N} \{L(2R/\lambda) - L(R/\lambda)^2\}$$

$$L(x) = \frac{6}{x^3} (1-e^{-x}) - \frac{3}{x} (1 + 2/x)e^{-x} \xrightarrow[x \to 0]{} 1 - \frac{3x}{4} + \frac{3}{10} x^2 + \cdots$$

(17)

where R is the nuclear radius and λ is the mean free path (the real part of the πN scattering amplitude makes λ complex). This crude calculation also disagrees strongly with the data but has the same qualitative behavior as do the cascade results: the cross section is maximized near the free resonance position and becomes very small at the low energies. It is important that such questions be resolved and, again, complementary photoreaction data would be most helpful. Because of the long photon mean free path, the behavior of the (π, π') and (γ, π) inclusive cross sections should be quite different at forward angles.

Another inclusive experiment which would be very important for understanding better the reaction mechanism is double charge exchange. This would be a rather direct measure of the multistep quasifree process and the energy dependence of this process as one goes above the resonance should be determined. Unfortunately, there is no analogous photoreaction in this case.

Another puzzling result is the total photoabsorption cross section on heavy nuclei in the resonance region. The experiment was actually an (e,e') scattering mechanism,[15] so that the cross section measured was for (virtual) spacelike photons. The response function, as a function of electron energy loss, is analogous to the pion total cross section as a function of energy. However, the photomeasurement should give a clearer picture of the single-particle Δ-propagator because of the long photon mean free path. The surprising result[15] shown in Figure 9 is the narrowing of the quasifree Δ-excitation peak with respect to the Fermi gas prediction.[16] For lighter nuclei † the experiments more or less agree with the Fermi gas calculations. It is very important that these measurements be re-

---

†See the contribution to these proceedings by J. Ahrens et al.

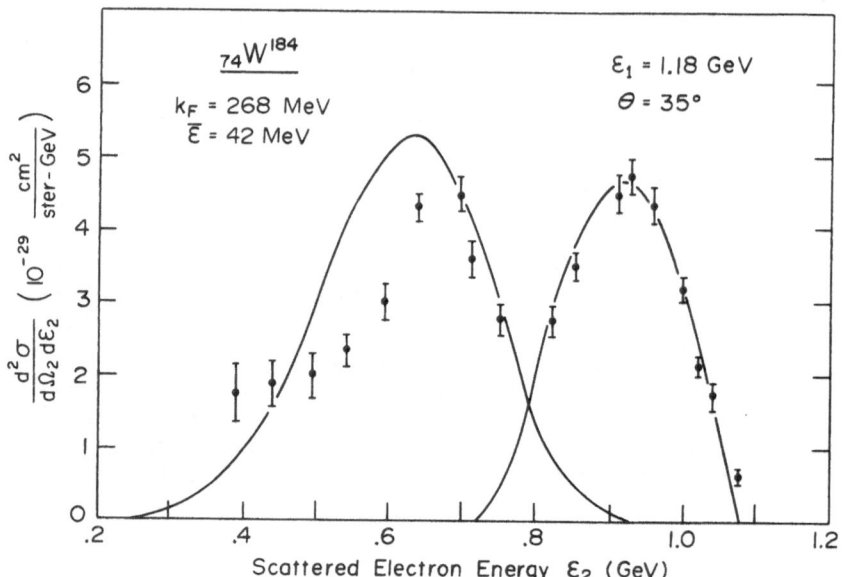

Figure 9. Inelastic electron scattering on tungsten.[15] Theoretical curves[16] are for quasi-free nucleon knockout and isobar excitation in the Fermi gas model.

peated on heavy nuclei; hopefully the measurements can be done
also with real photons (smallest momentum transfer for the
same energy transfer). If this result is correct, we are
missing some important aspects of isobar-nucleus dynamics.

IV.   CONCLUDING REMARKS

There clearly has been considerable progress in recent
years in understanding pion-nucleus interactions. In the isobar language, we saw that the same isobar dynamics led to a
good understanding both of elastic scattering and of total
inelastic and absorption cross sections for energies up to the
resonance. The physical interpretation that both quasi-free
scattering and pion absorption play important roles as "direct"
reaction mechanisms out of the elastic channel seems to be consistent (although not necessarily unique). We did not discuss
the Δ-hole calculations above the resonances. In fact, the
situation there is not nearly as good as at the lower ener-

gies.[3] This difficulty ties in directly to a lack of understanding of the underlying reaction dynamics. In fact, we
say that, even below the resonance, there may be a serious
problem in understanding the forward inclusive reactions.
Further, the total photon cross section in the resonance
region on heavy nuclei seems rather surprising. The same
remark applies to the old coherent $\pi°$ photoproduction data.
It may be that these discrepancies will disappear with a new
generation of experiments. In any case, it should be clear
that intermediate energy photoreactions can play a central
role in our attempt to understand better pion and isobar
dynamics in the nucleus.

REFERENCES

1. H. Feshbach, Ann. Phys. $\underline{5}$, (1958) 357.
2. M. Hirata, J. H. Koch, F. Lenz and E. J. Moniz, Phys.
    Lett. $\underline{70B}$ (1977) 281.
3. M. Hirata, J. H. Koch, F. Lenz and E. J. Moniz, to be
    published.
4. M. Hirata, F. Lenz and K. Yazaki, Ann. Phys. $\underline{108}$ (1977)
    116.
5. L. Kisslinger and W. Wang, Ann. Phys. $\underline{99}$ (1976) 374.
6. J. H. Koch and E. J. Moniz, contributed paper elsewhere in
    this volume and to be published.
7. J. P. Albanese et al., Phys. Lett. $\underline{73B}$ (1978) 119.
8. R. A. Eisenstein and F. Tabakin, Comp. Phys. Comm. $\underline{12}$
    (1976) 237; R. Landau and F. Tabakin, Phys. Rev. D$\underline{5}$ (1972)
    2746.
9. D. Ashery, private communication.
10. T. Bowles et al., Phys. Rev. Lett. $\underline{40}$ (1978) 97.
11. J. Ginocchio, Phys. Rev. C$\underline{17}$ (1978) 195.
12. J. Ginocchio, private communication.
13. F. B. Binon et al., Nucl. Phys. B$\underline{17}$ (1970) 168.
14. E. J. Moniz, unpublished.
15. Yu. I. Titov et al., Soviet Journal of Nuclear Physics
    $\underline{13}$ (1971) 660.
16. E. J. Moniz, Phys. Rev. $\underline{184}$ (1969) 1154; R. A. Smith and
    E. J. Moniz, Nucl. Phys. B$\underline{43}$ (1972) 605.

COHERENT π^0 PHOTOPRODUCTION IN THE

ISOBAR-HOLE FORMALISM

J.H. Koch

Institut voor Kernphysik Onderzoek
Amsterdam, The Netherlands

and

E.J. Moniz*

Center for Theoretical Physics
Laboratory for Nuclear Science and
Department of Physics
Massachusetts Institute of Technology
Cambridge, Massachusetts 02139

Coherent π^0 photoproduction is expected to provide a rather clean test of our understanding of the pion optical potential. As a production reaction, it is sensitive to the pion wavefunction inside the nucleus and not just to the phase shifts measured in elastic pion scattering. As a photoreaction, the reaction mechanism is comparatively well understood. Previous calculations have used the standard DWIA approach with a first order pion optical potential. These do not agree with the somewhat ancient data available for production in the resonance region. Recently, the isobar-hole formalism[1-4] has led to a detailed and quite successful description for $\pi^- {}^{16}O$ scattering in the resonance region[4], and our aim in this note is to show some results of applying the same formalism to the

coherent $^{16}O(\gamma,\pi^\circ)$ ^{16}O reaction. We shall discuss the way
in which the isobar-hole formalism goes beyond the DWIA
framework and the extent to which the (γ,π°) cross section
is sensitive to different phase-shift-equivalent descriptions
of the pion-nucleus dynamics.

The isobar-hole formalism is described elsewhere in this
volume[5]. Here, we restrict the πN interaction to the resonant $J = 3/2$, $T = 3/2$ channel and describe πN scattering as
mediated through intermediate Δ-isobar formation (Fig. 1a).
Pion-nucleus scattering is then determined by the appropriate
matrix element of the Δ-hole propagator $G_{\Delta h}$, as indicated in
Fig. 1b. In the projection operator language, we have the
elastic transition matrix

$$T_{PP} = H_{PD}\, G_{\Delta h}\, H_{DP}$$

$$= H_{PD}\, [E - E_R + i\Gamma/2 - H_{\Delta h}]^{-1}\, H_{DP}$$

$$H_{\Delta h} = H_{DD} + H_{D\overset{\uparrow}{D}} + H_{D\overset{\downarrow}{D}} \tag{1}$$

where P projects onto the states of pion plus nuclear ground
state and D onto the Δ-hole states. The escape interaction
$H_{D\overset{\uparrow}{D}}$ accounts for **intermediate** coupling to the P-space (pion
multiple scattering in the coherent approximation) and $H_{D\overset{\downarrow}{D}}$
is the spreading interaction (including coupling to the true
absorption channels). The diagonal interaction H_{DD} includes
isobar propagation, binding and Pauli effects. The coupling
interaction H_{DP} is just the πN-Δ vertex operator. For each
partial wave L, we define the doorway state

$$|D_0^L\rangle \equiv N_L\, H_{DP}\, |0;\, (\vec{q})_L\rangle$$

$$N_L = \langle 0;(\vec{q})_L | H_{PD}\, H_{DP} |0;(\vec{q})_L\rangle^{-1/2} \tag{2}$$

A complete set of basis states $|D_i^L\rangle$ is built up from $|D_0^L\rangle$ by
repeated application of $H_{\Delta h}$ with orthogonalization at each
step[3]. The partial wave elastic scattering amplitude becomes

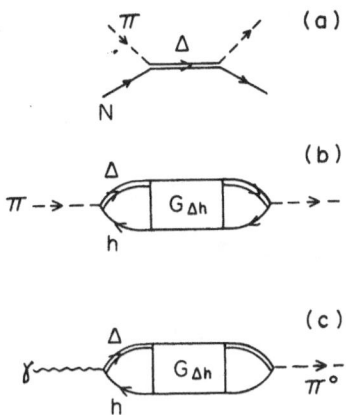

Figure 1. Role of the isobar in (a) πN scattering, (b) π-nucleus elastic scattering, and (c) coherent π^0 photoproduction.

$$T_L(q) = N_L^{-2} \langle D_0^L | G_{\Delta h} | D_0^L \rangle \tag{3}$$

Figure 2 shows the contributions to the imaginary part of $\langle D_0^L | G_{\Delta h}^{-1} | D_0^L \rangle$. We denote by D the sum of the various modifications: the elastic broadening due to $H_{DD}^\uparrow$ (labelled π), the spreading contribution (Δ), Pauli blocking (P) and binding corrections (B). An important feature is the strong L-dependence as seen by comparing the 1^+ and 4^- partial waves. This invalidates the phenomenological approach[6,1] in which an L-independent propagator is fitted to the π scattering data.

In the unified basis provided by the isobar-hole formalism, we now use the same $G_{\Delta h}$ to calculate π^0 photoproduction (Fig. 1c). With $H_{D\gamma}$ denoting the $\gamma N \to \Delta$ vertex, we obtain

$$T_{\gamma \pi^0} = H_{PD}[E-E_R+i\Gamma/2-H_{DD}-H_{DD}^\uparrow-H_{DD}^\downarrow]^{-1}H_{D\gamma} \tag{4}$$

which can easily be rearranged to

$$T_{\gamma \pi^0} = [T_{PP} \, G_P + 1]\{H_{PD}[E-E_R+i\Gamma/2-H_{DD}-H_{DD}^\downarrow]^{-1}H_{D\gamma}\}. \tag{5}$$

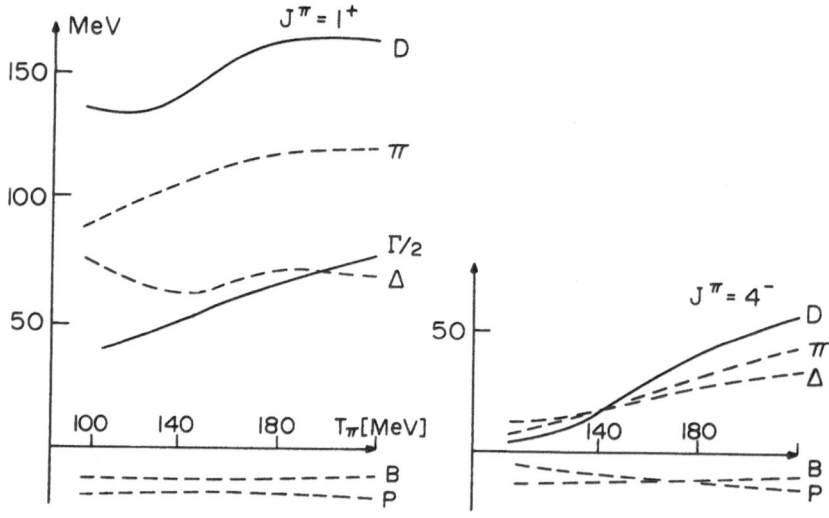

Figure 2.   Decomposition of $\langle D_0^L | G_{\Delta h}^{-1} | D_0^L \rangle$ for $L=1^+, 4^-$.

Equation 5 allows a direct comparison to the standard DWIA approach. The square brackets contain the distorted pion wave. The curly brackets contain the (γ, π) production operator. In a DWIA calculation this would be the free operator, but in our approach it contains the terms H_{DD} and $H_{DD}^{\perp}$, corresponding to multistep contributions. This is an important difference: multistep contributions are included in the Δ-hole framework with the same spreading interaction fit to elastic pion scattering. Figure 3 shows the results for the differential and total (γ, π^0) cross section for ^{16}O. There is a large reduction from the plane wave impulse approximation (IMP) to the full calculation (EX). A similarly strong reduction is found in DWIA calculations. In π- nucleus elastic scattering, it was shown[2,3] that the doorway state $|D_0\rangle$ exhausts most of the transition strength. The dashed curve shows the (γ, π^0) cross section if only this doorway state is kept. The result is remarkably close to the full result and emphasizes the compact unified approach provided by the isobar-hole formalism for different reactions. By inserting a complete set of states in Eq. 4 we obtain for the production

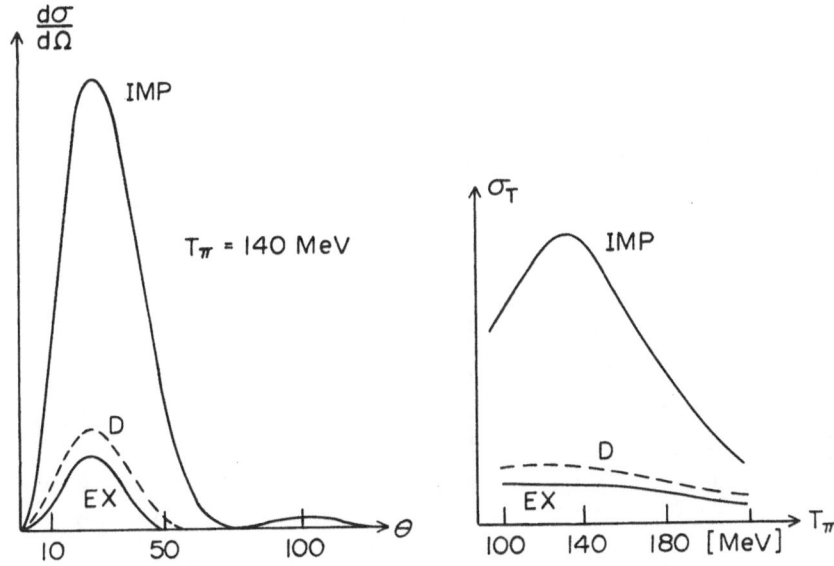

Figure 3. Relative cross sections for $^{16}O(\gamma,\pi^0)^{16}O$ in the impulse (IMP), one doorway state (D) and full isobar-hole (EX) calculations.

amplitude

$$T_{\gamma\pi^0} = (N^\gamma N_L)^{-1} \sum_{i \geq 0} <D_0^L|G_{\Delta h}|D_i^L><D_i^L|D_\gamma^L> \tag{6}$$

$|D_\gamma^L>$ is the photon doorway state defined analogously to Eq. 2. The dominant contribution to the sum from the doorway state, $|D_0^L>$, involves the same diagonal matrix element as probed in elastic scattering, Eq. 3. The new physics we can learn from the (γ,π^0) reaction is contained in the off-diagonal matrix elements, $i\neq 0$. Thus, the extent to which new information can be obtained depends on the overlap of the pion and photon doorway states, $<D_i^L|D_\gamma^L>$ shown in Fig. 4. Clearly, only for small L is the overlap very different from unity, allowing for significant contributions of other states with $i\neq 0$. Therefore the central partial waves will be most important to extract new information and to distinguish among

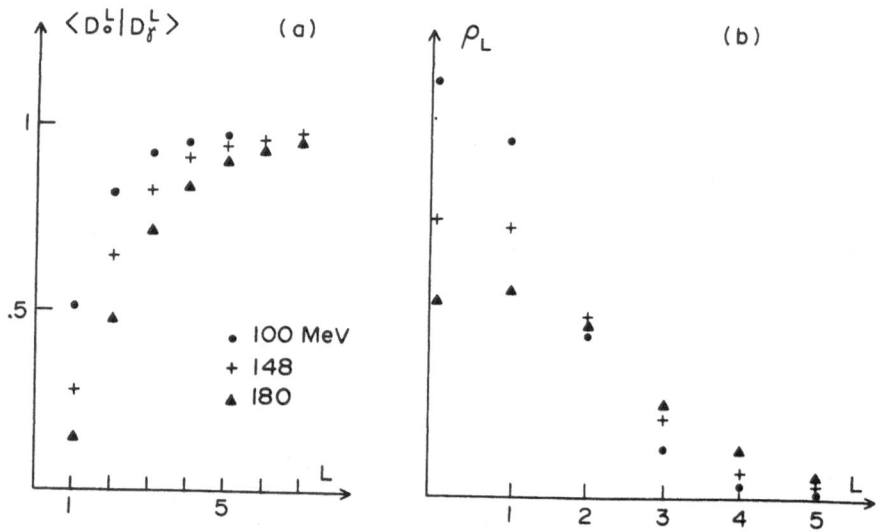

Figure 4. (a) Overlap of the pion and photon doorway states.
(b) The nuclear partial wave form factor (Eq. 7).

various models that fit elastic nuclear scattering equally
well. However, the central partial waves are strongly damped
(see Fig. 2) and most of the (γ, π°) cross section is due to
the semiperipheral and peripheral partial waves, thus reducing
the sensitivity of the (γ, π°) reaction to different on-shell
equivalent descriptions of π-nucleus scattering.

The behavior of the overlap shown in Fig. 4a is easily
understood. With $H_{\gamma D} \sim \vec{\epsilon} \cdot \vec{S} \times \vec{k}$ and $H_{PD} \sim \vec{S} \cdot \vec{q}$, the overlap is
proportional to

$$\langle D_0^L | D_\gamma^L \rangle \sim \sqrt{\frac{L(L+1)}{2L+1}} \; (\rho_{L+1}(k,q) - \rho_{L-1}(k,q))$$

$$\rho_L(k,q) = \int_0^\infty dr \; r^2 \; \rho(r) \; J_L(kr) \; J_L(qr) \qquad (7)$$

The ρ_L are shown in Fig. 4b. For $\rho_{L+1} \approx \rho_{L-1}$, we have poor
overlap, while for $\rho_{L+1} \ll \rho_{L-1}$ the overlap of the normalized

doorway states is good. The latter condition is achieved for low energies and smaller targets.

Further details will be published elsewhere[7].

NOTES AND REFERENCES

*This work is supported in part through funds provided by the U.S. Department of Energy (DOE) under contract EY-76-C-02-3069.
1. L. Kisslinger and W. Wang, Ann. Phys. 99 (1976) 374.
2. M. Kirata, F. Lenz and K. Yazaki, Ann. Phys. 108 (1977) 116.
3. M. Hirata, J.H. Koch, F. Lenz and E.J. Moniz, Phys. Lett. 70B (1977) 281.
4. M. Hirata, J.H. Koch, F. Lenz and E.J. Moniz, to be published.
5. E.J. Moniz, invited paper elsewhere in this volume.
6. R. Woloshyn, TRIUMF preprint.
7. J.H. Koch and E.J. Moniz, to be published.

UNIFIED DESCRIPTION OF PION-PHOTO-PRODUCTION AND ELASTIC PION

SCATTERING ON THE DEUTERON IN THE ISOBAR-DOORWAY MODEL

R. Händel, M. Dillig, M. G. Huber

Institute for Theoretical Physics

Univ. Erlangen, W. Germany

It has been recognized recently[1] that the strong dominance of the $\Delta(1236)$ isobar in nuclear reactions at medium energies advocates an explicit treatment of this degree of freedom in nuclei (in contrast to its implicit incorporation in an optical model description). Such a model, phrased commonly as isobar doorway or A^* model[1] has two particularly interesting features. (i) it forms a natural basis for a unified description of different isobar dominated scattering processes; (ii) it offers a more or less direct test of the isobar self energy in nuclei or, equivalently, of the ΔN interaction. An ideal testing ground for such a model seems to be deuteron: with all many-body complications being absent, predictions for scattering reactions on the deuteron should reflect and in this way allow direct investigation of the various ingredients of the model.

To test this conjecture a consistent analysis of elastic π-scattering (π,π) and of photo-production of neutral pions (γ,π°) on the deuteron was performed in the D^* model (Fig. 1). A unified description of both reactions is suggested by the structure of the resonant transition amplitude

$$T(\vec{k},\vec{k}') = \langle\vec{k}'|H'G_{\Delta N}(\omega)H|\vec{k}\rangle \tag{1}$$

where

$$G_{\Delta N}(\omega) = \frac{1}{\omega - H_{\Delta N}} = \sum_\mu \frac{|D_\mu^*\rangle\langle D_\mu^*|}{\omega - \varepsilon_\mu^*} \tag{2}$$

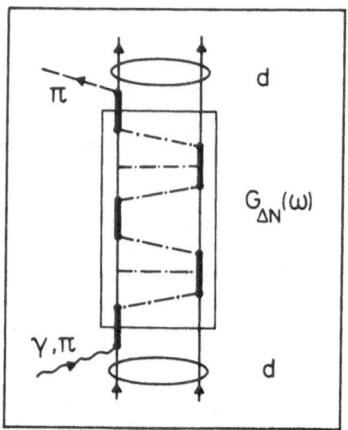

Figure 1. Schematical representation of elastic π scattering and π photoproduction on the deuteron in the D* model.

which differs for the two processes only in the transition operator $H_{\lambda N\Delta}$ into the (ΔN)-doorway state

$$H_{\pi N\Delta} = (f_\pi^*/m_\pi)\vec{S}^+\vec{k}_\pi(\vec{T}^+\vec{\phi}_\pi) \tag{3}$$

and

$$H_{\gamma N\Delta} = (g_u^*/m_\pi)\vec{S}^+(\vec{\epsilon} \times \vec{k}_j)T_o^+ \tag{4}$$

respectively. The spectrum of the $|D_\mu^*> = (\Delta N)_\mu$ states was calculated - in analogy to the NN interaction. - from a ΔN OBEP

$$V_\lambda(q) = -\frac{W_\lambda^+(q)W_\lambda(q)}{q^2 + m_\lambda^2 - \omega^2} \tag{5}$$

which includes contributions from $\lambda = \pi$, σ, ρ, ω^o exchange in the direct term and $\lambda = \pi, \rho$ in the exchange term (compare Fig. 2); corrections from vertex modifications and multiple meson exchange were included in a simple parametrization (for further details see Ref. 2).

The nonstatic correction in Eq. (5) is found to be of crucial influence: in contrast to the potential induced by the exchange of heavy mesons - which shows a radial dependence as

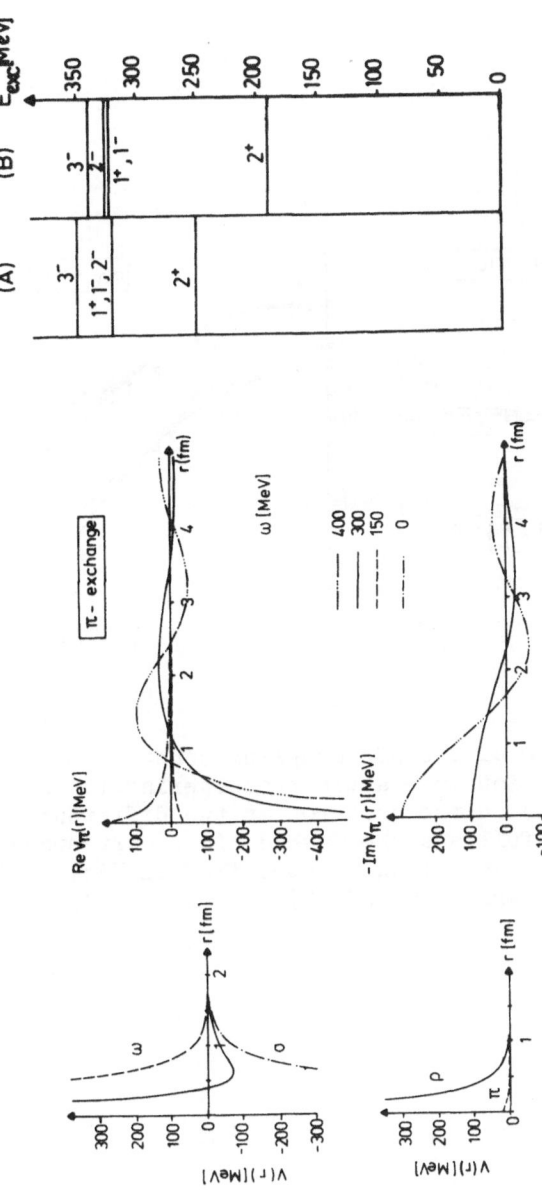

Figure 2. (a) Typical radial dependence of the central part of the ΔN potential induced by different agencies. (b) Energy dependence of the real and imaginary part of the π exchange term in the N potential. (c) Sensitivity of the D*(T=1) spectrum on the ρBB coupling constant. Spectrum (A) refers to $f_{\rho NN}/4\pi = 1.89$, spectrum (B) results from $f_{\rho NN}/4\pi = 3.32$ (for details see Ref. 2).

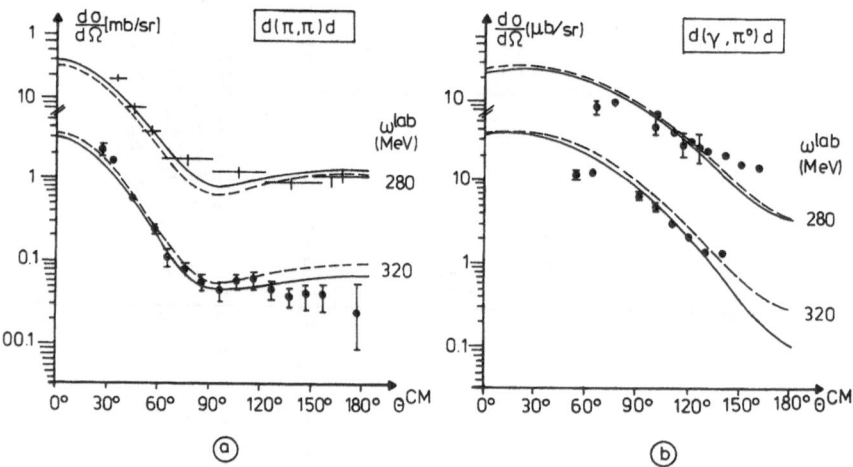

Figure 3.   Influence of the ρBB coupling constant on the differential cross section in elastic π scattering (a) and pion-
photo-production (b) on the deuteron at two different scattering energies.   The full and the dashed lines correspond to the
weak and the strong ρ-coupling as specified in Fig. 2c (the
experimental values are from Ref. 3).

familiar from the NN interaction (Fig. 2a) - the influence of
the scattering energy ω leads to an effective imaginary π-mass

$$m_\pi^* = i\sqrt{\omega^2 - m_\pi^2} \tag{6}$$

above the π production threshold, resulting in a strongly energy dependent, long-ranged and complex π- exchange potential
(Fig. 2b).

The resulting D^* spectrum for the T = 1 (ΔN) system is
shown in Fig. 2c (at a scattering energy ω = 300 MeV). While
the splitting of the D^* spectrum reflects directly the strong
spin dependence of the ΔN potential, the two spectra compared
exemplify the influence of different parametrizations (in
Fig. 2c two choices for the ρBB coupling constant are compared).
The differential cross sections for elastic pion scattering
and for coherent photopion-production have been calculated
using the same parameter sets as for the spectra (A) and (B),
respectively, of Fig. 2c. The results are compared with available experimental data in Fig. 3. The two main findings are
that the results show an appreciable sensitivity in particular
on the parametrization of the ΔN interaction (though the limited amount of data presently available does not allow to discriminate between the interactions (A) and (B); the isobar
doorway model allows an, at least qualitative, unified description of both scattering reactions.

Certainly the investigation of medium energy reactions in
the isobar model is only in its beginning; though the results
are rather encouraging further detailed studies are necessary,
an extension of OBE model to incorporate the true pion absorption as well as the double Δ(1236) excitation channels in a
TBE model is under way, to establish the model as a tool for
a quantitative and unified description of isobar dominated
medium energy scattering processes on nuclei.

References

1. L. S. Kisslinger, W. L. Wang, Phys. Rev. Lett. 30 (1973)
   1071; M. Dillig, M. G. Huber, Phys. Lett. 48 B (1974) 417;
   F. Lenz, Ann. Phys. 95 (1975) 348; G. E. Brown, W. Weise,
   Phys. Rep. 22 C (1975) 405.
2. R. Händel, M. Dillig, M. G. Huber, Phys. Lett. 73 B (1978)
   4; R. Händel, Dissertation (Erlangen 1978).
3. B. Bouquet et al., Nucl. Phys. 79 B (1974) 45; G. von
   Holtey et al., Z. Phys. 259 (1973) 51.

A SUM-RULE DESCRIPTION FOR THE INCLUSIVE ELECTROPRODUCTION

OF PIONS FROM NUCLEI

R. Rosenfelder

Institute of Theoretical Physics,
Department of Physics
Stanford University, Stanford, CA 94305

Angular distributions of low-energy pions (T_π = 21.6 MeV) have been measured recently at Mainz[1]. Due to the large excitation energy available (ϵ_0 = 280 MeV) the summation over the excited nuclear states can be done by the closure approximation. The double-differential cross section is then proportional to the spin-isospin correlation function of the nucleus $m_0 = <o|\sigma^+\sigma|o>$ where for s-wave production

$$\sigma = \sum_i \zeta_\pm(i)\vec{\sigma}(i)e^{-i\vec{k}\cdot\vec{\tau}_i} u_{\vec{q}}^*(\vec{r}_i).$$

Here $\vec{k}$ denotes the momentum transfer of the electron and $u_{\vec{q}}(\vec{x})$ is the pion wave function distorted by the optical potential between outgoing pion and nucleus. In the present calculation we have used pion wave functions (kindly provided by W. C. Haxton) which successfully describe low-energy elastic scattering[2], and retained four partial waves. The usual problem associated with the closure approximation - how to choose the mean excitation energy $\overline{E}$ of the nucleus - can be solved systematically by the evaluation of higher moments[3]. For simplicity we have restricted ourselves to the lowest approximation $\overline{E} = m_1/m_0$[4] where m_0 = 8(1-exp $(-\frac{1}{2} t^2 b^2)$ (1 + $\frac{1}{16} t^4 b^4$)) is the harmonic oscillator expression for ^{16}O when evaluated with plane pion waves ($\vec{t} = \vec{k} - \vec{q}$). It is well know that the classical TRK result for the first

367

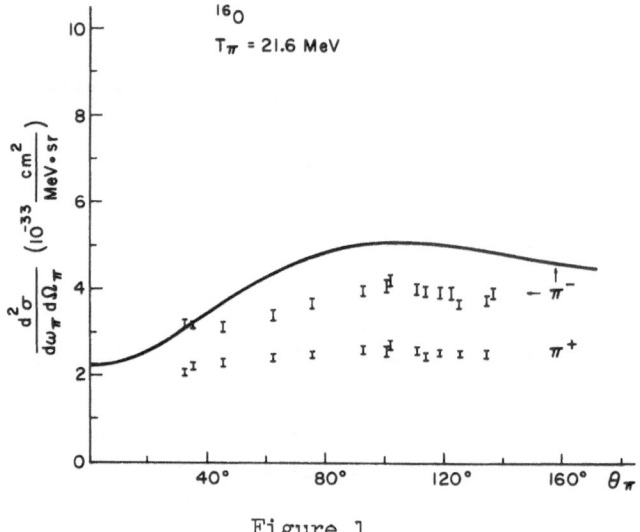

Figure 1

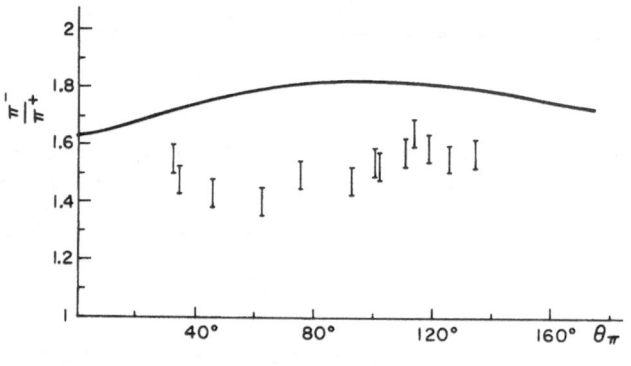

Figure 2

moment ($m_1 = 8 \cdot t^2/2M$) is enhanced by exchange parts of the
nuclear interaction. With effective interactions of the
Skyrme type this contribution can be incorporated in an
effective mass $M^{*[5]}$. Although this description is only
valid for small momentum transfers it partly compensates the
overestimate of the cross section at backward angles (large
t) given by the sum-rule calculation[4].

Figure 1 shows the numerical results using the Skyrme II force ($M^* = 0.63M$). Coulomb energy shifts and the interference term between s- and p-wave production have been included. Compared with previous calculations[1] which used pion wave functions derived from pionic atoms, we notice a considerable reduction of the theoretical curve due to a stronger absorption of the pion waves inside the nucleus. The simple nuclear model and the failure of the sum-rule description at backward angles probably account for the remaining discrepancy between theory and experiment. In fact, an elaborate shell-model calculation summing all kinematically allowed nuclear transitions[6] agrees quite well with the experimental cross sections. However, the π^-/π^+ ratio (shown in Figure 2) seems to be insensitive to details of the nuclear structure and thus can be used to test our understanding of the production operator and the pion final state interaction.

REFERENCES

1. F. Borkowski et al., Phys. Rev. Lett. **38**, 742 (1977).
2. N. J. Digiacomo et al., Phys. Lett. **66B**, 421 (1977).
3. R. G. Gordon, J. Math. Phys. **9**, 655, 1087 (1968).
4. R. Rosenfelder, Nucl. Phys. **A290**, 315 (1977).
5. R. Rosenfelder, Nucl. Phys. **A298**, 397 (1978).
6. W. C. Haxton, LASL preprint (1978).

INCLUSIVE ELECTROPRODUCTION OF LOW ENERGY PI MESONS

M. Pauli, A. M. Berstein, K. I. Blomqvist,

G. B. Franklin, J. LeRose, K. Min, N. Paras,

D. Rowley, B. Schoch, P. Stoler, E. J. Winhold,

and P. F. Yergin, MIT-RPI

INTRODUCTION

Because the electromagnetic interaction is well known and because the real or virtual photon can create a pion at any point inside the target nucleus, the study of gamma-pi type experiments can give important information concerning the final state interaction of the pion with the daughter nucleus.

The reaction that we wish to talk about today is inclusive electroproduction of pi mesons, namely $e^- + A \rightarrow e^- + \pi^\pm + X$ where only the pion is observed. There are several possibilities to generate low energy pions (see Figure 1). One of these is direct production of a low energy pion where the daughter nucleus remains in a specific final state. An example of this reaction is $^{12}C (\gamma, \pi^+)^{12}B$ which has been widely studied[1,2]. Another mechanism is the quasi-free process in which the pion is produced on a nucleon in the nucleus and both the pion and the recoil nucleon leave the nucleus. An example of this is $^{12}C (\gamma, \pi^+)^{11}B$ [3]. The quasi-free process is expected to dominate at higher energies. Another possible mechanism to produce low energy pions is direct production of high energy pions followed by energy loss as shown in figures 1c and 1d.

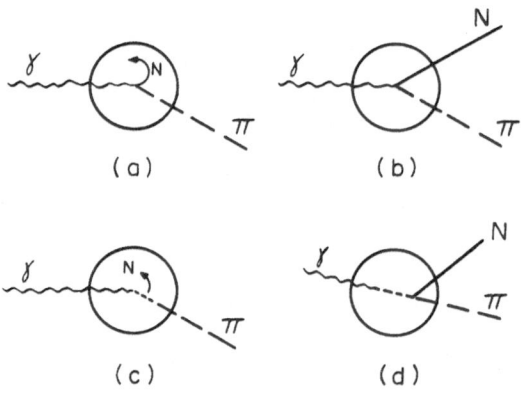

FIGURE 1

The π^- to π^+ ratio, R, is measured here for various elements and kinematics. This ratio will differ from the elementary ratio for various reasons; the most important of these being the different phase space available to π^+ and π^-. The elementary amplitude and the available phase space are well known and from this we will see the effect of the coulomb interaction and the strong final state interactions of the pion and the outgoing nucleon.

Our experimental setup has been explained in a previous talk by Winhold[4]. Our experimental program to date on inclusive pion electroproduction has been in the range of pion kinetic energies between 15 and 42 MeV. The electron energy has been varied between 200 and 280 MeV. We are fixed at a scattering angle of $90°$ in the lab frame. The targets we have measured so far include: ^1H, ^2H, Be, BeO (for oxygen), C, and Ca.

PI MINUS TO PI PLUS RATIO

The pi minus to pi plus ratio, R, is defined as:

$$R = (\frac{d^2\sigma}{d\Omega dE})_- \Big/ (\frac{d^2\sigma}{d\Omega dE})_+ \qquad (1)$$

It is interesting to investigate this ratio for a variety of reasons. Experimentally, many of the uncertainties that plague absolute measurements cancel out in this ratio. Theoretically, certain nuclear structure uncertainties are believed to cancel out[5] and the ratio is often a good way of isolating the dependence of the pion electroproduction on the coulomb interaction of the outgoing pion.

There are many ways to examine the quantity R. One of these is to examine the dependence of R on the beam energy (the end point energy of the virtual photon spectrum). While we have made many measurements at different pion energy on different targets, we shall consider here one representative example. Figure 2 is a plot of the ratio R versus the end point energy E_o for 29 MeV pions on ^{12}C. Since the threshold for ^{12}C $(\gamma, \pi^+)^{12}B$ occurs at 185 MeV and the threshold for ^{12}C $(\gamma, \pi^-)^{12}N$ occurs at 188 MeV, the ratio R will start at zero at 185 MeV. The ratio rises slowly to 1.38 at a beam energy of 280 MeV. At this energy, we are in close agreement with data from Mainz[6] and a theoretical calculation by Haxton[5]. Also shown is a point at 250 MeV which is the average of two measurements, one at T_π = 23 MeV by Watase et al[7]. An important feature to notice is that as we increase the beam energy, the ratio climbs up to roughly that of the free ratio as threshold differences become less important.

At this point, it should be mentioned that the free ratio $\frac{d\sigma}{d\Omega}(\gamma+n\rightarrow\pi^-+p)/\frac{d\sigma}{d\Omega}(\gamma+p\rightarrow\pi^++n)$ at $90°$ has been calculated

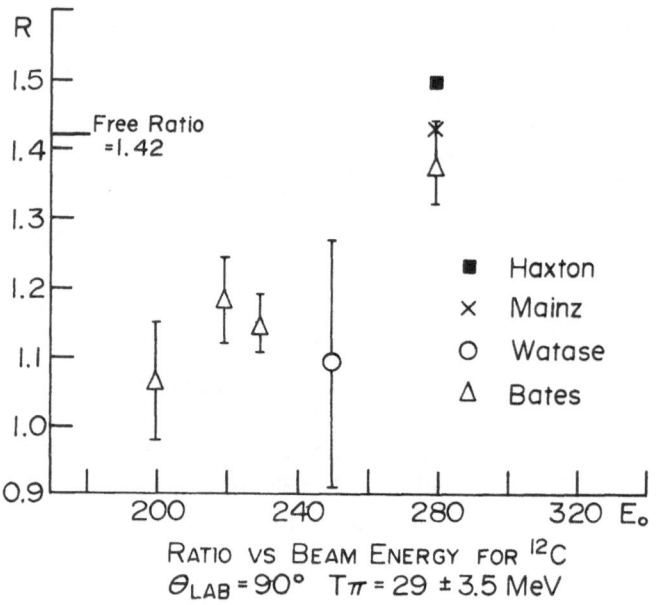

RATIO VS BEAM ENERGY FOR ^{12}C
$\theta_{LAB} = 90°$ $T\pi = 29 \pm 3.5$ MeV

FIGURE 2

by Blomqvist and Laget[8] and is between 1.4 and 1.44 for
pions of 10 to 50 MeV which is the kinematic range of our
studies so far. The ratio measured on deuterium at T_π =
17.4 MeV and E_0 = 230 MeV is 1.45 $\pm$ 0.04, which is close
to what one would expect since the deuteron is not a very
bound nucleus.

It is also useful to plot the π^-/π^+ ratio versus pion
kinetic energy. Such a plot is shown in Figure 3 for the
case of ^{12}C. Again, we see that the data is in agreement
with calculations of Haxton that sums over single particle
states and also contains quasi-free production but does not
have an energy loss mechanism in the final state interaction.
The triangles represent data taken at E_0 = 230 MeV at Bates.

One can see the rise in the ratio as one goes lower in pion
kinetic energy due to the effect of the pion coulomb interaction on the available phase space. We can estimate the
ratio for various T_π as:

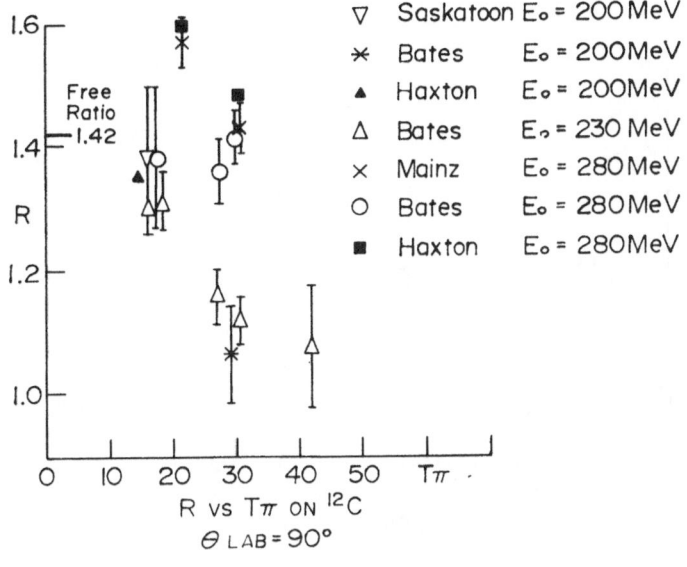

FIGURE 3

$$R = R_{free} \; \frac{N}{Z} \; \frac{\sqrt{T\pi + Vc}}{\sqrt{T\pi - Vc}} \qquad\qquad (2)$$

where V_c is the coulomb potential

The rise of R as one goes lower in pion kinetic energy is even more pronounced at 200 MeV[9]. We expect R to go to zero around 40 MeV since we will be hitting pion threshold.

A plot of R for 29 - 30 MeV pions at 90 degrees and a beam energy of 280 MeV is shown in Figure 4. We again confirm the basic agreement between our data and similar data from Mainz. The large fluctuations in R are due mainly to N≠Z and to threshold differences of daughter nuclei which give different phase space. The importance of the threshold differences was shown many years ago by Littauer and Walker[10].

DOUBLE DIFFERENTIAL CROSS SECTIONS

As stated earlier, we measured yields or double differential cross sections at various beam and pion energies.

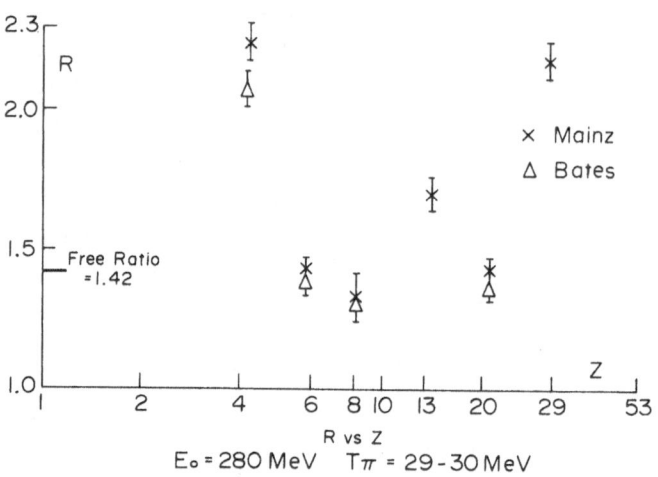

FIGURE 4

The absolute normalization on these yields are in question by an estimated 15%. The errors in the points themselves are due to systematic problems in the background separation and deadtime and are of the order of 5%.

One of the things we can examine is the yield versus beam energy. Figure 5 is a plot of the double differential cross section in units of 10^{-34} cm^2/str-MeV versus beam energy for 29-30 MeV pions at 90 degrees off ^{12}C. The isochromat of N. Paras, et al[1] would start at 185 MeV and go to about 195 MeV where it begins to flatten out[11]. This data is too small to show on this graph. The extension of the isochromat, including the giant resonance states, is shown (roughly) in Figure 5. We thus see that the direct production of low energy pions with the nucleus remaining in a definite final state can account for only a small amount of the pions that we detect at high energies.

Another interesting piece of physics that one may hope to extract from the yields is the A dependence on the total charged electroproduction yields. To remove the effects of fluctuations in the π^-/π^+ ratio as shown in Figure 4, we

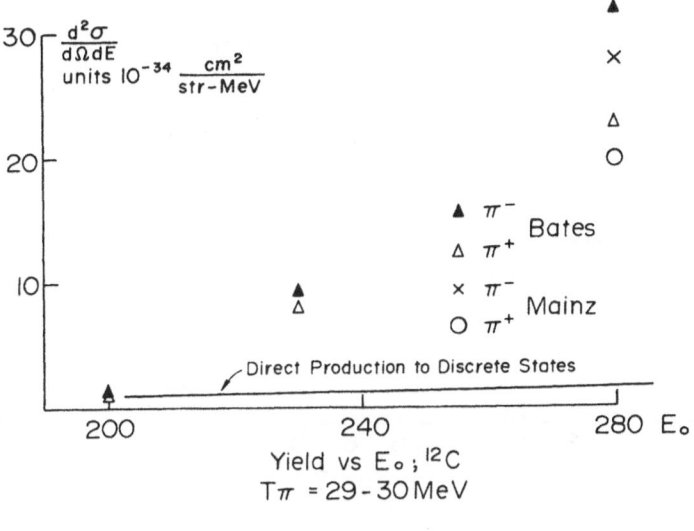

FIGURE 5

consider not the π^- or the π^+ yield itself but rather the
sum of these. Naively, we would expect the total charged
electroproduction yield to vary as A. There are two possible
reasons for departure from an A dependence. First of all,
in heavier nuclei, there are nucleons that are very deeply
bound. The photoproduction off these nucleons will be
suppressed by phase space and by the exclusion principle.
All things being equal therefore, we could expect that the
total yield per nucleon for a heavy nucleus such as lead
would be less than the total yield per nucleon for carbon.
One indication of this phenomena is what seems to be the
enormous contribution of the fifth (excess) neutron in
beryllium. This departure from A dependence does not involve
any assumptions about the pion final state interaction.

The other possible reason for departure from A dependence involves the absorbtion of the pion by the optical
potential in the nucleus. The pion mean free path, λ_π, is
a function of the pion energy. In the high pion energy
region where the mean free path is much smaller than the
nuclear radius, a pion produced in the nuclear interior is
very likely to be absorbed and we should see an $A^{2/3}$ dependence since the observed pions will be those from the nuclear
surface. For low pion energy where λ_π is large compared to

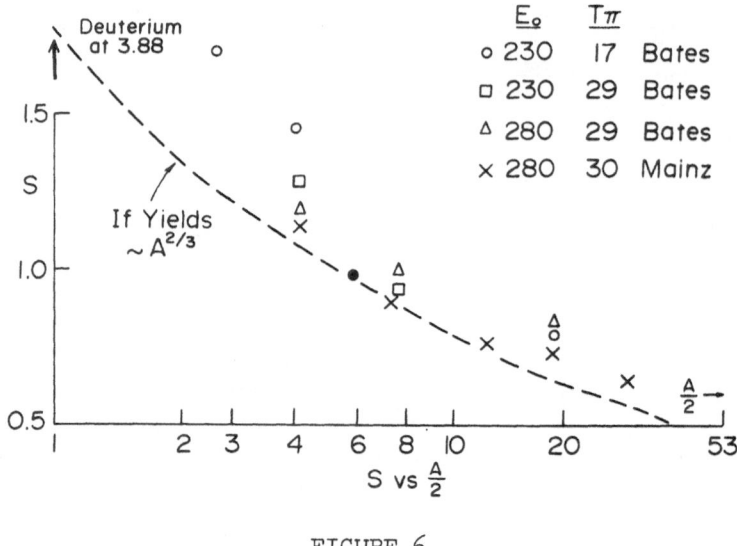

FIGURE 6

the nuclear size, the dependence may be closer to A. It will be interesting to study the A dependence for different kinematical conditions.

We have defined the quantity S as:

$$S \equiv \frac{\{\frac{1}{A}[(\frac{d^2\sigma}{d\Omega dE})_+ + (\frac{d^2\sigma}{d\Omega dE})_-]\}_{\text{NUCLEUS}}}{\{[(\frac{d^2\sigma}{d\Omega dE})_+ + (\frac{d^2\sigma}{d\Omega dE})_-]\}_{\text{CARBON}}} \qquad (3)$$

S is fixed at one for carbon. We have plotted S versus A measured for different E_o and T_π in Figure 6. If the total charged pion yields are proportional to A, we expect a horizontal line. The dashed line represents an $A^{2/3}$ dependence. Our data is somewhat in between the dashed line and being flat. Our one point on deuterium is far above the other points which is a manifestation of the fact that the deuteron is barely bound and its daughter nucleus is unbound. The slope for 17 MeV pions with 230 MeV beam energy is a bit different from that for 280 MeV beam energy and 29-30 MeV pions but more data is needed here.

FUTURE INCLUSIVE STUDIES

As already mentioned, we hope to extend the studies of the past to a wider range of beam energy, pion energy, and targets. One way to try to extract the nature and importance of the pion thermalization process is indicated by the work of Mecking, et al[12]. The Mecking experiment has measured the double differential cross section for real tagged photons with 9 MeV resolution on carbon at various angles as a function of kinetic energy. This work shows the effect of the pion final state interaction using a Monte Carlo program of Gabriel[13]. The data goes only to 50 MeV pion energy. It is our aim to fill in the data at 90 degrees for pion energies of 15 to 50 MeV (plus taking some higher points to get proper matching). We hope to be able to simulate a pseudo-monochromatic virtual photon beam through use of the virtual photon spectrum difference method.

We also believe that we can obtain better estimates on the amount of real photoproduction in our targets through better measurements of the real/virtual ratios. More measurements on hydrogen should give us a more reliable absolute calibration as well as ideas on the virtual photon shape. Improvements in our Cerenkov veto system, reducing deadtime, and using time of flight to sort out background may get rid of the systematic fluctuations in the yields that are presently at the 5% level.

REFERENCES

1. N. Paras et al, contributed paper to this conference.
2. H. Ohashi et al, contributed paper to this conference.
3. See for example; J. M. Laget, contributed paper to this conference; J. M. Laget Nuci. Phys. A194 (1972), 81; and references of these articles.
4. E. J. Winhold, contributed paper to this conference.
5. W. Haxton LASL preprint #LA-UR-78-393.
6. F. Borkowski, U. Mainz dissertation and F. Borkowski et al, Phys. Rev. Lett. 38 #14 (1977).
7. Y. Watase et al, J. Phys. Soc. Japan 40 (1976), 1531.
8. I. Blomqvist and J. M. Laget Nucl. Phys. A280 (1977) 405.
9. Saskatoon group private communication.
10. Littauer and Walker, Phys. Rev. 86, #6 (1952), 838.

11.   Alder et al, contributed paper to this conference.
12.   Arends et al, contributed paper to this conference.
13.   T. A. Gabriel, Phys. Rev. <u>C13</u> (1976), 240.

PHOTOPION PRODUCTION IN THE FERMI GAS MODEL

J.S. O'Connell, W.M. MacDonald*, and E.T. Dressler+

National Bureau of Standards, Washington, D.C.

In order to understand some of the general properties of pion photoproduction from a nucleus, we have studied the photoproduction process from a Fermi gas. In this calculation, we used the non-relativistic pion photoproduction operator of Blomqvist and Laget[1] which reproduces the $\gamma + p \rightarrow \pi^{+} + n$ data through the 3-3 resonance region. One of the advantages of using this operator is that the momentum dependence of the struck nucleon is given explicitly. We studied the importance of the momentum dependent terms by calculating the Fermi gas cross section exactly and compared it with the results with these terms neglected, i.e., the frozen nucleon approximation. We found this to be a poor approximation leading to errors of about 25% in the resonance region. However, fixing the nucleon's momentum at the Fermi momentum and in the direction $\phi_p = \frac{\pi}{2}$ (measured with respect to the plane formed by the photon and pion) the results only differed by up to 7%. Fixing the magnitude and direction of the struck nucleon's momentum at some optimum value enables one to factor the pion production operator out of the nuclear overlap integral and greatly simplifies the calculation.

Using this factored approximation, we considered the effect on the cross section from the interaction of the target

*Permanent Address: Department of Physics and Astronomy, University of Maryland, College Park, Maryland 20742.

+NAS/NRD Postdoctoral Fellow

nucleon with the rest of the nucleus both before and after
pion production. This effect was tested by letting the nucleon energy be

$$P_0 = \sqrt{\vec{p}^2 + \vec{m}^2} + V(\vec{p}^2)$$

where $V(\vec{p}^2)$ is the real part of the nucleon-nucleus optical
potential[2]. Curve A of Figure 1 shows the results for the
Fermi gas (γ, π^+) total cross section without the nucleon-nucleus interaction. Including the nucleon-nucleus interaction
(Curve B, Figure 1) reduced the cross section in the region
below the resonance and increased it above.

The effect of the interaction of the pion with the nuclear medium is taken into account by introducing[3] a pion
index of refraction, n, where the pion momentum in the nuclear
medium is given by

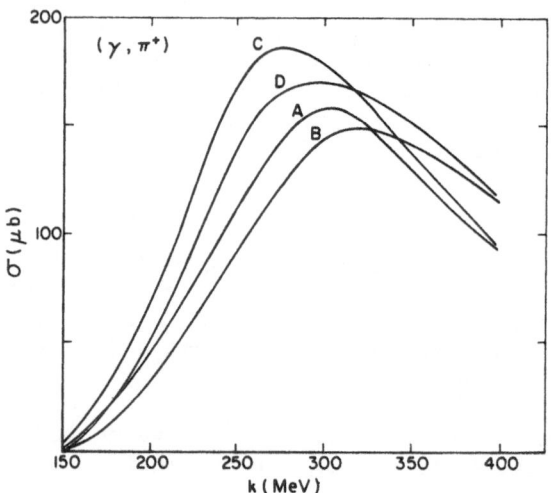

Figure 1. Curve A is the total (γ, π^+) cross section using the
Fermi gas model in the factored approximation. Curve B includes nucleon-nucleus interactions, Curve C includes π-nu-
cleus interactions, and Curve D includes both.

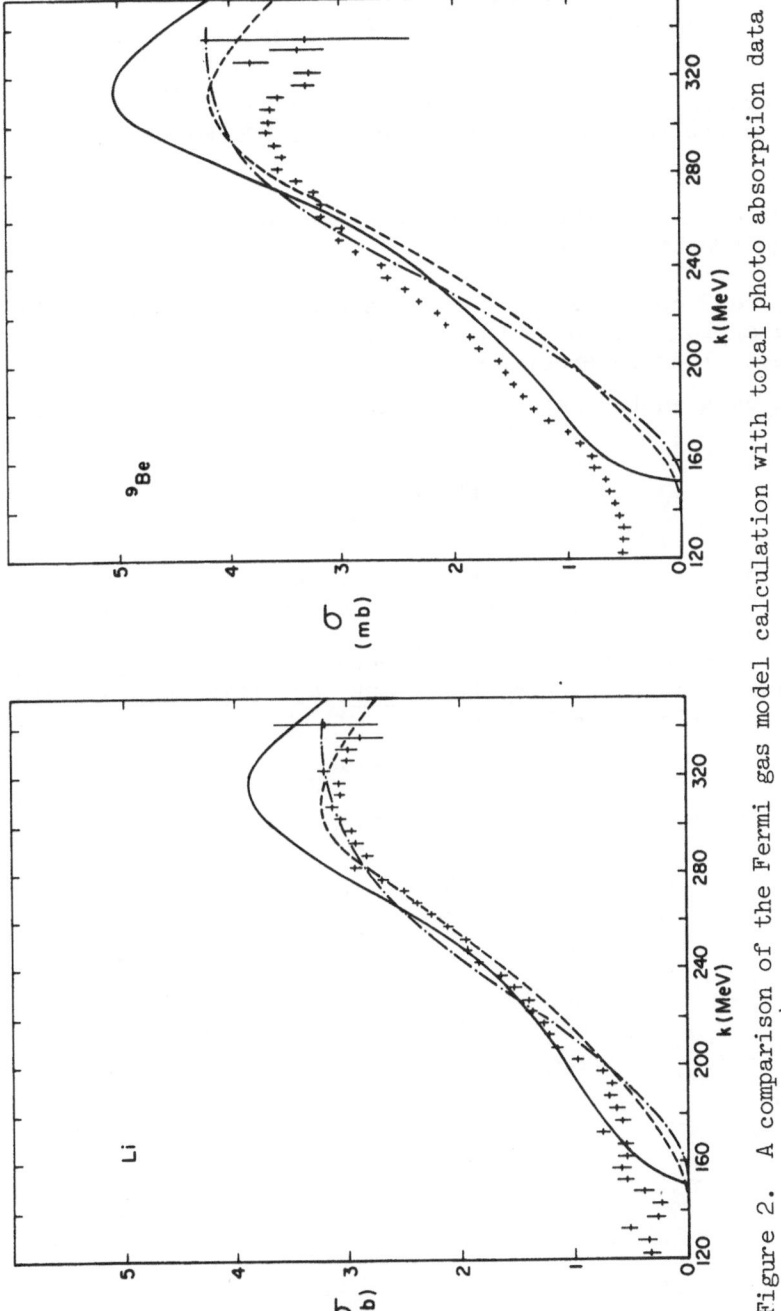

Figure 2. A comparison of the Fermi gas model calculation with total photo absorption data on natural Li and ^{9}Be in the resonance region. The dashed curve is for the free Fermi gas in the factored approximation. The dashed-dot curve has N-nucleus and π-nucleus interactions. The solid curve is for free nucleons not in a Fermi gas.

$$q_m = Re\ (n) \cdot \sqrt{q_o^2 - m_\pi^2}, \quad q_o = \text{total pion energy}.$$

This effect by itself gave a large increase to the cross section in the region below the resonance (Curve C, Figure 1).

Including both the pion and nuclear optical potentials gives the results in Curve D.

Figure 2 compares the Fermi gas cross section using the factored approximation (with and without the nucleon-nucleus and pion-nucleus interactions) with recent total photo absorption data in the resonance region for natural Li and $^9Be^4$.

REFERENCES

1. I. Blomqvist and J.M. Laget, Nucl. Phys. A280, 405 (1977).
2. F.A. Brieva and A. Dellafiore, Nucl. Phys. A292, 445 (1977).
3. T.E.O. Ericson and J. Hufner, Phys. Lett. 33B, 601 (1970).
4. H. Ahrens et al., Phys. Lett. 52B, 43 (1974) and private communication from H. Gimm.

TOTAL PHOTONUCLEAR ABSORPTION CROSS SECTIONS FOR Li AND Be UP TO 350 MeV

J. Ahrens, H. Gimm, R.J. Hughes, R. Leicht,

P. Minn, A. Zieger and B. Ziegler

Nuclear Physics Group, Max-Planck-Institute for

Chemistry, Mainz, Germany

The knowledge of the total photonuclear absorption cross-section in composite nuclei may help to understand nuclear matter effects in photopion production. This cross section has therefore been measured up to 350 MeV in the Mainz Linac Laboratory.

In Fig. 1 the final results for natural Be- and Li-targets and a preliminary result for a ^{6}Li-target are given. These values were obtained by subtracting a "Quasideuteron" background, that had been found by fitting the total nuclear absorption cross sections of various light nuclei in the region 50 to 100 MeV. The results for the rising part of the (3/2, 3/2) resonance can be compared with the incoherent sum of the photopion production cross sections for the free nucleons. The values experimentally found for the composite nuclei are smaller than the incoherent sum by a factor .7 to .85. This reduction factor is due to the binding of the nucleons. The main source of error in the determination of this factor arises from the uncertainties in the photo-pion production cross sections for the free nucleons.

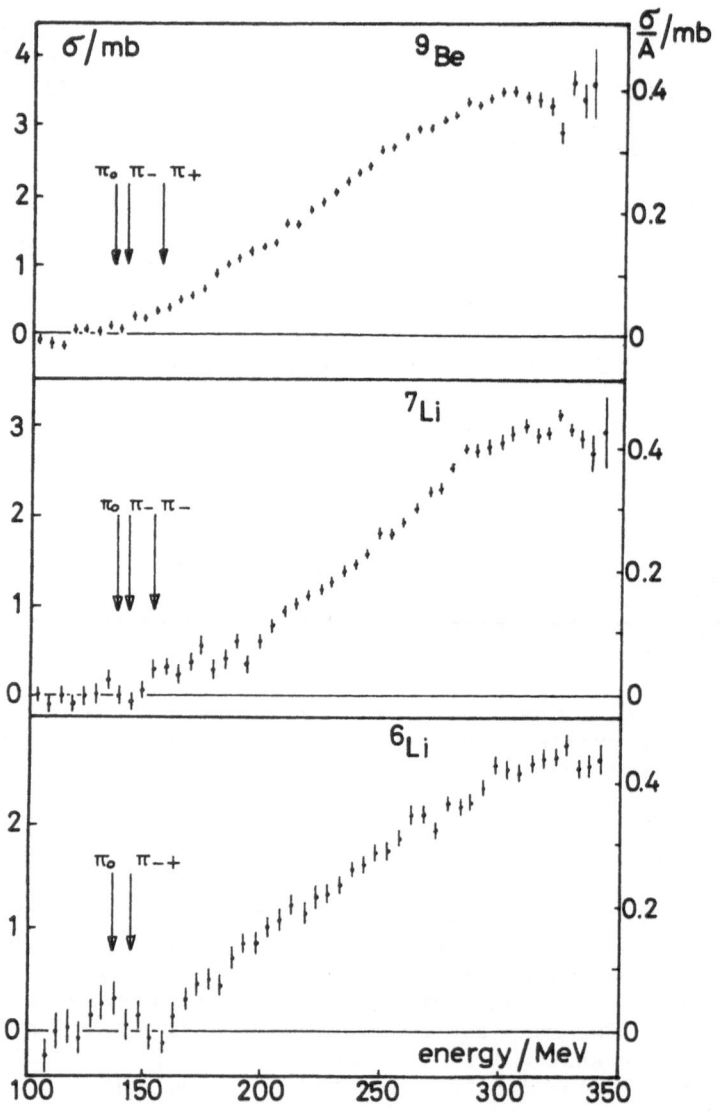

Figure 1.   Total photonuclear cross sections minus a "Quasi-deuteron" cross section for Be, ^{7}Li and ^{6}Li (preliminary). For direct comparison the right hand ordinate is given in cross section per nucleon.   Also given are the thresholds for the production of the different pions.

ISOMERIC AND GROUND STATE FORMATION IN ^{197}Hg BY THE ^{197}Au(γ,π^-)^{197} Hg REACTION NEAR THRESHOLD[*]

P. H. Ballentine[+], J. K. Hersh[+], H. A. Medicus,

S. Planeta, E. Potenziani[+], and S. Rossdeutscher

Rensselaer Polytechnic Institute

Troy, New York 12181

In $\gamma,\pi^{\pm}$ reactions near threshold that populate in the product nucleus an isomeric state and a radioactive ground state, the ratio of the cross-sections leading to these two states, or isomer ratio, can yield information about the production mode of the emitted photopions. In such reactions, the isomeric and the ground state may be reached directly when the pion is emitted. However, these states can also be reached indirectly via gamma cascades originating from highly excited states in the product nucleus. If the target nucleus is excited to an energy only little above meson emission threshold, one can expect that the fraction of reactions giving rise to such gamma cascades is smaller than at higher excitation energies of the target nucleus.

We have studied the isomer ratio in ^{197} Hg resulting from the ^{197}Au(γ,π^-)^{197}Hg reaction. Gold foils were irradiated for approximately 3 hours in the bremsstrahlung beam of the Bates Linear Accelerator. In order to minimize as much as possible the frequency of the secondary p,n reactions in gold leading to the same product nucleus as the photomeson reaction, the targets had a thickness of 0.1 to 0.2 g/cm^2 and were freely hanging. Runs were made at several bremsstrahlung endpoint energies between 134 and 187 MeV. After irradiation the mercury was separated by heating the

gold foils in a testtube-like device, and the vaporized
mercury was collected on a cooled silver foil. The decay of
the activity was then observed with a Ge(li) detector for
approximately one month.

Several runs were made below threshold to assess the
contribution of the p,n reaction. We also observed the
radioactivity in ^{195}Hg from the p,3n reaction. For our data
reduction we assumed that the ratio of these two proton-
induced reactions was constant near threshold. In this
way we corrected for the p,n contribution. At the higher
energies this contribution was negligible relative to that
from the γ,π^- reaction.

The isomeric state lies 0.3 MeV above the ground state
and has a halflife of 24 h. Its decay can easily be
observed. The 64-h ground-state decay, however, cannot be
determined from the observation of the 77-keV gamma line
alone, because the energy of this line coincides with that
of the K_β X-ray lines emitted in the electron-capture decay.
Furthermore, the same or similar X-rays are also present
from the decay of ^{195}Hg which has a 40 h halflife, the
isomeric transition, etc. Therefore, considerable corrections
have to be made to deduce the number of nuclei formed in
the ground state. A very careful determination of the
detector efficiency and the knowledge of the pertinent
nuclear data is therefore necessary. We acknowledge the
great help of the Oak Ridge National Laboratory Nuclear
Data Center it gave us by furnishing us with their newest
data on the decay of ^{197}Hg, ^{195}Hg, ^{195}Au, and ^{133}Ba.

In the present study we found the isomer ratio
σ(isomer): σ(ground state) lower than reported in an
earlier communication[1]. The cause for this discrepancy is
due to an improved efficiency determination of the Ge(li)
detector. The isomer ratios are given in the table below.

Table 1

Bremsstrahlung Energy (MeV)	Isomer Ratio σ(isomer): σ(ground)
148	1.4 ± 0.4
153	0.7 ± 0.3
155	0.8 ± 0.2
165	1.0 ± 0.2
180	1.0 ± 0.2
187	1.0 ± 0.2

The spin of the ground state of the target nuclide ^{197}Au is 3/2, that of the product nuclide ^{197}Hg is 1/2, that of the isomeric state 13/2, and below the isomeric state are two states with spins 3/2 and 5/2, respectively. Considering these conditions, the relatively high fraction of transitions ending up in the isomeric state indicates that direct transitions to the isomeric state seem to play a considerable role, in spite of the considerable change in angular momentum with respect to the target nucleus. This can be understood because the neutron configuration of the isomeric state is the only one among the low lying states in ^{197}Hg that differs in only one neutron from that of the neutron ground state configuration in ^{197}Au. This comes about because the $p_{1/2}$ neutron subshell in ^{197}Au is empty and the two lowest excited levels are more characteristic of core-excited states than of single-particle states.

---

*Supported by the National Science Foundation (Grant SPI77-26207) and by the Department of Energy.

+NSF Undergraduate Research Participation program participant.

REFERENCES

1. H. A. Medicus and R. Smalley, Proc. 6th Intern. Conf. on High-Energy Physics and Nuclear Structure, Los Alamos 1975, p. 287.

INVESTIGATION OF THE CHARGE EXCHANGE PROCESS $\pi^- d \to \pi^0 nn$

Il-T. Cheon

Department of Physics
Yonsei University
Seoul, Korea

Five years ago, on the basis of the dispersion theory, an effective interaction was proposed to analyse the pion absorption and production at low energy[1]. With this interaction, the absorption rate for the process $\pi d \to nn$ at rest is given as

$$\hbar\omega(\pi^- d \to nn) = \frac{16g^2 \hbar^3 k}{3a_o^3 M_\eta \tilde{\mu}_\pi c} \left[F(k)\right]^2$$

where the coupling constant $g^2 = 0.08$, a_o and c are the Bohr radius and the speed of light, M_η and $\tilde{\mu}_\pi$ are the neutron mass and the pion reduced mass and $k = 1.854$ fm^{-1} is the momentum of relative motion of the two out-going neutron. The matrix element $F(k) = 0.3642$ $fm^{1/2}$ is given in ref. 1. After some numerical calculations, we obtain

$$\hbar\omega(\pi^- d \to nn) = 0.682 \text{ eV}.$$

By the same method we have evaluated the transition rate of the charge exchange; $\pi^- d \to \pi^0 nn$.

$$\hbar\omega(\pi^- d \to \pi^0 nn) = \frac{8M_\eta \hbar^2}{(2\pi)^3 \tilde{\mu}_\pi^2 a_o^3} \left[\frac{4\pi}{3}(a_1 - a_3)\right]^2 G(F)$$

where

$$G(F) = \int_0^\xi \left[\frac{9}{25} F_d^2(k,q) + \{F_s(k,q) + \frac{\sqrt{2}}{5} F_d(k.q)\}^2\right] kq^2 dq$$

with

$$F_{s,d}(k,q) \equiv \int_0^\infty g_1^{NN}(kr) f_1^\pi\left(\frac{q}{2} r\right) \exp\left(-\frac{r}{a_o}\right) \psi_{s,d}(r) r^2 dr.$$

$\psi_{s,d}(r)$ are the radial parts of the deuteron wave function and $g_1^{NN}(kr)$ and $f_1^\pi\left(\frac{q}{2} r\right)$ are the radial wave functions of

the p-state for relative motion of the two out-going neutrons and the neutral pion relative to the CM of the two neutrons, respectively. The upper limit of the integral is

$\xi = \left[2 \tilde{\mu}_\pi \hbar^{-2} \times 1.0878 \text{ MeV}\right]^{1/2}$. With $a_1 - a_3 = 0.291\lambda_\pi$ and $a_1 + 2a_3 = 0.055\lambda_\pi$, we obtain

$$\hbar\omega(\pi^- d \rightarrow \pi^\circ nn) = 0.695 \times 10^{-4} \text{ eV}.$$

Contribution from the d-state of the deuteron wave function is only 4.2%. The higher order terms are negligibly small. Then,

$$\omega(\pi^- d \rightarrow \pi^\circ nn)/\omega(\pi^- d \rightarrow nn) = 1.02 \times 10^{-4}$$

With the value calculated by Gibbs et al.[2] for the radiative pion capture rate, $\hbar\omega(\pi^- d \rightarrow \gamma nn) = 0.283$ eV, we obtain

$$\omega(\pi^- d \rightarrow nn)/\omega(\pi^- d \rightarrow \gamma nn) = 2.41$$

$$\omega(\pi^- d \rightarrow \pi^\circ nn)/\omega(\pi^- d \rightarrow \gamma nn) = 2.46 \times 10^{-4}$$

$$\omega(\pi^- d \rightarrow \pi^\circ nn)/\omega(\pi^- d \rightarrow \text{all}) = 0.720 \times 10^{-4}$$

for which the experimental results[3] are (2.97 ± 0.07), $(5.76 \pm 0.87) \times 10^{-4}$ and $(1.45 \pm 0.19) \times 10^{-4}$, respectively.

REFERENCES

1. Il-T. Cheon et al., Nucl. Phys., B66, 412 (1973).
2. W. R. Gibbs et al., Phys. Rev., C16, 327 (1977).
3. R. MacDonald et al., Phys. Rev. Letters, 38, 746 (1977).

PION-ELECTROPRODUCTION FROM SINGLE NUCLEONS

Howard B. Miska

Physics Department
The Pennsylvania State University
University Park, Pennsylvania 16802

We are currently examining pion-electroproduction from
various nuclei. To do so, we are employing the formalism
developed by L.M. Saunders[1] which treats electroproduction
and photoproduction on the same footing. This formalism
neatly divides the problem into single nucleon production
amplitudes, standard nuclear physics, and pion-nucleus interactions in the final state. Thus, it was natural to begin
by looking at the single particle production amplitudes.

We are using the amplitudes developed by Fubini, Nambu,
and Wataghin[2] (FNW) from dispersion theory. They provide
two forms: a complete form that should be good for incident
beams up to 500 MeV and a simplified form for lower energy.
The simplified form consists of an electric and magnetic dipole term and a term arising from the pion current graph.
Most people doing electroproduction using the FNW amplitudes
employ the latter with further simplifications. We examined
these amplitudes for the case of pion production from a single
nucleon with different forms for the amplitudes. The results
of the investigations are shown in Fig. 1. We looked at the
full amplitudes (curve one) and the amplitudes as simplified
in FNW (curve two). We also looked at the cases where the
amplitudes consisted only of the electric dipole and resonant
magnetic dipole terms (curve three) and only the electric
dipole term (curve four). We looked at the electron energies
of 200 and 280 MeV.

From the graphs we see that the simplifications (curves
two-four) differ from the full FNW curve both in magnitude

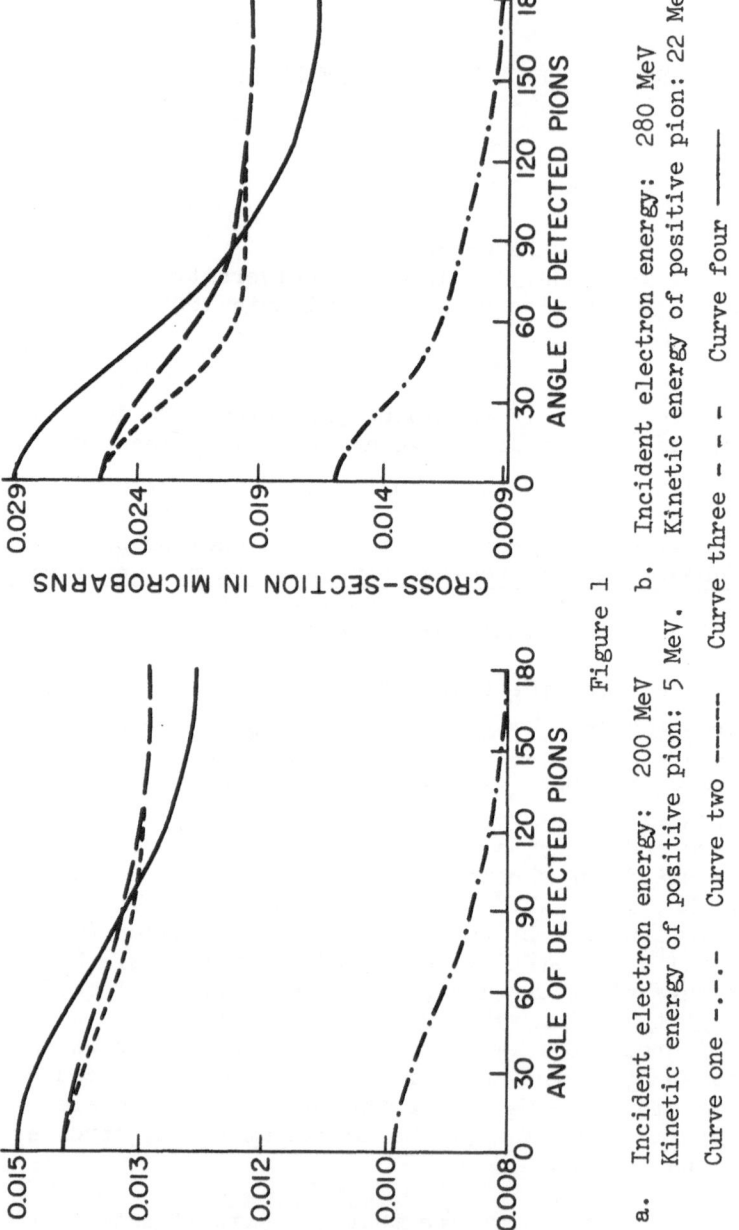

Figure 1

a.   Incident electron energy: 200 MeV   b.   Incident electron energy: 280 MeV
     Kinetic energy of positive pion: 5 MeV.   Kinetic energy of positive pion: 22 MeV

     Curve one -·-·-   Curve two ——   Curve three - - -   Curve four ——

and in structure. The difference in magnitude comes from the dropping of a factor $(1 + \omega/M)^{-1}$ that was present in the full form and multiplies the electric dipole there and the pion current graphs. The difference in structure comes from the dropping of a term that breaks the charge degeneracy of the pion, namely $k^2/2\omega$. This term is important since it is proportional to k, the photon momentum and even for threshold production k can be large. M is the mass of a nucleon and ω is essentially the pion energy.

The conclusion one must draw is that these terms must be retained in any quantitative prediction of pion-electroproduction on nuclei in order to be consistent with the FNW dispersion results.

REFERENCES

1. L.M. Saunders, Nucl. Phys. B7 (1968) 293.
2. S. Fubini, Y. Nambu, and V. Wataghin, Phys. Rev. 111, (1958) 329.

DEEP INELASTIC ELECTRON SCATTERING

NEAR AND ABOVE PION THRESHOLD

Peter D. Zimmerman

Louisiana State University

Claude F. Williamson

Massachusetts Institute of Technology

Recent experiments by two collaborations, the LSU-MIT group[A] and the Saclay-LSU-Sanitá-Clermont-Frascati-Basel effort[B], have greatly multiplied the amount of data available on deep-inelastic electron scattering. This paper reviews some of the more recent results from those experiments[1,2].

The work at the Bates Laboratory was intended, first of all to provide data on the transverse response of ^{40}Ca. This could in turn, give information on the contribution of transverse processes other than "quasi-elastic" (one nucleon knockout) scattering. In particular, meson exchange currents were expected to be important even well below pion threshold[3]. Above pion threshold, diagrams formally similar to some of those considered in the MEC calculation (such as real $\Delta(1232)$ production) appear. The contributions of these, however, are proportional to the Δ production amplitude, and are, therefore, quite small.

In Figure 1 are shown spectra from calcium at a scattering angle of 160° and bombarding energies of 150, 200 and 250 MeV. The data were first fitted with a zero-temperature Fermi gas momentum distribution[4] treating k_F (the Fermi momentum) and $\bar{e}$ (the average separation energy) as variable parameters.

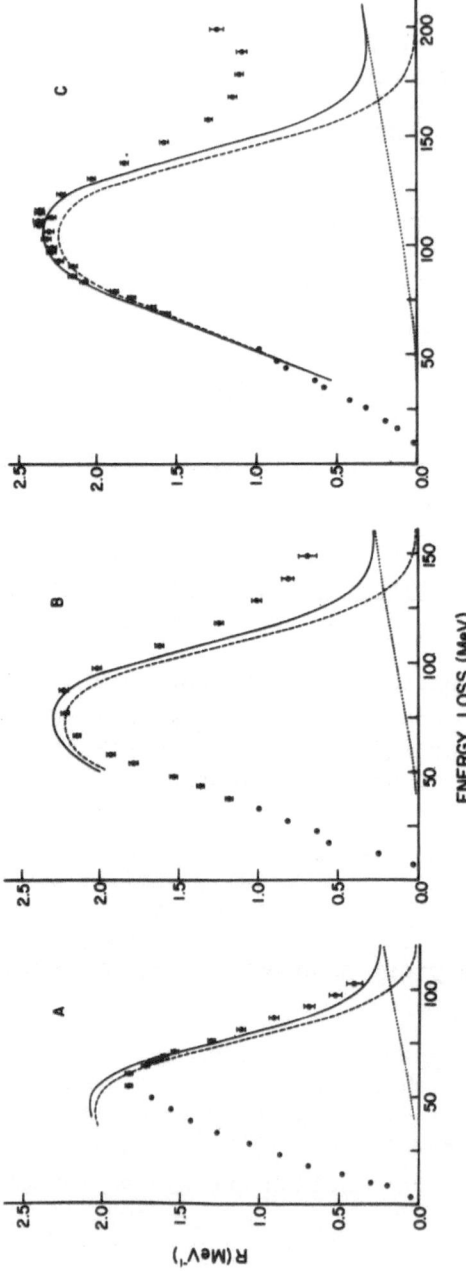

Figure 1: Response functions for ^{40}Ca at 160° scattering angle vs. energy loss, at three bombarding energies (a:150, b:200 and c:250 MeV). Dashed lines represent the harmonic oscillator shell model contribution, the dotted lines the MEC contribution, and the solid lines are the sums of these two terms.

Values of k_F and $\bar{\varepsilon}$ for four incident energies are shown in Table I. The large variation in effective k_F, even at these relatively low values of momentum transfer, was considered to be serious, even with this simple version of the Fermi gas model. If the parameter k_F is fixed at any value, then the energy dependence of the response function $R(q,\omega)$ is incorrectly given. The function, R, is defined as the measured cross section divided by the Mott cross section for a singly-charged particle.

Table I

Fermi Model Parameters as Functions of Incident Electron Energy

E (MeV)	k_F (MeV/c)V/c	$\bar{\varepsilon}$ (MeV)
150	330 ± 1.6	14.6 ± 0.8
200	285 ± 6.7	31.6 ± 1.7
250	245 ± 1.4	33.0 ± 0.5
300	232 ± 1.5	32.0 ± 0.5

A more elaborate calculation was made by DeForest and Hermans[1] using a shell model description of the nucleus for the quasi-elastic scattering[5,6] and treating two body knockout mediated by meson exchange currents[3]. These curves are shown in Figure 1 (dashed line: one body; dotted line: MEC and solid line: sum). This simple shell model treatment does not include Pauli blocking, and therefore overestimates the cross-section at low values of the energy loss, ω. The "filled in dip region" first noted by Whitney, et al[7] between the exhaustion of the quasi-elastic cross section and the onset of isobar production is clearly in evidence in these data. Although the MEC contribution goes in the right direction, it is insufficient to bring about agreement between theory and experiment. It should be emphasized that no parameters used in this calculation were adjusted to improve agreement; all were determined from independent experiments.

At Saclay the "600" spectrometer in the HE1 experimental
hall, with its very large momentum acceptance ($\Delta P/P \sim 36\%$)
has been used to obtain a very large number of spectra from
^{12}C at bombarding energies between 160 MeV and 520 MeV at
40 MeV intervals. Data were obtained at two scattering
angles, 60° and 130°. The sets of response functions for
the highest eight beam energies at the two angles are shown
in Figures 2 and 3. (The response function is defined the
same as it was for calcium). The forward angle data look
much as anticipated, with moderately filled in "dips",
particularly at the higher energies.

The data at the larger energies and at 130° are, however, truly surprising. From a bombarding energy of 440 MeV
on up the quasi-elastic "peak" all but vanishes. At 520 MeV
only a slight shoulder remains. We have tested simple one
body, MEC, coherent pion production and isobar production

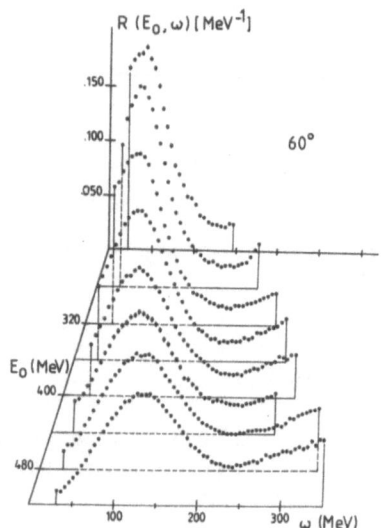

Figure 2: Response functions for ^{12}C at 60° and electron
bombarding energies between 240 MeV (topmost curve) and
520 MeV in steps of 40 MeV. The data have been averaged at
5 MeV intervals and the points for $\omega \leq 30$ MeV have been
deleted for clarity in the figure. Where not indicated,
the error bar is smaller than the plotted point.

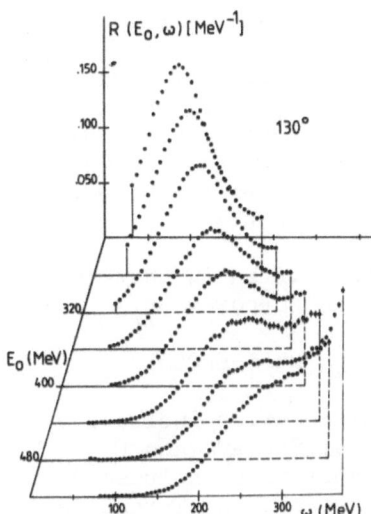

Figure 3:   Response for ^{12}C, as in Figure 2 but for a 130°
scattering angle.

calculations against these data^2 but with very poor success.
We find, as we have reported in Reference 2, that an additional reaction mechanism may be needed to provide strength
between the quasi-elastic and Δ peaks. One candidate may be
other types of two-nucleon emission not explicitly calculated,
although this is expected to be small. We tend, therefore,
to conclude that a yet unknown mechanism is responsible for
the pronounced strength between the quasi-elastic and Δ-peaks
which was systematically observed in this experiment.

     The techniques used to make radiative corrections were
similar in both experiments and are based on the ideas of
Maximon and Isabelle[8], Borie[9] and G. Miller[10]. Pion-rejection
at MIT was provided by lucite Cerenkov counters. This technique was inadequate at the larger energies used at Saclay.
A specially designed Cerenkov detector using silica aerogel
(index of refraction ~ 1.06) was built to replace the lucite
counters for the high energy part of the experiment.

ACKNOWLEDGEMENTS

Conversations with Drs. T. W. Donnelly, P. N. Kirk,
L. C. Maximon, T. Kawazoe, R. R. Whitney and E. Borie were
very helpful to me. The calculations provided by Drs.
DeForest and Hermons, and T. Kawazoe were most useful. I
thank all of my collaborators on these two experiments,
without whose work there would be little to report.

Work at LSU is supported in part by the National Science
Foundation and Research Corporation.

REFERENCES

A. The LSU-MIT collaboration includes P. D. Zimmerman,
    J. M. Finn, now at Saclay, and C. F. Williamson.
B. The Saclay collaboration includes J. Mougey, M. Bernheim,
    D. Royer, D. Tarnowski, S. Turck (Saclay); P. D.
    Zimmerman, J. M. Finn (LSU); S. Frullani (Sanitá and
    INFN, Rome), D. B. Isabelle (Clermont-Ferrand); G. P.
    Capitani, E. De Sancticis (Frascati Laboratory, INFN);
    and I. Sick (Basel).
1. P. D. Zimmerman, J. M. Finn, C. F. Williamson, T.
    DeForest, Jr., and W. C. Hermans, "Deep Inelastic
    Electron Scattering from ^{40}Ca in the Transverse Region",
    Physics Letters B (in press).
2. J. Mougey, et al., "Deep Inelastic Electron Scattering
    from ^{12}C", submitted to Physical Review Letters.
3. T. W. Donnelly, J. W. Van Orden, T. DeForest, Jr., and
    W. C. Hermans, Physics Letters 76B, 393 (1978).
4. R. A. Smith and E. J. Moniz, Nucl. Phys. B43, 605 (1972).
5. T. DeForest, Jr., Nucl.Phys. A132, 305 (1969).
6. A.E.L. Dieperink and T. DeForest, Jr., Phys. Rev. C 10,
    543 (1974).
7. R. R. Whitney, et al., Phys. Rev. C 9, 2230 (1974).
8. L. C. Maximon and D. B. Isabelle, Phys. Rev. 133, B1344
    (1964).
9. E. Borie, Lett. Nuovo Cim. 1, 106 (1969).
10. G. Miller, SLAC Report No. 129 (1971).

SUMMARY AND OUTLOOK

Aron M. Bernstein

Physics Department and Laboratory for Nuclear

Science

Massachusetts Institute of Technology

I. GENERAL REMARKS

This conference has signified the coming of age of a new era in photopion physics in complex nuclei. The size, enthusiasm, and young average age of the participants indicates the recent origins of this field, which in fact is just several years old. The conference has been characterized by the enthusiasm and high professional standards of both speakers and participants, with lively discussions, so that a review paper would be superfluous. Therefore, I will present a status report of where we are and where we are heading. In a sense I will try to anticipate the agenda of the next Photopion Symposium which we anticipate will be as successful as the present one. In particular, I would like to try to assess the range of possible potentialities of this field with some educated guesses about new physics to learn. This is done in a sceptical spirit, keeping in mind what Gerry Brown concluded in his introductory review - it is not clear what new physics surprises are in store for us or what new things we may learn. However, it is clear that just based on the things we know about we can anticipate that we have very interesting things to study.

The field of photopion physics is intimately related to the study of the internal structure of the nucleon. The first (γ,π) reaction on the nucleon was performed about thirty years years ago and was subsequently utilized to confirm the quantum numbers to the first excited state of the nucleon, the Δ.[1]

Subsequently the higher excited states of the nucleon were explored.[2] In more recent times elementary particle physicists have turned their attention to (e,e'π) coincidence experiments and to deep inelastic electron scattering, and have discovered scaling which is related to the parton (or quark) structure of the nucleon.[3] It is interesting that deep inelastic electron scattering, properly scaled, shows a similar scaling behavior for atoms, nuclei, and nucleons.[4]

This symposium was held at the present time because of the large amount of activity in photopion physics in complex nuclei. The reason that this field is so interesting is that it is basic, involving the quanta of the electromagnetic (EM) and nuclear interactions. In general terms the interest here is to use the precisely understood nature of the EM interaction to study the pionic properties of nuclei. In electron scattering to discrete nuclear states these properties manifest themselves as a result of virtual pion emission and absorption which effect the current and charge distributions in nuclei. In the photopion reaction the pions are real, rather than virtual, and one can study their propagation and interaction with the nucleus. In this way it can be seen that there are deep connections between photopion physics and electron scattering on the one hand and with pion physics on the other hand.

One of the most interesting features of the pion nucleus interaction is the rapid variation in strength with pion energy. Because of this I shall divide the following discussion into three energy regimes:

1. low energy 0 to 10 MeV (pion mean free path[†] $\geq$ 15 fm for interaction with a single nucleon)
2. intermediate energy 10 to 70 MeV (pion mean free path[†] varies from $\sim$ 15 to 4 fm)
3. the Δ region 70 to 400 MeV (pion mean free path $\sim$ 4 fm to 1 fm)

Of course, the precise dividing line between the three regimes is fuzzy. The total cross section for photopion production on the nucleon is shown in Fig. 1. The Δ peak is quite prominent.

[†]not including true absorption, e.g., $\pi NN \rightarrow NN$.

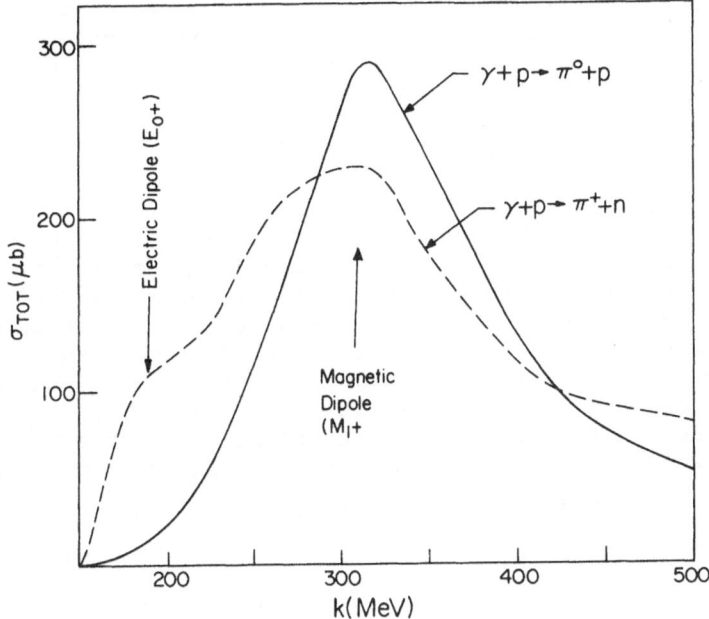

Figure 1.   Total cross section for (γ, π) reaction on the proton.

I will conclude the general comments by noting that the present experimental activity is a direct consequence of the improvements in experimental techniques. This includes the following items:

1.   the new generation of electron linear accelerators (linacs) with their high intensity and quality beams;

2.   new detector technology which includes magnetic spectrometers of large solid angle and momentum acceptance as well as high energy resolution so that discrete nuclear final states can be resolved. In addition fast timing and wire chamber technology is an important part of present experiments;

3.   the progress in computer technology including CAMAC interfacing between the experimental apparatus and the computer which allows for event by event recording of experiments.

In this connection one can anticipate in the not too distant future that computer control of experimental parameters, which will further increase the precision of experiments, will come into operation.

II. INGREDIENTS OF THE PHOTOPION REACTION

The application of the photopion reaction to complex nuclei raises some interesting and fundamental problems. One takes the elementary amplitude as calculated for the free nucleon[6] which is in good agreement with experiment,[6] and applie it to bound nucleons. For the kinematic variables this means an extension off the mass shell. At present one usually neglects one of the more interesting effects - namely the dynamical corrections to the amplitude due to the fact that it is imbedded in the medium. At the present time the usual procedure is to take the relativistic amplitude and to perform a non-relativistic expansion to order $(P/m)^2$ where P and m are the final nucleon momentum and mass.[5] It clearly would be of interest to apply the fully relativistic operator, particularly to the deuteron. Probably the most serious problem is that the Δ has been added in a phenomenological way. This means that the prescription for taking the amplitude off shell has no firm theoretical foundation. Since the amplitude represents the data for the elementary process quite accurately one can be confident that it is correct only when the off shell extrapolation is not too large.

The second major ingredient of the photopion process is the nuclear matrix element. At threshold the $\vec{\sigma} \cdot \vec{\epsilon}$ term dominates (for charged pion production) and the nuclear matrix element is a spin flip, which is the Gamow-Teller matrix element at the appropriate momentum transfer. At higher energies there should also be the analogue of the Fermi matrix element, and as Tabakin* showed in his paper, matrix elements with derivatives. One should note that these matrix elements either do not occur in electron scattering or do not occur in the same combinations. For example magnetic electron scattering

---

* The appearance of an asterisk will signify that reference is being made to an earlier paper in this volume.

contains both a spin and an orbital contribution. This fact
has led some people to think that the (γ,π) reaction can be
used for spectroscopic purposes. This may be true in the long
run but at the present time one should treat this point of
view with caution. However, this will not preclude the dis-
covery of certain "giant" spin-flip states and some interesting states have been observed in radiative pion capture (e.g.,
see contributions of Alder et al.[*] and Perroud[*]). The photopion reaction should enable one to learn more about these excitations whose properties are presently quite uncertain by
studying their differential cross sections.

The last main ingredient in the photopion reaction is the
final state interaction of the outgoing pion. The hope is
that one can learn some things that cannot be obtained from
pion scattering. For example: 1) low energy pions can be detected and therefore this technique is a bridge between pionic
atom studies and low energy meson nucleus scattering experiments; 2) the photo can produce pions in the interior so that
pion interactions can be studied in the interior as well as in
the surface region (this is of particular importance in the Δ
region; 3) the pion interaction is generally in a different
nucleus than in pion scattering, e.g., in the $^{12}C(\gamma,\pi^-)^{12}N$
reaction the pion interacts with ^{12}N while in pion scattering
or the (γ,π^0) reaction it interacts with ^{12}C, so that information on the isovector part of the pion-nucleus interaction
can be obtained; and 4) there is some sensitivity to the pion
wave functions in the (γ,π) and $(e,e'\pi)$ reactions.

In the Δ region there are some specific features which
are of interest. Here one can picture the mechanism as proceeding by the creation of a Δ-hole state. In this picture
one then has the Δ-hole interaction and Δ propagation and
decay as new and interesting features of the (γ,π) reaction.
One should note that this model also is closely connected to
other reactions such as pion scattering and the (γ,π) reaction.

I would like to conclude this section with some remark
on the present experimental state of the elementary interaction[6,7] and the agreement between theory and experiment.[5]

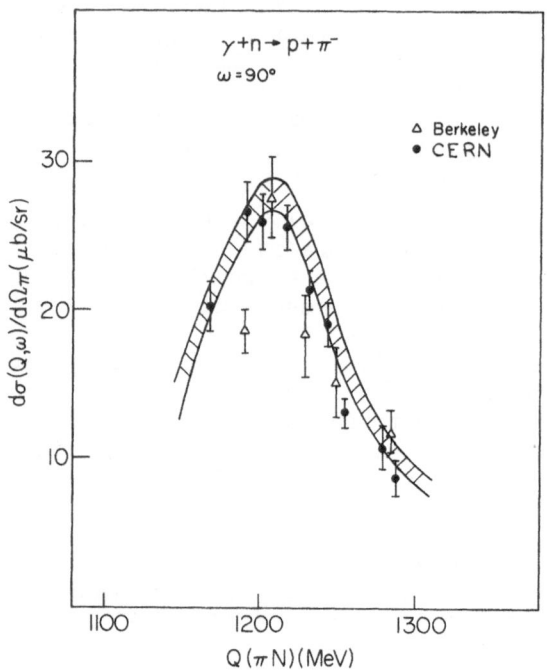

Figure 2. Total cross section for the $\gamma n \rightarrow p\pi^-$ reaction in the Δ region. The cross hatched area is from a coincidence experiment $\gamma D \rightarrow \pi^- pp$.[8] The points with error bars are from the inverse reaction $\gamma^- p \rightarrow n\gamma$.

In general terms the $\gamma + p \rightarrow n + \pi^+$ reaction is well studied, with accuracies generally $\sim$ 10% for energies starting around 50 MeV above threshold and continuing to about 500 MeV. For the first 50 MeV above threshold the accuracy is $\sim$ 20%.

For the $\gamma + n \rightarrow p + \pi$ amplitude the experimental situation is less certain since experiments on the deuteron are involved. The impulse approximation is made and the neutron amplitude is obtained from the data. There is also the possibility to use the inverse reaction $\pi^- p \rightarrow \gamma n$ and then apply detailed balance. The problem with this approach is that these later experiments are difficult. Recently at Saclay a $\gamma D \rightarrow pp\pi^-$ experiment has been performed with the $p\pi^-$ pair detected in coincidence. The neutron amplitude obtained

from this experiment is compared to the value obtained from
the inverse reaction in Fig. 2 for the Δ region and the agreement is reasonable. Unfortunately for the threshold region
the experimental situation is quite sparse. A recent experiment on the $\pi^-p \to n\gamma$ reaction has recently been performed at
the Saclay linac but the results are not yet available. Here
is an area in which we as nuclear physicists might make a
significant contribution by disentangling the deuteron structure and the elementary neutron amplitude.

For the $\gamma + p \to p + \pi^0$ amplitude the situation is again
well studied for photon energies above $\sim$ 200 MeV but for the
first $\sim$ 50 MeV above threshold the data is skimpy. For the
$\gamma + n \to n + \pi^0$ amplitude the situation is poorly studied. In
this case the threshold values are of special interest as discussed by N. de Botton.

In all cases in which there is good data the level of
agreement between theory and experiment is at the level of

10 to 20 percent[5] and the general feeling is that the experiments must be improved before any statements about discrepancies could possibly be made. Therefore at the present time
we believe that we are in good shape as far as the elementary
amplitude is concerned, but further experimental work, particularly on the neutron amplitudes and in the threshold
region is recommended.

III. REACTION MECHANISM

The usual method to calculate the (γ,π) process is in
first order distorted wave impulse approximation. This represents the simplest way to approach the problem. The major
ingredients of this approach have been discussed in the introductory section by Brown[*] and in more detail at this symposium by Tabakin[*]. On the other hand there are already cases
for which the DWIA is not believed to be sufficient. In the Δ
region Moniz[*] and others have discussed the Δ-hole model. In
the threshold region for the (γ,π^0) reaction, for which the
cross section is small, Koch and Woloshyn[9] have proposed that
the dominant reaction mechanism could be two-step, i.e., (γ,π^+)
followed by (π^+,π^0). Since the (γ,π^+) cross section is much

larger than the (γ,π^o) cross section it is clear that the second order process could dominate. Detailed calculations show that this is indeed true.[10] A comparison of theory and experiment for (γ,π^o) just above threshold in d and ^{3}He is given by de Botton* and Vincent* along with a quantitative discussion of the rescattering effect.

An interesting case for coupled channel effects has been discussed for the nucleon by Laget and in these proceedings by de Botton*. Here the reaction $\gamma p \rightarrow \pi^o p$ was calculated by considering the reactions $\gamma p \rightarrow \pi^+ n$ and $\pi^+ n \underset{\leftarrow}{\rightarrow} \pi^o p$. Since at low energies the pions are produced in s waves, and the $\gamma p \rightarrow \pi^+ n$ threshold is higher than for $\gamma p \rightarrow \pi^o p$ one expects a cusp in the $\gamma p \rightarrow \pi^o p$ cross section at the $\gamma p \rightarrow \pi^+ n$ threshold. Figure 3

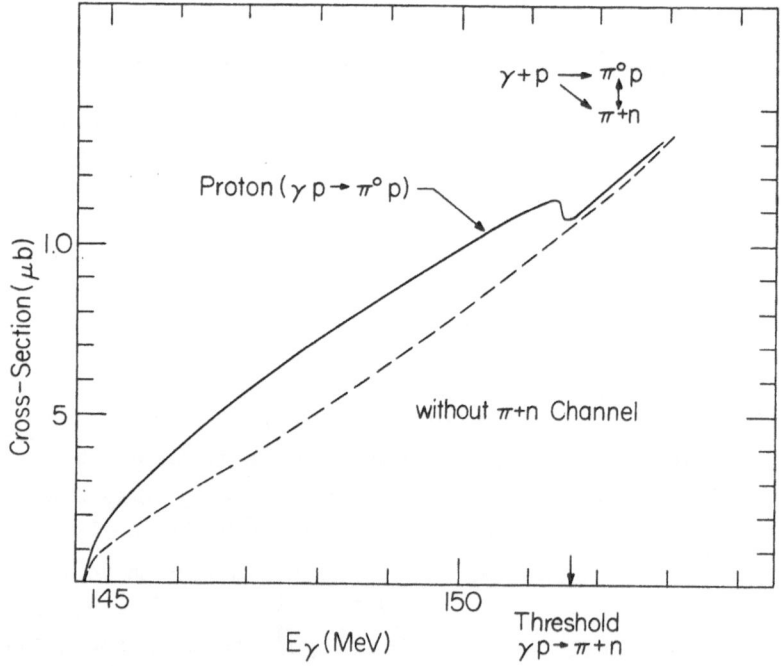

Figure 3. Cross section for the $\gamma p \rightarrow \pi^o p$ reaction including the channel coupling to the $\pi^+ n$ channel as calculated by Laget.

shows the results of a coupled channels calculation by Laget
in which this cusp appears. It will be interesting to see if
this can be observed experimentally.

Since the (γ,π) field is in its infancy and already several second order processes have been discussed one might suspect that more will surface. In the history of nuclear reactions with hadronic probes we have gradually discovered
more and more evidence for higher order processes. Hopefully
since the photon part of the reaction is relatively weak there
is good reason to expect that the situation will be simpler
than with hadronic probes in both channels. However, there
are still possibilities that have never been discussed, e.g.,
the coherent (γ,π^{o}) reaction followed by pion charge exchange
to compete with the $(\gamma,\pi^{\pm})$ reaction. Clearly the reaction
mechanism must be quantitatively examined before we can know
how much information can be obtained from (γ,π) reactions.

IV.   THRESHOLD REGION AND RADIATIVE PION CAPTURE

The $(\gamma,\pi^{\pm})$ reaction just above threshold and the study of
the $(\pi^{-}\gamma)$ reaction from bound pionic atomic states has been
the most active part of this field. There have been several
experiments whose purpose it is to check the basic reaction
theory, which as Gerry Brown*has told us, is now on a firm
footing. I might add that this was not the case when the
present generation of experiments started. These experiments
are (see the papers by E. C. Booth and E. J. Winhold for more
detail):

$$d(\gamma,\pi^{+})nn$$

$$^{3}He(\gamma,\pi^{+})^{3}H$$

$$^{6}Li(\gamma,\pi^{+})^{6}He \qquad ^{6}Li(\pi^{-}_{1s},\gamma)^{6}He$$

$$^{12}C(\gamma,\pi^{-})^{12}N$$

For all of these cases we are considering the cross section
for the first few MeV above threshold to the ground state of
the final nucleus except for the deuteron in which the strongly correlated nn final state gives the most important contribu-
tion to the cross section. For the $^{6}Li(\pi^{-}_{1s},\gamma)^{6}He$ case we are
considering the new SIN experiment in which the high energy

gamma ray was detected in coincidence with the π-mesic x-ray
in order to be certain that the capture took place from the
pionic 1s state. The agreement between theory and experiment
is at the 10% level for all of the above cases which indicates
that we understand the basic reaction mechanism (see the paper
of Booth[*] for more detail). I believe that this is a healthy
situation for the field because we can now go on to tackle
more difficult problems with confidence. In particular we
note that at low energies the final state pion-nucleus interaction is relatively weak, and therefore does not contribute
to the cross section as much as it will in the Δ region. In
addition the major contribution comes from the $\vec{\sigma} \cdot \vec{\varepsilon}$ term in the
Hamiltonian. At higher energies one will have to deal with a
more complicated elementary interaction photoproduction
operator.

We should note that these experiments have now been
carried out at the present state of the art and large improvements in accuracy cannot be anticipated in the near future.
In fact the level of activity in the threshold region studies
is diminishing rapidly.

The (π^-, γ) experiments have yielded a large body of data.
One special and very beautiful experiment was discussed by
Perroud and that is the $\pi^- d \rightarrow nn\gamma$ reaction which was analyzed
for information on the nn interaction by examining the high
energy end of the γ ray spectrum. The results are:

$$a_{nn} = -18.3 \pm 0.55 \text{ fm}$$

$$r_{nn} = 2.35 \pm 0.34 \text{ fm}$$

These are to be compared with:

$$a_{pp} = -17.1 \pm 0.2 \text{ fm}$$

$$r_{pp} = 2.84 \pm 0.3 \text{ fm}$$

where a is the scattering length and r is the effective range.
This indicates a clear difference between the nn and pp interactions characterized by a difference of scattering lengths of
approximately 1 fm. The implications for the nucleon-nucleon
interaction and for the violation of isospin symmetry in complex nuclei need to be explored. One should note the change in

a_{nn} from previous values and should be cognizant of the difficulties in obtaining this number from experiment.

The bulk of the (π^-,γ) data are in complex nuclei and
consist of a series of strongly excited states and a continuum.
For nuclei heavier than ^{40}Ca only a continuum is observed. For
many nuclei the "continuum" is really not smooth but shows signs
of intermediate structure which to my knowledge has never been
discussed. The entire phenomenology is now well explored experimentally but lacks explanation. A summary of calculations
is presented by Perroud.[*] For example the question of which
states in the final nucleus are being excited and their quantum
numbers is not well known. One can anticipate that the (γ,π^+)
reaction may give some insight here because the angular distribution of the pions may enable us to assign quantum numbers
to these states. The dynamics of these states and their possible relation to states seen in other charge exchange reactions
has yet to be explored. There is need for charge exchange
experiments with other projectiles, and for theoretical calculations to further elucidate this subject. The question of
why no discrete structures are seen for nuclei heavier than
^{40}Ca also need explanation. The question of the continuum
could use more experiments. In particular one could attempt[*]
$(\pi^-,n\gamma)$ coincidence experiments as is discussed by Eramzhyan
and Balashov.[*]

The (γ,π^0) reaction has also been studied near threshold.
N. de Botton[*] presented the very nice results of the Saclay
group for d, ^{3}He, and ^{4}He relative to the proton, for the first
few MeV above threshold. As was discussed in Sec. III the two
step contributions to this reaction are dominant. This has
several unfortunate consequences. First, the original goal of
obtaining information about the elementary $\gamma n \to \pi^0 n$ cross section from this experiment will not work. Second, a definitive
comparison between theory and experiment cannot be made on the
basis of the relative cross sections. Therefore we shall have
to wait until the absolute cross section measurement for the
proton is performed, which I understand is under active con-
sideration at this time.

I anticipate that interesting results will be forthcoming
in the (γ,π^0) reaction in the next few years. In this connec-

tion one should note that a group at Boston University is
starting measurements at Bates, as reported by Milder*, et al.

V.   INTERMEDIATE ENERGY REGION

Measurements of the pion spectra for the $(\gamma, \pi^{\pm})$ reaction
in the region of 10 to 70 MeV above threshold have begun at
Sendai and also at Bates by an MIT-RPI collaboration. Several
reports of this work were presented at this symposium. These
measurements are actually being made by electroproduction in
which the final electron is not detected so that the gamma ray
is virtual rather than real. Since the electrons go through
the target they are also scattered into the detector which
makes observations of the (γ, π^{-}) reaction experimentally difficult. For this reason the Sendai group has concentrated on
the (γ, π^{+}) measurements. The work at Bates is being performed
at 90° and and at this large angle both π^{+} and π^{-} mesons can
be separated from the backgrounds of positrons and electrons
respectively. At forward angles the separation of negative
pions from electrons will be impossible with this technique.

The first experiments have just been completed. At Sendai
angular distributions have been measured for the $^{12}C(\gamma, \pi^{+})^{12}B$
reaction for the ground and sum of the two excited states, as
well as for several other targets. At Bates the $^{12}C(\gamma, \pi^{\pm})$
reaction has been measured. This experiment while fixed at
90° has obtained better energy resolution and statistics and
has examined the excitation spectra to approximately 10 MeV
in the final nucleus. In addition, the $^{12}C(\gamma, \pi^{-})$ spectra to
discrete final states has been measured for the first time.
The results have been compared with preliminary calculations
of Nagl and Uberall and are in reasonable agrement. In addition the strongly excited 2^{-}, 4^{-} doublet at 4.5 MeV has been
seen in this and in the $^{12}C(\pi^{-}, \gamma)$ reactions. This gives an
indication of the kind of data one can expect in the next
several years from these groups.

There are several transitions in light nuclei for which
the initial and final state wave functions can be considered
well enough known by a combination of shell model calculations
and magnetic (e,e') studies to be used as test of the reaction

theory. Accurate data for these transitions will enable us to test this theory. Although the ingredients of the calculations are well known we do not know at present how sensitive they are to each specific input ingredient such as the pion optical potential or to specific terms in the elementary interaction. These studies have just commenced and were discussed by Tabakin[*]. This is an area in which we can anticipate rapid progress.

Another topic for which we may obtain new information is in the assignment of quantum numbers to the strongly excited states by making use of the angular distributions in the (γ, π) reaction. The Sendai experiment on ^{12}C gives a clear indication of this. This should help our understanding of this subject.

From the experimental point of view one anticipates a lot of progress as both the Sendai and Bates groups are building new, much improved, spectrometers. (A more detailed discussion of this point is made by Winhold[*] and in the conclusion of this paper.)

Finally I would like to comment on the fact that these experiments are actually being performed by electroproduction without detection of the final electron, and then often interpreted by virtual photon theory. It would be quite useful if experimentalists would give the original electroproduction data and if this is what theorists would calculate. One then would avoid the pitfalls of virtual photon theory. In this connection Shoda showed us several curves of the same data which were almost a factor of two difference because different formulas for the virtual photon spectra were utilized. A study of the cross sections for both real and virtual photons may lead to some interesting information. The idea was originally tried for the $ep \rightarrow en\pi^+$ reaction by Panofsky and coworkers[11] and examined theoretically by Dalitz and Yennie[12]. In that case the results are not too encouraging although some interesting dynamic effects did show up in the real to virtual ratio. It is quite possible that for higher Z targets these effects will be larger and will give us some information. The Bates group is presently working on this subject and it is a completely open question as to what will come out of it.

VI. THE Δ REGION

Some of the most beautiful work that has been done on the (γ,π) reaction has been the study of the $\gamma D \to PP\pi^-$ reaction at Saclay which was reported by Laget[*] at this symposium. They have performed coincidence experiments in which the kinematics were selected to emphasize certain higher order diagrams, particularly those in which there is a Δ-N interaction in the intermediate state. In that way they have been able to increase the sensitivity of their results to the Δ-N interaction, and perhaps to measure the angular momentum and scattering length of the dominant Δ-N intermediate state. As pointed out by Laget the two pion photoproduction becomes important above 400 MeV. An interesting exchange effect can occur when one of these pions is subsequently re-absorbed.

It will be interesting to see that if some other kinematic situations can be measured and compared to theoretical calculations in which there are no more free parameters to see if the scheme has predictive power. It would be misleading not to point out the difficulty of performing these coincidence experiments at high momentum transfer in which the dominant quasi-free production mechanism is small. This means that the count rates are also small. The extension of the experiments will depend in part on higher energy and higher duty cycle accelerators, although improvements in the detection techniques could make some differences in the short run. The Saclay group has also begun to employ a monochromatic photon beam from their energetic positron beam, and experiments are just beginning. This is of particular importance for the study of heavier nuclei than the deuteron for which some energy resolution is required to ascertain the final state. The extension of such experiments to heavier nuclei will be quite interesting, as was discussed by Laget[*] In particular, we note the earlier Saclay work of the ^{4}He$(\gamma,p\pi^-)$ reaction which was quite difficult to interpret, although the situation has been resolved (see the paper by J. M. Laget[*])

The total cross section for photo-absorption in the Δ region is of considerable interest. This has been studied for light nuclei at Mainz and reported at this symposium by Ahrends.[*] Unfortunately their technique is applicable only to light nuclei·and a new method must be found to make these important measurements in heavier nuclei. Groups in Bonn and Saclay are

presently thinking about experiments to measure the total
hadronic production cross sections. It should be pointed out
that this type of measurement has been performed for the proton and deuteron at high energies but that there the problem
is much simpler because the background electronic effects are
relatively small.

Another interesting measurement is to look at the pion
spectrum produced by quasi-monochromatic γ rays with energies
from 250 to 400 MeV. An experiment of this type using tagged
photons was reported here by Mecking*, who obtained results
which are consistent with quasi-free photo-production followed
sometimes by a quasi-free πN collision which further reduced
the pion energy. A similar experiment using the bremsstrahlung difference technique has been started at Bates and was
reported here by Pauli*. These experiments are complimentary
to a study of the pion spectra emerging from nuclei when monoenergetic pions are incident; the latter reactions have just
been started both at SIN and LAMPF. Such studies, taken as a
whole, should provide some insight into the behavior of pions
in nuclei. They go beyond the optical model which removes
flu by the action of the imaginary potential but which does
not specify the physical mechanism for flux removal. The
photo-induced reactions should be of special importance in
this field because one can produce pions in the nuclear interior while energetic pion beams do not penetrate into the
nuclear interior. Therefore a comparison of the pion and
photon results can possibly distinguish between surface and
volume phenomena. It will be of interest to see how this
develops.

One should not discuss the Δ region without mentioning
the effects of virtual as well as real pions. In this connection we have the rise in the cross section for the deuteron
photodisintegration cross section and also the relatively
large cross sections observed in the $^{16}O(\gamma,p)$ and $^{4}He(\gamma,p)$
reactions, one of the interpretations that has been put
forward for the large size of the cross section in the photoproduction of the Δ in an intermediate state. It should be
clear that at this time it is just an interesting hypothesis
and will require further testing.

An extensive theoretical program to understand the region
has been carried out by Moniz and his co-workers and was

reported at this symposium by Moniz[*]. The model which is used involves the production of the Δ, with subsequent Δ-N interactions, and finally Δ decay. This approach leads to a non-local angular momentum dependent potential. One of the nice features of these calculations is that once the Δ-nucleus potential parameters have been determined from elastic pion-nucleus scattering, all other processes can then be predicted in a unified way without further free parameters. An example of the application of this the (γ, π^{o}) reaction was presented by Koch[*] and Dillig[*] et al. With the experimental advances in pion induced reactions and in photo-reactions that we anti-cipate in the next few years we can be sure that this will be a lively arena.

VII. CAN WE MEASURE THE PION WAVE FUNCTION?

One of the most interesting aspects of electron scatter-ing experiments is the ability to measure wave functions in momentum space, i.e., form factors. This ability depends upon the known properties of the EM interaction and on the kine-matic ability to independently vary $\vec{q}$ and ω the momentum and energy transfer to the nucleus. The same possibility exists for the $(e, e'\pi)$ reaction. For the case of $(e, e'\pi)$ reactions for low energy pions, as well as the (e, e') reaction to dis-crete nuclear states, the cross section can be written as:

$$\frac{d^2\sigma}{d\Omega_e d\epsilon_e} = \frac{4\pi\sigma_{Mott}}{M_T} \left(\frac{q_u}{q}\right)^2 S_L(\vec{q}, \omega) + \left[\frac{q_u}{2q^2} + \tan^2\frac{\theta}{2}\right]S_T(\vec{q}, \omega)$$

where Ω_e and ϵ_e are the solid angle and energy for the detec-ted outgoing electron, σ_{Mott} is the Mott cross section, M_T is the target mass, q_u is the four momentum transfer $(\omega, \vec{q})$ and S_L and S_T are the longitudinal and transverse invariant struc-ture function which includes the transition form factors and the pion wave functions. We note that the photon point is purely transverse. The new information that comes from the $(e, e'\pi)$ reaction, in contrast to the (γ, π) reaction is the inclusion of a longitudinal form factor, as well as the abili-ty to independently vary $\vec{q}$ and ω. It should be stressed that these experiments are extremely difficult.

By constrast we should discuss the amount of information which is available from the more easily measured (γ, π) reaction.

As was discussed by Gerry Brown[*] the cross section for the (γ,π°) reaction is proportional to the square of the amplitude $A^\circ(k,\theta)$, where:

$$A^\circ(k,\theta) = <0|\nabla\phi_\pi^*(r) \cdot (\vec{k}\times\vec{\epsilon})e^{i\vec{k}\cdot\vec{r}}|0>$$

$$= \int\rho(r)\vec{\nabla}\phi_\pi^*(r) \cdot (\vec{k}\times\vec{\epsilon})e^{i\vec{k}\cdot\vec{r}}d^3r$$

where $\vec{k}$ and $\vec{\epsilon}$ are the photon momentum and polarization. The gradient of the pion wave function $\nabla\phi_\pi$ enters because of the p wave nature of the Δ resonance which dominates the (γ,π°) reaction. Assuming that $\rho(r)$, the nuclear ground state density, is well known one can obtain information about $\phi_\pi(r)$.

Similarly for the $(\gamma,\pi^\pm)$ reaction the amplitude is given by:

$$A^\pm(k,\theta) = <J_f|\phi_\pi^*(r) \sum_{j=1}^{A} f_j\tau_j^\pm e^{i\vec{k}\cdot\vec{r}_i}|J_i>$$

$$= \int\rho_{fi}^\tau(r)\phi_\pi^*(r)e^{i\vec{k}\cdot\vec{r}}d^3r$$

where f_j is the elementary photopion operator on the jth nucleon and $\rho_{ti}^\tau(r)$ is the appropriate isospin transition density. For example just above threshold we have $f = \vec{\sigma}\cdot\vec{\epsilon}$ and then ρ_{fi}^τ is the spin-isospin transition density (i.e., the Gamow-Teller transition density). Again, if one considers ρ_{fi}^τ to be well known then one can obtain information about $\phi_\pi(r)$.

For both the positive and neutral pion photoproduction, there is a unique relationship between the pion and photon energy (for a specific final nuclear state). Therefore the amplitudes cannot be used to map out the pion wave function (in momentum space) because if k is varied, so is the pion energy, and therefore the pion-nucleus interaction. Therefore one has sensitivity to ϕ_π but cannot map it out as in $(e,e'\pi)$ reaction.

VIII. CONCLUSIONS

In my opinion this symposium has indicated the interest and vitality of photopion physics. The low energy region just above threshold is in good shape which indicates that our

theoretical tools are in reasonable shape. As is befitting a
field in an early stage of development there are more open
questions than there are answers. Therefore it seems to me
to be appropriate to end the summary paper with a brief list
of open problems in the field. As was discussed in Sec. IV
the systematic data from the (π^-,γ) reaction are largely un-
explained, particularly the nature of the strongly excited
states and the fact that none are seen above for heavier
nuclei than ^{40}Ca, and also the apparent appearance of intermediate structure. For the (γ,π) reaction it would be nice
to see some improvements in the basic approach to the calculations: for example the inclusion of fully relativistic
amplitudes in light nuclei such as the deuteron (as opposed
to a non-relativistic reduction), a better description of
the off shell elementary amplitude and an evaluation of the
medium effects like virtual meson current contributions and
the effects of binding on the elementary operator. We look
forward to realistic estimates of the sensitivity of the (γ,π)
reaction to each of the major ingredients so that we can have
a more realistic assessment of what we may learn in this field.

A challenge to both experimentalists and theorists is the
apparent large discrepancy of the old (γ,π^0) results. In particular it is of importance to perform a new series of experiments in this field.

Since the field has been limited by experimental difficulties, advances in experimental techniques will have a de-
cisive role in the future development of the field. I have
saved these open problems for the end. A very nice example
was the proposal by Schoch[*] to measure (γ,π^-) total cross
sections just above threshold by observing the $\pi^-d \to nn$ reaction of the outgoing negative pions. This technique can work
with accelerators with very short bean pulses or with 100%
duty cycle. It should open up an interesting class of measurements.

An entire open class of measurements is the total hadronic cross sections, particularly in the Δ region. `Another ob-
jective which is important to achieve is good absolute accuracy of cross sections. Experiments on the $\gamma p \to n\pi^+$ process
have claimed accuracies as good as 3% in absolute value using
older accelerators. Assuming the accuracy of this claim we
should be able to duplicate this accuracy with present tech-

nology, even for much smaller cross sections since the limits
are the accuracy of the incident flux, target thickness, and
detector efficiency and solid angle, and not counting statistics.

Finally we have the problem of energy resolution. Most
people believe that the achievements of good resolution is
important only for the purposes of spectroscopy. This is only
partially true since one does need good energy resolution to
do spectroscopy. However, this isn't the entire story since
good resolution enables one to accurately measure the cross
sections to discrete states which have different quantum numbers. This increases the selectivity and sensitivity of the
tests that one can perform, and will open a larger number of
states that we can examine. There is another common miscon-
ception that one cannot achieve high resolution in photoreactions because the end point is too "mushy". In fact with high
intensity accelerators this is no longer true, but the limitation in energy resolution comes from the fact that photons do
not lose energy in the target while the emerging charged particles do. At Bates we have solved this problem by proposing
to perform the experiments via electroproduction so that the
energy loss of electrons in the target balances the energy
loss of the outgoing pions as discussed in the review by K.
Shoda. Using this idea and the energy loss mode one can
achieve a fractional energy resolution $\sim 10^{-4}$. An instrument
with these specifications is presently being built and calculations of the spectra were shown by Winhold.[*] For that case
assuming no other experimental uncertainties, separation of
the low lying quartet of levels in the ^{16}N can easily be
achieved. I believe that this is just one example of an
experimental improvement which will move the field forward
in the new few years. I look forward with anticipation to
the next Photopion Symposium to see how our specifications
will work out in the future. I also anticipate the inclusion
of photopion physics into the main stream of intermediate
energy physics.

References

1. For a bibliography of the early papers on pion and photopion physics see H. A. Bethe and F. de Hoffman, Mesons and
   Fields, Vol. II, Dow, Peterson & Co., Evanston, Ill.
   (1955).

2. See e.g., K. H. Althoff in Proceedings of the June Work-
    shop in Intermediate Energy Electromagnetic Interactions,
    MIT, A. M. Bernstein, editor, to be published, and Bonn
    Report HE 76-4 (1976).
3. J. I. Friedman and H. W. Kendall, Ann. Rev. Nucl. Sci.
    22, 203 (1972).
4. G. B. West, Phys. Report C 18, 263 (1975).
5. K. I. Blomqvist and J. M. Laget, Nucl. Phys. A280, 405
    (1977).
6. G. Von Holtey, Springer Tracts 59, 3
7. M. I. Adamovich, Proceedings (Trudy) of the P. N. Lebedev
    Physics Inst., Vol. 71, Photonuclear and Photomesic
    Processes (1976).
8. P. E. Aragan et al., to be published.
9. J. H. Koch and R. M. Woloshyn, Phys. Lett. 60B, 221 (1976).
10. P. Bosted and J. M. Laget, Nucl. Phys. A296, 413 (1978).
11. W. K. H. Panofsky and ·E. A. Allton, Phys. Rev. 110, 1155
    (1958).
12. R. H. Dalitz and D. R. Yennie, Phys. Rev. 105, 1598 (1957).
13. T. W. Donnelly in Proceedings of the June Workshop in
    Intermediate Energy Electromagnetic Interactions, MIT,
    A. M. Bernstein, editor, to be published and T. W. Donnelly
    and J. B. Camerata, to be published.

LIST OF CONTRIBUTORS

Ahrens, J., Max-Planck Institut, Mainz, W. Germany
Alder, J. C., University of Lausanne, Lausanne, Switzerland
Arends, J., University of Bonn, W. Germany
Argan, P., CEN, Saclay, France
Audit, G., CEN, Saclay, France
Baba, K., Hiroshima University, Japan
Balashov, V. V., Moscow State University, USSR
Ballentine, P. H., Rensselaer Polytechnic Institute
Bernstein, A. M., Massachusetts Institute of Technology
Bloch, A., CEN, Saclay, France
Blomqvist, K. I., Massachusetts Institute of Technology
Booth, E. C., Boston University
Brown, G. E., SUNY-Stony Brook - Nordita
Cheon, Il.-T., Yonsei University, Seoul, Korea
Comuzzi, J., Massachusetts Institute of Technology
Dahme, W., University of Munich, W. Germany
de Botton, N., CEN, Saclay, France
DeCarlo, V., Pennsylvania State University
Dillig, M., University of Erlangen, W. Germany
Donnelly, T. W., Stanford University
Dressler, E. T., National Bureau of Standards
Endo, I., Hiroshima University, Japan
Epstein, G. N., University of Pittsburgh
Eramzyhan, R. A., JINR, Dubna, USSR
Eyink, J., University of Bonn, W. Germany
Faure, J. L., CEN, Saclay, France
Franklin, G., Massachusetts Institute of Technology
Fujii, H., University of Tokyo, Tanashi, Japan
Fujisaki, M., Hiroshima University, Japan
Furui, S., University of Tokyo, Japan
Gabioud, B., University of Lausanne, Lausanne, Switzerland
Gimm, H., Max Planck Institute, Mainz, W. Germany
Gmitro, M., JINR, Dubna, USSR
Händel, R. University of Erlangen-Nurnberg, W. Germany
Hartmann, H., University of Bonn, W. Germany
Hegerath, A., University of Bonn, W. Germany
Hersh, J. K., Rensselaer Polytechnic Institute
Huber, M. G., Universität Erlangen-Nurnberg, W. Germany

425

Joseph, C., University of Lausanne, Lausanne, Switzerland
Kadota, S., Hiroshima University, Japan
Kissener, H. R., Zentral. für Kernforschung, Rossendorf,
    E. Germany
Klein, F., University of Mainz, W. Germany
Koch, J. H., IKO, Amsterdam, Holland
Laget, J. M., CEN, Saclay, France
Leicht, R., Max Planck Institute, Mainz, W. Germany
LeRose, J., Rensselaer Polytechnic Institute
Loude, J. F., University of Lausanne, Lausanne, Switzerland
Lührs, G., University of Mainz, W. Germany
MacDonald, W. M., University of Maryland
Mecking, B., University of Bonn, W. Germany
Medicus, H. A., Rensselaer Polytechnic Institute
Milder, F. L., Boston University
Min, K., Rensselaer Polytechnic Institute
Minn, P., Max Planck Institute, Mainz, W. Germany
Miska, H., Pennsylvania State University
Moniz, E. J., Massachusetts Institute of Technology
Morel, N., University of Lausanne, Lausanne, Switzerland
Murakami, A., Saga University, Japan
Murata, Y., University of Tokyo, Tanashi, Japan
Nagl, A., Catholic University
Nakahara, K., Tohoku University, Sendai, Japan
Noguchi, S., University of Tokyo, Tokyo, Japan
Nöldeke, G., University of Bonn, W. Germany
O'Connell, J. S., National Bureau of Standards
Ohashi, H., Tohoku University, Sendai, Japan
Panke, H., University of Munich, W. Germany
Paras, N., Massachusetts Institute of Technology
Pauli, M., Massachusetts Institute of Technology
Perrenoud, A., University of Lausanne, Lausanne, Switzerland
Perroud, J. P., University of Lausanne, Lausanne, Switzerland
Planeta, S., Rensselaer Polytechnic Institute
Potenziani, E., Rensselaer Polytechnic Institute
Rao, K. S., Matscience, Madras, India
Renker, D., University of Munich, W. Germany
Roberts, B. L., Boston University
Rosenfelder, R., Stanford University
Rossdeutscher, S., Rensselaer Polytechnic Institute
Rost, H., University of Bonn, W. Germany
Rowley, D., Rensselaer Polytechnic Institute
Schoch, B., University of Mainz, W. Germany
Schuhl, C., CEN, Saclay, France
Shoda, K., Tohoku University, Sendai, Japan

Singham, M. K., University of Pittsburgh
Sridhar, R., Matscience, Madras, India
Stoler, P., Rensselaer Polytechnic Institute
Strassner, G., University of Zurich, Switzerland
Sumi, Y., Hiroshima University, Japan
Sung, B. N., Seoul National University, Korea
Susila, S., Matscience, Madras, India
Tabakin, F., University of Pittsburgh
Tamas, G., CEN, Saclay, France
Tosunjan, L. A., JINR, Dubna, USSR
Tran, M. T., University of Lausanne, Switzerland
Truöl, P., University of Zurich, Switzerland
Tzara, C., CEN, Saclay, France
Überall, H., Catholic University
Vincent, E., CEN, Saclay, France
Werntz, C., Catholic University
Williamson, C. F., Massachusetts Institute of Technology
Winhold, E. J., Rensselaer Polytechnic Institute
Winkelmann, E., University of Lausanne, Switzerland
Wünsch, R., JINR, Dubna, USSR
Yamazaki, M., Tohoku University, Sendai, Japan
Yergin, P. F., Rensselaer Polytechnic Institute
Zeigler, B., Max Planck Institute, Mainz, W. Germany
Zieger, A., Max Planck Institute, Mainz, W. Germany
Zimmerman, P. D., Louisiana State University

LIST OF PARTICIPANTS

Ahrens, J., Max Planck Institute, Mainz, W. Germany
Audit, G., CEN, Saclay, France
Baer, H., Los Alamos Scientific Lab
Bergstrom, J. C., University of Saskatchewan, Canada
Bernstein, A. M., Massachusetts Institute of Technology
Blomqvist, K. I., Massachusetts Institute of Technology
Booth, E. C., Boston University
Bosco, B., University of Florence, Italy
Cammarata, J. B., Virginia Polytechnic Institute
Cannata, F., Catholic University of America
Cottam, R., Pennsylvania State University
Cottman, B., Massachusetts Institute of Technology
Crannell, H., Catholic University of America
Dahme, W., University of Munich, W. Germany
de Botton, N., CEN, Saclay, France
De Carlo, V. A., Pennsylvania State University
de Forrest, T., IKO, Amsterdam, The Netherlands
de Witt-Huberts, P., IKO, Amsterdam, The Netherlands
Donnelly, T. W., Stanford University
Dressler, E. T., National Bureau of Standards
Dubach, J., Massachusetts Institute of Technology
Edge, R. D., University of South Carolina
Fabian, W., University of Mainz, W. Germany
Franco, W., Brooklyn College
Franklin, G., Massachusetts Institute of Technology
Friar, J. L., Los Alamos Scientific Laboratory
Hackman, R. H., Virginia Polytechnic Institute
Hadjimichael, E., Fairfield University
Hammen, M., Max Planck Institute for Chemistry, Mainz,
    W. Germany
Haxton, W., Los Alamos Scientific Laboratory
Homma, S., University of Tokyo, Tanashi, Japan
Hotta, A., University of Massachusetts, Amherst
Itoh, K., University of Saskatchewan, Canada
Kawazoe, Y., Tohoku University, Sendai, Japan
Knuepfer, W., University of Erlangen, W. Germany
Koch, J. H., IKO, Amsterdam, The Netherlands
Koester, W., University of Erlangen, W. Germany

Laget, J. M., CEN, Saclay, France
LeRose, J. J., Rensselaer Polytechnic Institute
Lindgren, K., University of Illinois at Champaign
Maas, R., IKO, Amsterdam, The Netherlands
Mecking, B. A., University of Bonn, W. Germany
Medicus, H. A., Rensselaer Polytechnic Institute
Milder, F. L., Boston University
Min, K., Rensselaer Polytechnic Institute
Miska, H., Pennsylvania State University
Miska, H. H.,Johannes Gutenberg University, Mainz, W. Germany
Moniz, E. J., Massachusetts Institute of Technology
Nagl, A., Catholic University of America
Negishi, T., McMaster University, Canada
Neuhausen, R., Universität Mainz, W. Germany
O'Brien, J. T., Catholic University of America
Paras, N., Massachusetts Institute of Technology
Pauli, M. R., Massachusetts Institute of Technology
Perroud, J. P., University of Lausanne, Switzerland
Pitthan, R., Naval Postgraduate School
Prats, F., George Washington University
Quinn, B., Massachusetts Institute of Technology
Rao, K. S., Matscience, Madras, India
Roberts, B. L., Boston University
Rosenfelder, R., Stanford University
Rowley, D., Rensselaer Polytechnic Institute
Sanzone, M., Istituto Nazionale de Fisica Nucleare, Italy
Sapp, B., Massachusetts Institute of Technology
Sapp, W., Massachusetts Institute of Technology
Schoch, B. H., Johannes Gutenberg University, Mainz, W. Germany
Sealock, R. M., University of Saskatchewan, Canada
Shoda, K., Tohoku University, Sendai, Japan
Singham, M., University of Pittsburgh
Skopik, D. M., University of Saskatchewan, Canada
Sober, D., Catholic University of America
Speth, J., Institut für Kernphysik, Julich, Germany
Stoler, P., Rensselaer Polytechnic Institute
Sung, B. N., Seoul National University, Korea
Tabakin, F., University of Pittsburgh
Thies, H. H., University of Western Australia
Tomusiak, E. L., University of Saskatchewan, Canada
Truöl, P., University of Zurich, Switzerland
Überall, H., Catholic University of America
Vincent, E., CEN, Saclay, France
Werntz, C., Catholic University of America
Winhold, E. J., Rensselaer Polytechnic Institute

Yergin, P. F., Rensselaer Polytechnic Institute
York, R., University of Massachusetts, Amherst
Zieger, A., Max Planck Institute, Mainz, W. Germany
Zimmerman, P., Louisiana State University

SUBJECT INDEX

Absorption, photon,
   total, 385
Absorption, pion, 301, 316
$^{27}Al(e,\pi^+)$, 186
$^{27}Al(\gamma,\pi^+)$, 140
$^{197}Au(\gamma,\pi^-)$, 387
Angular correlation
   function, 126

Bates linac, 211, 245
Bates magnetic spectrometer
   29, 36, 38, 212
$^{11}B(\pi^-,\gamma)$, 107
$^{10}B(\gamma,\pi^+)$, 146
$^{10}B(\gamma,\pi^-)$, 137
$^{9}Be(\pi^-,\gamma)$, 83
$^{9}Be(e,\pi^+)$, 186, 194
$^{9}Be(\gamma,\pi^+)$, 151, 384
$^{9}Be(\gamma,\pi^-)$, 137
$^{209}Bi(\pi^-,\gamma)$, 94
Blomqvist-Laget Model, 302,
   303-308, 328, 408
Bonn Synchroton, 275
Bonn magnetic spectrometer
   379
Bonn tagged photon
   facility, 43

Cascase model, 275, 285
   for pion nuclear
   interactions, 347

$^{11}C(\gamma,\pi^-)$, 133
$^{12}C(\gamma,\pi^0)$, 235, 245
$^{12}C(\gamma,\pi^+)$, 150, 211, 276, 287
$^{12}C(e,\pi^+)$, 182, 186, 194, 200
$^{12}C(\gamma,\pi^-)$, 8, 27, 63, 131,
   137, 140, 155, 211, 327,
   416
$^{12}C(\pi^-,\gamma)$, 34, 83, 107
$^{13}C(\gamma,\pi^+)$, 146, 416
$^{13}C(\gamma,\bar{\pi})$, 135, 137, 144
$^{13}C(\pi^-,\gamma)$, 117
$^{40}Ca(\gamma,\pi^0)$, 245

Cerenkov detectors, 239
CGLN amplitude, 220, 247, 250,
   307
Closure approximations, 112,
   367
Cluster model, 249
Cohen-Kurath wave
   function, 108, 111
Coherent photoproduction, 8,
   226, 247
   and isobar doorway
   model, 335
Continuum shell mode, 125
Correlation function, γ-n, 126
$Cu(\gamma,\pi^0)$, 235
$Cu(\gamma,\pi^+)$, 295
Current algebra predictions, 3

Delta-hole
   See isobar hole

Delta-nucleon interaction, 257, 264

Delta resonance, 7, 16, 41, 256, 276, 290, 292, 306, 361, 382, 409, 428

Density matrix, nuclear, 51, 56

Double pion photoproduction, 257, 265

DWIA, 199, 411

Elementary electrophoto-production, 395, 410

Elementary production amplitude, Born terms, 259

Electromagnetic processes, 51

Electron scattering, 1
   deep inelastic, 399
   quasi-elastic, 400

Electroproduction, 199, 211

Ericson-Ericson effect, 155, 315

Fermi gas model, 381

Few body operator, 54

Final state interaction, 290, 409

Form factor, nuclear, 420
   and π^o production, 11

Gamow-Teller matrix element, 71

Gauge invariance, 1

Giant resonances, 89, 117

Giant M2 resonance, 87

$^1H(e,e'\pi^+)$, 48

$^1H(\gamma,\pi^o)$, 411, 412

$^1H(\gamma,\pi^+)$, 146

$^1H(\pi^-,\gamma)$, 411

$^2H(\pi^-,nn)$, 391

$^2H(\pi^-,\gamma)$, 73

$^2H(\pi,\pi)$, 361

$^2H(\gamma,\pi^o)$, 25, 227, 231, 239, 361

$^2H(\gamma,p\pi^-)$, 262, 418

$^2H(\gamma,\pi^+)$, 24, 147

$^2H(\gamma,\pi^-)$, 42, 171

$^3H(\pi^-,\gamma)$, 78

$^3He(\gamma,\pi^+)$, 148

$^3He(\gamma,\pi^o)$, 228, 239

$^4He(\gamma,\pi^+)$, 146

$^4He(\gamma,\pi^o)$, 239

$^4He(\gamma,p\pi^-)$, 262

Helm model, 155, 185, 214

$^{165}Ho(\pi^-,\gamma)$, 93

Hodoscope, 276, 280

Inclusive reactions, 38, 367, 371

Incoherent photoproduction and isobar doorway model, 336

Isochromat, 198

Isobars
   see delta resonance

Isobar doorway model, 12, 353, 361
   connection to DWIA, 356

Isobar-hole, 337, 353, 361

Isomer ratio, 387

Interaction, Δ-N
   see delta-nucleon interaction

Kroll-Ruderman amplitude, 1

$^6Li(e,\pi^+)$, 186, 194

$^6Li(\gamma,\pi^-)$, 132, 149

$^6Li(\gamma,\pi^+)$, 21

$^6Li(\gamma,\pi^+)$, 58, 80, 149, 384

$^6Li(\gamma,\pi^0)$, 245, 249
$^6Li(\gamma^-,\pi)$, 78
$^7Li(\gamma,\pi^+)$, 146
$^7Li(\gamma,\pi^0)$, 245
$^7Li(\pi^-,\gamma)$, 83
Low energy theorems, 1

Mainz linac, 171, 385
Meson exchange currents,
    54, 60
$^{27}Mg(\gamma,\pi^+)$, 140
Monochromatic photons, 419
    Saclay, 43, 269
    Bonn, 275
Monte Carlo calculation, 240
Multiple scattering
    expansion, 259
Muon contamination, 283

$n(\gamma,\pi^0)$, 221
$n(\gamma,\pi^-)$, 410
n-n scattering length, 69, 73
$^{14}N(\pi^-,\gamma)$, 85, 107
$^{14}N(\gamma,\pi^-)$, 165
$^{14}N(\gamma,\pi^+)$, 151
Neutral pion production 8,
    219, 239, 245
    elementary amplitude, 220
Neutrino reactions, 51
Nuclear structure
    and photopion inter-
    actions, 407
$^{16}O(\gamma,\pi^-)$, 155
$^{16}O(\gamma,\pi^+)$, 34, 151
$^{16}O(e,\pi^+)$, 186, 194
$^{16}O(\pi^-,\gamma)$, 93, 121
$^{18}O(\pi^-\gamma)$, 93
$^{16}O(\pi^-,\gamma n)$, 125

$^{16}O(\pi^+,\pi^+)^{16}O$, 342
Off-shell effects, 382
One body operator, 54
Optical potential, pion, 156,
    165, 301, 314-321, 328,
    335
    absorption, 377
    phase shift equivalent, 320

$\pi \rightarrow \mu \rightarrow e$ technique, 21, 144
Panofsky ratio, 72, 231
$^{208}Pb(\pi^-,\gamma)$, 94
$Pb(\gamma,\pi^+)$, 295
PCAC, 2
Peaking approximation for
    electroproduction, 199
Phase velocity, pion, 321
Photoproduction amplitude
    momentum-dependence, 382
Pionic atoms, 301
Pion-nucleus interaction, 301,
    335
    cascade model, 347
Pion wave function, 12
Positron annihilation, 276
Primakoff equation, 112
PWIA, 298

Quasifree production, 42, 255,
    259, 290, 295

Radiative capture, 23, 34, 69,
    107, 111, 121, 414
    direct, 125
    R-matrix theory, 121
    ratio to absorption, 111
    resonance, 125
Ratios π^-/π^+, 371
Reflection coefficient, 318
Residual radioactivity,
    method of, 134

$^{32}S(e,e')$, 108

^{32}S(π^-,γ), 107
Saclay linac, 239
Saskatchewan spectrometer, 38
Scattering length, Δ-N, 259
   n-n, 73,,414
Scattering, pion, 301, 391
   elementary amplitude, 226
   charge exchange, 227
Scintillation counters, 280
Semi-leptonic reactions, 51
^{28}Si(π^-,γ), 85
^{28}Si(γ,π^-), 161
^{28}Si(γ,π°), 245
SIN pair spectrometer, 34, 69
Spectator model, 260
Spectroscopy, pion, 27, 175, 193, 211
Spin isospin giant resonances, 186, 204
Spin-isospin transitions, 106, 108, 130, 211, 213
Spreading interaction, 339, 344

Virtual photons, 212
   spectrum, 181, 184, 194, 198, 201

Wave function, pion, 11, 314-321, 420
^{89}Y(e,π^+), 186
Structure function, 420
Surface production model, 162
SU(4), 71

Tagged photons, 275, 277
Threshold production, 1, 135, 144, 145, 413
Tohoku linac, 175, 194, 205
Tohoku spectrometer, 32
Tokyo synchroton, 43, 295
Total cross section, pion, 129, 349
   near threshold, 20
   reference list, 139
Transition densities, nuclear, 308-314
Two-body operator, 57

^{51}V($\gamma,\pi^\pm$), 140